Lecture Notes in Mobility

More information about this series at http://www.springer.com/series/11573

Gereon Meyer · Susan Shaheen
Editors

Disrupting Mobility

Impacts of Sharing Economy and Innovative Transportation on Cities

Springer

Editors
Gereon Meyer
Department Future Technologies and Europe
VDI/VDE Innovation + Technik GmbH
Berlin
Germany

Susan Shaheen
Transportation Sustainability Research Center
University of California
Berkeley, CA
USA

ISSN 2196-5544 ISSN 2196-5552 (electronic)
Lecture Notes in Mobility
ISBN 978-3-319-51601-1 ISBN 978-3-319-51602-8 (eBook)
DOI 10.1007/978-3-319-51602-8

Library of Congress Control Number: 2016961328

Printed on acid-free paper

This Springer imprint is published by Springer Nature
The registered company is Springer International Publishing AG
The registered company address is: Gewerbestrasse 11, 6330 Cham, Switzerland

Preface

Disruptive innovations, in contrast to just revolutionary ones, not only substitute existing solutions, but they also create new markets and change society. It was not the revolutionary invention of the automobile alone but rather it is a cheap mass manufacturing platform that replaced the carriage a century ago. Today, none of the much discussed revolutions in transportation, be it electric cars or self-driving pods, alone has a truly disruptive potential. Each of them faces high financial and legal hurdles for mass adoption. However, if these innovations are combined with each other into a vs. system approach with game-changing business and operational models, true synergies can be unlocked. The sharing economy, which includes models such as ridesourcing, carsharing, and pooled rides, offers the opportunity to unlock the potential of other technologies. Combined, an innovative mobility system could arise that makes travel cheaper, cleaner, and more accessible, particularly in cities.

At times when society is deeply divided over the impacts of globalization and digitization and doubts are voiced about opportunities for individuals, it is not sufficient to extol and sell novel solutions. The needs of users cannot only be anticipated but also have to be heard and taken into account in the design and engineering process from the start so that societal ownership of the results can manifest. Otherwise, disruptive solutions may fail, as citizens rely on conventional mobility in response to global concerns regarding exclusion, depreciation, labor supply, and safety. An integrated and sustainable urban mobility system that is developed in a participatory way, however, can acknowledge the various potentials and interdependencies for a more equitable, resilient, and transparent transportation system. A community and customer-centered approach could help to balance out the root causes of reservations against this disruption but also actively foster the disruptive process, readying the transportation infrastructure and the built environment for a transition to shared electric connected and automated mobility.

Choosing the right technologies, service concepts, and policy measures requires a dialogue among regional and city planners, public transit operators, engineers, entrepreneurs, and scholars of all fields. The Disrupting Mobility Summit held in Cambridge, MA (USA), in November 2015, was the first global occasion that brought together notable thought leaders from across the globe for this critical exchange. Organizers included leading academics from the City Science Initiative at MIT Media Lab; the Transportation Sustainability Research Center at the University of California, Berkeley; LSE Cities at the London School of Economics and Politics; and the Innovation Center for Mobility and Societal Change (InnoZ) in Berlin.

We invited speakers and poster presenters from the Disrupting Mobility Summit to submit chapters for this book to continue this dialogue. The chapters were peer-reviewed prior to publication. This process helped us to identify some the best pieces from the vast amount of knowledge shared at the Disrupting Mobility Summit and to make these contributions available to key stakeholders, including cities, to facilitate informed decision making on how to best prepare and plan for future disruption.

This book is divided into three parts: public sector activities (I), sharing economy and multimodal mobility (II), and innovative transportation technologies and city design (III).

In the first chapter in Part I, titled "Beyond Traffic: Trends and Choices 2045—A National DialogueAbout Future Transportation Opportunities and Challenges," Deputy Secretary Victor M. Mendez and his colleagues from the US Department of Transportation analyze trends that shape the mobility of people and goods, and they describe options for public policy. This is followed by a chapter written by Dewan M. Karim, a city planner from Toronto (Canada), on a comprehensive mobility ecosystem model that combines technology development and public transit planning in light of shared mobility.

In Part II, Francesco Ciari and Henrik Becker from the Swiss Federal Institute of Technology ETH Zurich present results from a study that simulates the costs and benefits of bikesharing, carsharing, and shared-ride services, indicating that the latter mode can fill a gap between the other two. Next, Joseph Iacobucci and colleagues from the US-based Sam Schwarz Consulting firm report on the outcomes of a study commissioned by the civic foundation TransitCenter on the relationship between conventional public transit and shared mobility systems that lead to a set of tangible recommendations for policy makers and city governments. Referring to traditional and innovative, digital shared-ride schemes in the city of Manila, Philippines, Katja Schechtner from MIT and the Vienna Technical University in Austria and Melinda Hanson from the Urban Project Collective in New York City conclude that Transportation Network Companies (TNC) or ridesourcing services in Asian megacities only serve the mid- to high-class income market, whereas traditional sharing schemes reach a far larger part of the population, mainly due to lower costs. In their chapter, Christopher Lisson from the Karlsruhe Institute of Technology in Germany and his colleagues summarize findings regarding behavioral response to Intelligent Traveler Information Systems; they find that usage decisions for such devices are determined less by cost and time savings but by pleasant design and convenient interaction. Mark-Philipp Wilhelms and his colleagues from the EBS University of Business and Law in Germany are also concerned with customer acceptance issues, as they analyze what motivates customers to use peer-to-peer carsharing, namely economic rather than environmental considerations. In their paper on multimodal transportation payments, Michael Dinning and Timothy Weisenberger from the US Department of Transportation's Volpe National Transportation Research Center argue that people expect easy-to-use apps for mobility planning and booking. Wolfgang Gruel and Joseph Stanford from MIT discuss the role of carsharing in mitigating the undesired long-term effects of self-driving cars as price transparency may discourage unnecessary tripmaking due to the ease and convenience of privately owned automated vehicles. This is followed by a chapter on smartphone applications in transportation written by Susan Shaheen and her colleagues from the University of California, Berkeley in the USA. According to the study findings, mobility aggregators can lead to more public transport use and less driving. In the second chapter on the influence of smartphone apps on travel behavior, Adam Davidson from the City University of New York in the USA explains that apps provide users with an elevated level of reliability that allows them to reach more active and better controlled mobility decisions. Next, Susan Shaheen and her colleagues from UC Berkeley and France report on the results of a survey among the users of a carpooling platform in France that indicates that individuals from lower-income groups tend to use the system as passengers, whereas higher-income users act as drivers, implying an equity balancing effect. Finally, Alejandro Henao and Wesley Marshall from the University of Colorado in Denver in the USA explore the impacts of ridesourcing on travel behavior.

In the first chapter of Part III, Florian Lennert and Robert Schönduwe from the Innovation Center for Mobility and Societal Change in Berlin, Germany, present an extensive comparison of scenario studies on the mitigation potential for greenhouse gas emissions when they noticed that modal shift, demand, and land-use management are not sufficiently covered. They see great potential, however, from the convergence of electric propulsion, automated and connected vehicle technology, and on-demand mobility services. In addition, Philipp Rode and colleagues from the London School of Economics and Politics in the UK review the pathways that cities can take to become more accessible; they conclude that cities either choose more sprawling, car-intense developments or more dense, public transit-related options that result in a ten-fold less in fuel consumption. Furthermore, Nicole Ronald from the Swinburne University of Technology and colleagues from the University of Melbourne, Australia, predict human behavioral shifts toward shared, connected, and automated vehicle services in the urban mobility system. Bern Grush and John Niles from the USA introduce a deployment concept for shared self-driving vehicles that could help avoid disruption failures. Victoria Adams and her colleagues from Booz Allen Hamilton in the USA discuss the role of electric, connected, and smart bicycles in encouraging more biking, and they explain how transportation agencies can support this. Luis E. Ferreras from the Parsons Transportation Group in the USA presents a concept for a drone-based ITS that could enable the observation of a city's traffic from the sky. Finally, Lino Vital García-Verdugo, an independent design researcher from Spain, introduces the concept of mobilescapes where automated vehicles transform into meaningful dynamic environments.

We hope that the readers of these chapters gain many new and useful insights from this volume of the Lecture Notes in Mobility series. We also wish that the contents stimulate fruitful discussions, proactive choices, and informed decision making around the future of mobility that yields more sustainable, equitable, and accessible transportation choices for all citizens across the globe. We invite scholars, managers, planners, and engineers to continue the academic discourse in light of upcoming challenges and opportunities for mobility disruption in the areas of aviation, goods movement, and suburban and rural transport at both national and international levels.

Finally, we thank the authors for their notable efforts to write the chapters of this book. We also acknowledge the many of contributions of our peer reviewers. Thanks also go to the organizers of the Disrupting Mobility Summit, Ryan Chin and Wolfgang Gruel of MIT's Media Lab, Susan Shaheen and Adam Cohen of the Transportation Sustainability Research Center at UC Berkeley, Florian Lennert of InnoZ, and Philipp Rode and Pia Laube of LSE Cities of the London School of Economics. Special thanks also go to Diana Tobias from VDI/VDE-IT for back office support during the editorial process and to Jan-Philip Schmidt from Springer for the accelerated publication timeline for this volume.

We also express our deep appreciation to the summit partners: the Transportation Research Board of the National Academy of Sciences, Stiftung Mercartor, and the Alfred Herrhausen Gesellschaft Das Internationale Forum der Deutschen Bank. Thanks also go to our many summit sponsors: Zipcar (city sponsor); Enterprise CarShare (diamond sponsor); RideScout (platinum sponsor); BCycle and Bridj (gold sponsors); Lyft, Hubway, Munich RE, the Shared-Use Mobility Center, TransitCenter, and Via (silver sponsors); and Transportation for America (bronze sponsor).

Berlin, Germany Gereon Meyer
Berkeley, CA, USA Susan Shaheen
November 2016

Contents

Part I
Public Sector Activities

Beyond Traffic: Trends and Choices 2045 —A National Dialogue About Future Transportation Opportunities and Challenges

Victor M. Mendez, Carlos A. Monje Jr. and Vinn White

Abstract This paper summarizes the findings of the U.S. DOT's comprehensive report on the current and future conditions of America's transportation system, *Beyond Traffic: Trends and Choices 2045*. The full report, and in turn this paper, discusses long-term and emerging trends in passenger and freight travel and the potential impacts of technological advances, climate change, and evolving governance institutions and funding sources. This analysis provides a framework for a fact-based discussion about the critical transportation policy choices we are faced with to address these long-term challenges. *Beyond Traffic* is intended to stimulate a national dialogue about the nation's future transportation opportunities and challenges.

1 Introduction: *Beyond Traffic: Trends and Choices 2045*

The enabling legislation that gave rise to the U.S. Department of Transportation was signed by President Lyndon B. Johnson in 1966. That legislation calls on the Secretary of Transportation to periodically report on the current and future conditions of our transportation system. In 1977 Secretary William T. Coleman, Jr. honored this responsibility by publishing National Transportation Trends and Choices. The report examined the conditions of the national transportation system

V.M. Mendez (✉)
Deputy Secretary of Transportation, U.S. Department of Transportation, 1200 New Jersey Ave. SE, Washington, DC 20590, USA
e-mail: Deva.Tucker@dot.gov

C.A. Monje Jr.
Transportation for Policy, Washington, DC, USA

V. White
U.S. Department of Transportation, Washington, DC, USA

G. Meyer and S. Shaheen (eds.), *Disrupting Mobility*,
Lecture Notes in Mobility, DOI 10.1007/978-3-319-51602-8_1

across all modes and estimated the long-term consequences of various policy options, and served as a model for subsequent Transportation Secretaries.

Inspired by Secretary Coleman's report, Secretary Foxx assembled a team of experts to develop *Beyond Traffic: Trends and Choices 2045*, a comprehensive analysis of the latest data and trends shaping transportation to lay the framework for a fact-based discussion about the critical policy choices that need to be made. A draft of the report was published in February 2015 and stakeholder input was solicited over the course of the year. A final version will be released in the spring of 2016. The key findings of this draft report are described below.

2 How We Move

Our growing population and economy will continue to increase demand for passenger travel and freight across nearly all modes of transportation. Over the next 30 years our population is expected to grow by nearly 70 million [1] and our economy will nearly double.

2.1 Growing Travel Demand in Metropolitan Areas

If long-term historic trends hold, the vast majority of this growth will occur in metropolitan areas, and particularly in suburban areas outside of primary cities. Between 1980 and 2010, the population living in suburban and urban areas of the United States increased by 50%, while the overall rural population declined [2].

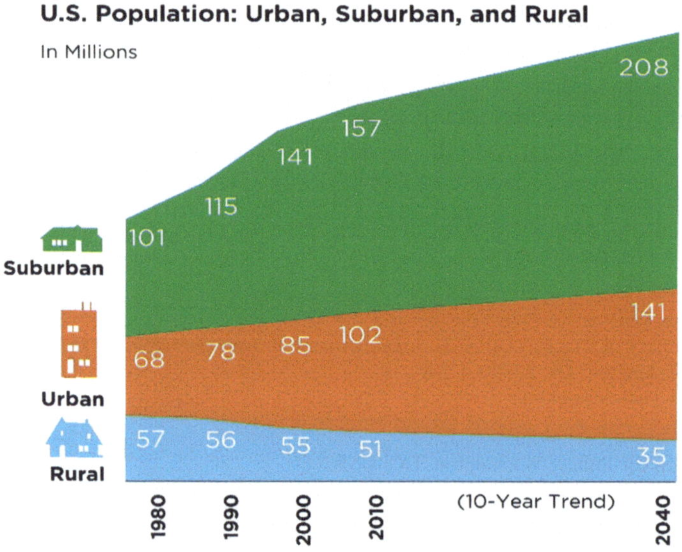

Fig. 1 U.S. population distribution: 1980–2010 with projection to 2040

Despite evidence of growing demand for urban housing and workplaces, the majority of our nation's population growth continues to be in the suburban areas surrounding cities. The suburbanization of population and employment has increased demand for vehicle travel, exacerbated congestion, strained existing roadway infrastructure, and increased household transportation costs (Fig. 1).

Population growth is also distributed unevenly across regions of the country. The fastest growing states and metropolitan areas are in the South and West. Florida, Texas, and California alone account for nearly 40% of population growth in the past decade [3]. The fastest growing metropolitan areas are predominantly young, auto-oriented Sunbelt cities such as Austin, Charlotte, Orlando, Phoenix, and Houston that have rapidly expanding suburbs, lower than average urban population densities and developing transit systems. The continued expansion of metropolitan areas in the South and West is likely to spur continued demand for auto-travel and increase pressure on already congested highways in those areas.

2.2 *Slowing Demand for Vehicle Travel*

While population growth will continue to drive overall demand for travel and freight, particularly in growing metropolitan areas, there are several countervailing demographic, cultural, and economic trends that will likely dampen demand for vehicle travel. Indeed, the most recent Federal Highway Administration (FHWA) forecasts predict long-term growth in vehicle miles traveled (VMT), a measure of how much people drive, will be slower than it has been over the past 30 years, increasing at an annual rate of less than 1% [4].

Between 2006 and 2013, VMT per capita declined each year. Despite recent increases, VMT per capita remains approximately 6% off its 2005 peak [5, 6]. High gas prices and a deep recession clearly played a major role in suppressing demand for vehicle travel, as has severe congestion in metropolitan areas.

As the growth in VMT slowed over the past decade, we saw increased use of other modes of travel including transit, walking, cycling, and telecommuting. Cycling and walking make up a substantial proportion of local trips that people take for nonwork purposes, particularly in urban areas. Together, they account for approximately one-half of all trips taken under one mile and more than 10% of all trips of any length [7]. Overall, however, shifts to nondriving modes have had a relatively minor impact on overall travel patterns for which auto-use remains the dominant mode [8].

As the economy has recovered and fuel prices have dipped in more recent years, VMT itself has rebounded. In 2015, total VMT grew by 3.5% and hit a record high [5]. However, long-term demographic and socioeconomic trends still point to slower VMT growth over the next 30 years.

2.3 *The Travel Preferences of Older and Younger Americans*

One important factor affecting long-term demand for vehicle travel is the age of our population. The percentage of the population that is over 65 is expected to increase by 77%, from 48 to 85 million people [1]. Today, on average, Americans over the age of 65 drive half the amount of Americans aged 25–64 [7]. As the percentage of the population over retirement age increases, workforce participation and demand for commute travel may decline.

In addition, evidence suggests that young adults are driving less today than they did a decade ago. License rates for young Americans are declining. Young adults are more likely to live in urban areas and use transit than older age cohorts. They are also more likely to shop or socialize online as alternatives to driving [9].

Socioeconomic forces likely play a role in the changing travel preferences of Millennials. Many came of age during the Great Recession and have high levels of student debt and limited employment opportunities which affect their ability to afford a vehicle or purchase a home. Recent Census data shows that a third of Americans aged 18–34 are living with their parents [10]. Household size is also decreasing as many young adults are waiting longer to form families. Together these trends may contribute to reduced demand for vehicle travel among younger Americans.

2.4 *Increasing Travel Choices*

Technology is expanding the choices for how we travel or if we travel at all. For example, Americans are increasingly likely to shop online rather than go to a store. Within 5 years, online purchases could account for 10% of all retail sales [11]. This trend could reduce personal travel associated with shopping, which currently accounts for one in five household trips, and reduce demand for private vehicle ownership.

Travel preferences, flexible schedules, "hoteling" (unassigned office seating), teleworking, and improvements in communications technologies are all changing how many people work and commute to work. Twenty-eight percent of workers in a recent Gallup survey said that they typically telecommute at least once a month. The number of Americans who work from home at least 1 day a week increased by 43% between 1997 and 2010 [12]. The fastest growing "mode" for commuting is, in fact, telecommuting.

New business models and technologies are also creating new ways for Americans to access vehicles. Car sharing is growing at a rapid rate. In 2015, there were 1.2 million members in 23 active programs in the United States, marking a more than tenfold increase in membership over the past decade [13]. Bicycle sharing has also emerged in more than 30 cities. Bike-share systems allow people to conveniently travel short distances and connect to other modes, particularly public transit. Finally, ride sourcing services, such as Lyft and Uber, are disrupting and augmenting traditional taxi service using mobile apps to connect for-hire drivers to riders.

2.5 *Improving Safety*

Transportation safety is a critical societal issue and the U.S. DOT's top priority. Among Americans aged 1–34, motor vehicle crashes are the leading cause of death. Americans spend more than 1 million days in the hospital each year as a result of crash injuries. In 2014, motor vehicle crashes resulted in 32,675 deaths.

Fortunately, passenger safety is rapidly improving across all transportation modes. Since 1990, the total number of transportation fatalities has decreased by 27% despite increasing passenger travel on all modes [14]. The rate of fatalities per vehicle mile traveled has decreased by more than 37%. Fatalities in commercial aviation have become exceedingly rare, while fatalities in rail and on waterborne transportation have steadily declined.

These changes are the result of a wide range of factors including changing demographics, improving vehicle technologies, safer infrastructure, increased enforcement, and higher safety standards. In addition to frontal and side air bags that help to prevent injuries in crashes, a number of crash avoidance technologies are now featured in passenger cars and trucks including automated emergency braking systems, lane-departure and forward-collision warning systems, and electronic stability control. In the future, connected and automated vehicle technologies could help drivers avoid a significant portion of the type of vehicle crashes that occur today.

2.6 *Key Policy Options*

These trends point to a number of possible policy responses:

- Increase infrastructure capacity: build new roads, bridges, and other facilities; maintain existing facilities more effectively; use existing facilities more effectively by implementing better designs and technologies; or use some combination of these methods
- Expanding and/or improving the quality of transit services
- Reduce congestion through land use, telework and flextime work schedules, smaller and automated vehicles, and pricing
- Supporting services that promote public transit, biking, and walking.

3 How We Move Things

Efficient freight movement is critical to our nation's economic health and competitiveness. Today, increasing demand for freight and changing freight patterns—strained infrastructure, increased trade, new technologies—present major challenges for government and industry decision makers.

3.1 Growing Freight Demand and International Trade

Over the next 30 years, our growing population and economy will lead to increased demand for freight. FHWA expects truck and rail freight movements to increase by nearly 45% by 2045 and air freight is expected to triple [15]. The volume of imports and exports is expected to double significantly increasing demand for capacity at ports of entry and intermodal facilities.

Total exports and imports of goods reached $4 trillion in 2014, accounting for 23% of U.S. GDP [16]. International trade will grow in importance and put increasing pressure on our ports, border crossings, airports, and intermodal facilities to efficiently move imports and exports to market. Increasing imports and exports could lead to greater congestion at container ports and border crossings, resulting in delayed shipments, increased transportation costs, and intensifying pollution.

Containerized shipments are concentrated in the top 10 to 12 deepwater ports with the requisite infrastructure. Eighty-five percent of America's imported and exported containerized freight flows through just 10 ports [17]. The concentration of shipments in these ports makes our international freight system vulnerable to disruption. If security incidents were to lead to heightened inspection requirements, they could further slow goods movement at ports of entry. Labor disputes and natural disasters also have the potential to impact operations at key ports and disrupt the national economy.

3.2 Impacts of Freight on Our Communities and Environment

Many of the worst freight highway bottlenecks are located on the roads surrounding major urban areas and near coastal container ports and large intermodal terminals, where freight traffic and passenger traffic compete for capacity. Congestion along truck corridors decreases the reliability of truck deliveries affecting the industry's ability to respond to customer requirements and raising the costs of goods. Limited capacity on rail increases freight costs and increased demand for higher value goods is limiting the supply of transportation available for lower value bulk goods, causing delays in passenger rail and vehicle travel.

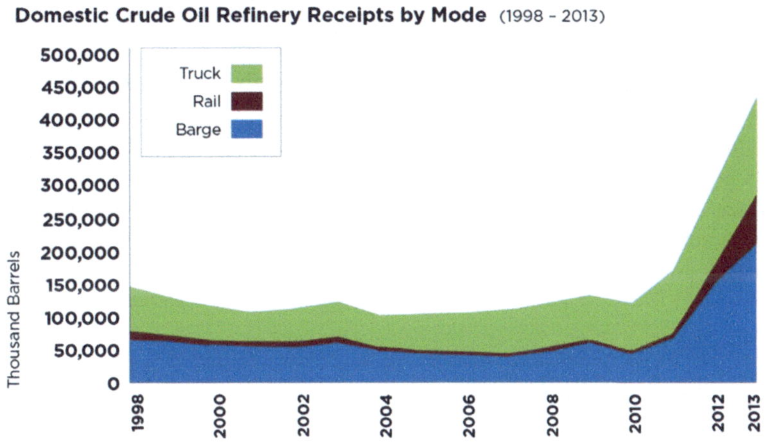

Fig. 2 Domestic crude oil refinery receipts by mode (nonpipeline)

Increasing freight movement may contribute to noise, air pollution, and safety hazards that negatively affect local populations. For example, transportation of oil by rail has increased dramatically since 2008, when less than 1% of oil was transported by rail. Today, more than 10% of all crude oil is now shipped by rail [18]. Recent derailments of tank cars highlight rising safety and environmental risks associated with increasing transportation of oil by rail. As more oil has moved by rail, accidents involving oil spills have increased [19] (Fig. 2).

Freight is also a major source of greenhouse gas emissions. Trucking accounts for about 9% of all highway miles traveled, but it is the source of 22% of all transportation-sector greenhouse gas emissions. Trucking is the single largest contributor to freight-related air pollution nationally. Since 1990, greenhouse gas emissions from trucking have increased five times faster than emissions from passenger travel [20].

3.3 Innovations in Freight

The freight industry is quietly going through a technological revolution as information and communications technologies are applied to optimize global supply chains. Advances in information and communications technologies will improve data collection and analysis capabilities of logistics firms and freight planners, enabling faster and more accurate analysis of freight routes, travel times, and infrastructure capacity.

Automation will transform the freight industry. Automated driver-assist features are improving the safety of new trucks. Connected, partially automated truck technologies could soon enable truck fleets to travel more closely together to improve fuel efficiency and safety and reduce travel times. With respect to ports, the process of transferring containers from ships to docks, trucks, and trains is becoming highly automated, reducing reliance on human operators. On rail, positive train control, is gradually improving the safety of rail operations.

Freight planners and policymakers are increasingly looking to intermodal logistics hubs to improve the efficiency of transfers between multiple freight modes and reduce the negative impacts of freight on local communities. Increased use of practices, such as double-stacking of containers on railway cars, will also increase the capacity and efficiency of the freight system.

3.4 Workforce Challenges

The transportation industry employs nearly 5 million people. Growing demand for transportation coupled with an expected wave of retirements will result in more than 4 million job openings in the transportation sector over the next decade.

As a generation of transportation workers passes into retirement, a more diverse workforce will take its place. Transportation employers need to find ways to attract and retain a younger generation of workers in an evolving and competitive economy. New technologies are also changing the nature of transportation work. The increasing use of information technology and computerized equipment necessitate new, more advanced skill sets in many transportation industry jobs.

3.5 Moving Fuel

Over the past 3 years, domestic oil production has increased by nearly 60%. In 2013, the United States surpassed Saudi Arabia to become the largest oil-producing country in the world [21]. Domestic shale gas production increased from 2 trillion cubic feet in 2007 to 13 trillion cubic feet in 2013. High-value energy products already account for more than 30% of the domestic ton-miles of freight moved each year. Should U.S. energy production continue to grow, it will have profound implications for our transportation system.

Demand for transportation in energy production regions is straining regional road and rail capacity and creating demand for additional pipeline capacity. Industrial traffic—heavy trucks, and drilling and other production equipment—is overwhelming many roads in states such as Pennsylvania, West Virginia, Ohio, and North Dakota. The current pipeline network, oriented toward imports arriving at Gulf Coast refineries, may need to be built up to accommodate growing gas and oil supply and demand.

3.6 *Key Policy Options*

These trends point to a number of possible policy responses:

- Improve freight planning at national, regional, and local levels
- Target policies and investments aimed at resolving freight congestion
- Encourage innovative strategies to address last-mile freight issues.

4 How We Move Better

Advances in data collection, computing, navigation systems, communication and mobile technologies, and robotics have the potential to dramatically change the way we travel and deliver goods and services. Many technology applications developed in other sectors, including satellite-based navigation systems, advanced data collection and analytic tools, smart phones, automation and 3-D printing, have applications in transportation.

4.1 *Nontransportation Technologies Transforming Transportation*

Used in various combinations, rapidly advancing technologies from across sectors are changing travel. For example, GPS-based navigation combined with real-time traffic information and mobile platforms have allowed individuals to optimize their routing and schedule choices. Similar mobile phone applications can allow public agencies and transportation providers to monitor traffic and infrastructure conditions. Sensors on transit, taxi, and truck fleets can monitor where vehicles drive, how fast they are being driven, and when maintenance is required.

These technologies are allowing new methods of payment that are changing how user fees are collected. Transponders for toll roads have become commonplace and open road tolling technologies are now the default option for new toll lanes. Over the past decade many transit systems have shifted to smart card payment systems and some are now experimenting with open payment systems that allow direct fare payment with standard credit and debit cards. Ride sourcing companies are revolutionizing taxi services by changing how the service is paid for.

Advances in robotics and sensor technology are enabling increased automation of transportation tasks, including advanced driver and operator assist features in vehicles, aircraft, trains, and marine craft. Robots are being deployed to improve how transportation agencies perform operations and maintenance for functions such as inspecting aviation radio towers, railroad tracks, bridges, and pipelines. Lidar systems, which use lasers to survey objects, are being used to map and assess conditions across entire road networks.

4.2 Advances in Aviation Technology

In aviation, advances in navigation systems and the advent of unmanned aircraft systems are creating great opportunities and challenges. The Next Generation Air Transportation System (Next Gen) will modernize the air traffic control system, which currently relies on decades-old radar technology. Shifting to satellite-based navigation technology is expected to make aviation safer and more efficient by improving the precision of air traffic control operations.

Widespread commercial use of unmanned aircraft systems (UAS) is imminent. Authorization has recently been granted for limited commercial UAS uses, including Arctic pipeline inspection and moviemaking. Many of the early missions for civilian UAS will be for such purposes as remote sensing and surveillance. Larger vehicles capable of carrying large payloads will enable more uses, such as spraying farm fields, surveying lands, and infrastructure, and providing radio coverage to broad areas as an alternative to satellites. Remotely piloted drone deliveries could be used to provide high-value and urgent cargo to remote and hard-to-reach locations within a decade.

4.3 Connected and Automated Vehicles

We are at the beginning of a period of dramatic change in the capabilities of, and expectations for, the vehicles we drive. Connected and automated vehicle technologies are rapidly advancing and promise great safety and mobility gains.

Vehicle-to-vehicle (V2V) and vehicle-to-infrastructure (V2I) technologies currently under development will allow nearby vehicles to communicate with each other and surrounding infrastructure. V2V technologies are capable of warning drivers of potential collisions that are not visible, such as a stopped vehicle blocked from view, or a moving vehicle at a blind intersection. V2I systems could be used to coordinate signal timing, improving parking information systems, and warn drivers of safety hazards. Data from connected vehicle systems could provide traffic management centers with detailed, real-time information on traffic flow, speeds, and other vehicle conditions, and allow more rapid response to traffic incidents.

Automated vehicles use GPS extensive mapping data, wireless communications, and sensor systems, including cameras, lasers, and radar, to "see" and to navigate through their environment and assist drivers. Partial automation of driving functions, such as lane guidance, active cruise control, and automatic braking, have been available in luxury vehicles for more than a decade, and are becoming more widespread. Automated driving on limited-access highways could be an option on luxury vehicles in several years. Fully automated vehicles, where a driver no longer has to steer or adjust speed, could be commercially available within the next 10–20 years.

Despite the technical feasibility and potential benefits, there are a number of barriers to more widespread adoption of automated vehicles. First and foremost, public agencies will need to determine how to regulate automated vehicles to ensure their safety. New performance standards will be needed to ensure that automation systems are safe and secure. Standards may also be needed to assuage driver privacy concerns regarding the generation, ownership, and sharing of automated vehicle travel data.

4.4 Key Policy Options

These trends point to a number of possible policy responses:

- Address regulatory barriers to deployment of new technology
- Collect and manage data and transition to a data-driven investment system, while protecting individual privacy
- Support research on technological developments and deployment
- Maintain a paramount focus on safety.

5 How We Adapt

Our transportation facilities—our roads, bridges, tunnels, rails, airports, waterways, and ports—are uniquely vulnerable to damage caused by severe storms, rising sea levels, drought, and extremes of temperature associated with climate change.

5.1 Increasingly Vulnerable Infrastructure

Sea level rise and increased frequency of severe weather caused by climate change will have a major impact on infrastructure, as low-lying infrastructure particularly in coastal watersheds becomes increasingly vulnerable to flooding from storm surges [22]. Flooding caused by rising sea levels and storms could force tunnels, subway stations, low-lying roads, rail lines, and marine cargo facilities to be relocated or even abandoned. There are 60,000 miles of coastal roads in America that are exposed to flooding from heavy rain and storm surges. Thirteen of the 47 largest airports in America are within reach of moderate-to-high storm surges, including all three major New York area airports [23]. Increased frequency of hurricanes could have a major impact on our nation's ports. In fact, 7 of the 10 largest ports (by tonnage) are located on the Gulf Coast.

5.2 *Limiting Transportation Emissions*

Transportation is a major contributor to carbon emissions that cause climate change. In 2013, transportation sources directly accounted for 27% of total U.S. greenhouse gas emissions [20]. Next to the generation of electricity, which contributes 31% of greenhouse gases, the transportation sector is the second largest source in the United States. However, emerging technologies and newly established policies are helping to stem the growth of transportation emissions: new types of fuels that promise to dramatically reduce emissions for automobiles, trains, planes, and vessels are emerging and higher fuel efficiency standards for cars and trucks are helping to reduce fuel consumption.

Corporate Average Fuel Economy (CAFE) standards have had a significant positive impact on the fuel efficiency of light-duty vehicles. CAFE standards were updated in 2011 and they are now scheduled to rise from 32.7 miles per gallon in 2012 to 47.2 miles per gallon in 2025. Greenhouse gas and fuel efficiency standards have recently been expanded to cover medium- and heavy-duty trucks. By 2018, new combination trucks will be required to achieve a 20% reduction in fuel consumption and greenhouse gas emissions.

5.3 *Supporting Hybrid, Electric, and Alternative Fuel Vehicles*

In recent years, a number of "alternative" fuels, many of which emit fewer pollutants than do petroleum-based fuels, have emerged as economically competitive alternatives. These include biomass fuels, natural gas, and hydrogen.

Hybrid vehicles now account for approximately 3% of all vehicles sold and the sale of plug-in electric vehicles is increasing rapidly [24]. Hybrid vehicles use regenerative braking and an internal combustion engine to charge a battery and are significantly more fuel efficient than are nonhybrid vehicles. Electric vehicle motors are more efficient than internal combustion engines and produce no emissions at the tailpipe. While improvements in battery storage capacity will almost certainly expand the market for electric vehicles, plug-in electric vehicles require investments in recharging infrastructure to be a widely viable transportation mode. Transportation agencies at all levels of government are acting to accelerate the adoption of electric vehicles by supporting research, infrastructure, and fleet and consumer adoption (Fig. 3).

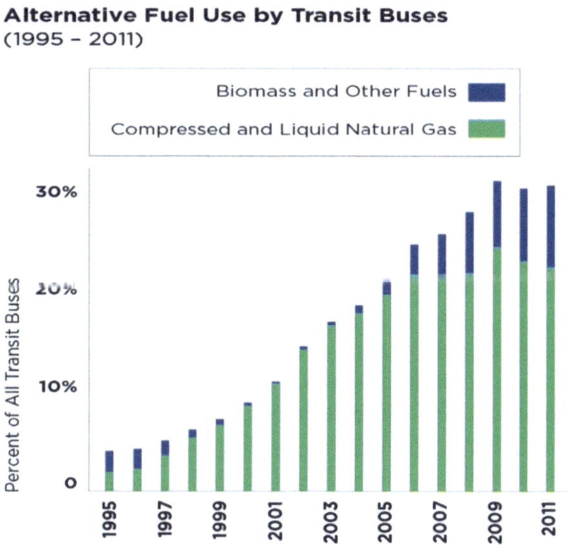

Fig. 3 Alternative fuel transit buses as percent of fleet (1995–2011)

5.4 Key Policy Options

These trends point to a number of possible policy responses:

- Reduce transportation emissions by improving fuel efficiency and increasing the use of alternative, cleaner fuels
- Align costs and incentives to encourage development patterns, and research into new technologies, that can aid in reducing greenhouse gas emissions and energy use
- Design and build better infrastructure that is more resilient to anticipated climate change effects
- Avoid developments in vulnerable locations.

6 How We Align Decisions and Dollars

As the transportation system has grown and become more complex, transportation decision-making has become more difficult. Greater coordination across stakeholder groups and generation of sufficient revenues to bolster the increase in infrastructure costs are imperative.

6.1 Coordinating Transportation Decisions

Federal support for transportation modes has created constituents with a distinct interest in preserving federal support, even while some have promoted more devolution of authority to state and local levels. With the increased local responsibility over transportation decision-making, the number of local governments and independent authorities has grown. Responsibilities for planning, financing, permitting, constructing, and operating infrastructure have become more and more fragmented and it has become increasingly difficult to reconcile local goals while ensuring transportation investments are efficient at regional and national levels. The necessary process of developing a consensus among numerous transportation agencies, local governments, and community stakeholder groups with varying objectives often leads to delays and inefficiencies in delivering projects.

6.2 Constrained Transportation Revenues

Since the mid-1990s, inflation has eroded the purchasing power of federal transportation funds by nearly 40% and the balances of most dedicated transportation trust funds have declined as outlays have exceeded revenues. Federal fuel taxes per gallon have not been increased since 1993. Compounding surface transportation revenue shortfalls is the slow growth in vehicle travel over the past decade and increasing fuel efficiency that have led to declining fuel consumption (Fig. 4).

In 2000, user charges accounted for more than 95% of all federal highway revenues. By 2010, less than half of all federal highway revenues were derived from user charges as a result of the use of General Funds to cover Highway Trust Fund

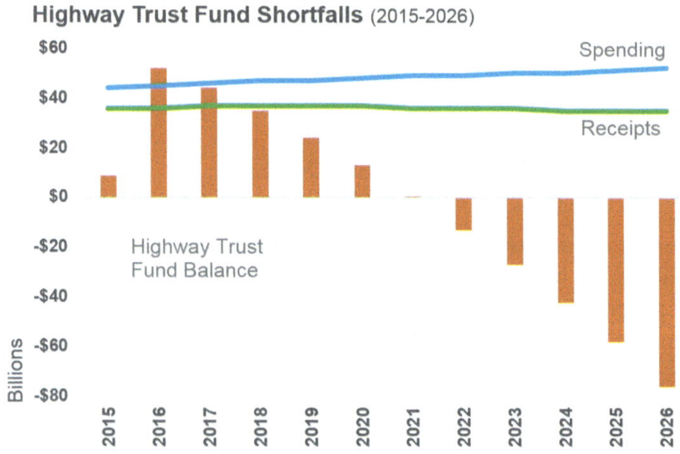

Fig. 4 Highway trust fund balance, 2015–2026 (CBO projections)

shortfalls. The portion of state highway revenues derived from user charges also declined—from 74% in 2000 to 56% in 2010 [25]. In 2010, public transit systems recovered 38% of operating expenditures from system-generated revenues, down from 46% in 2000 [25].

As public revenues have become increasingly scarce relative to the costs of maintaining, operating, and expanding infrastructure assets, public agencies at all levels have had to find ways to do more with less and, in some cases, scale back services. Adjusted for inflation, federal, state, and local spending on surface transportation fell by $29 billion, or 12%, between 2002 and 2012. Often, funding scarcity constrains options in ways that are ultimately detrimental and inefficient. For example, sometimes public agencies have to defer maintenance such that the ultimate costs associated with repair increase.

Funding limitations have come at a time of rising need. Underinvestment in transportation infrastructure has created a massive maintenance backlog, which has increased maintenance costs and reduced transportation system performance. Of the 607,000 public road bridges one-quarter in our transportation system are not meeting today's standards [26]. Public transit systems in our nation face an estimated $86 billion backlog in preservation investments [25]. Aging infrastructure affects waterways as well—over half of the system of locks and dams are over 50 years of age and out of service.

In response, some states have raised gas taxes. Other states, such as Pennsylvania and Virginia, have transitioned from a traditional motor fuels tax levied as a flat amount per gallon to a sales tax at the wholesale level. Others have dedicated a portion of the state sales tax to transportation funding or have raised license, registration, and excise fees on vehicles. Local governments have also demonstrated success raising taxes for transportation, often in exchange for a dedicated program of projects. A number of states have turned to tolling and priced express lanes, in particular, to deliver projects that expand roadway capacity while managing congestion.

6.3 Innovative Financing

As interest rates have reached historic lows, many states have increased their use of debt to fund transportation projects. States' use of debt to fund highway projects tripled over the past decade. Beyond the municipal bond market, innovative financing for infrastructure investment is becoming increasingly important as public budgets continue to tighten at all levels of government. Federal credit assistance programs, such as the Transportation Infrastructure Finance and Innovation Act (TIFIA) and the Railroad Rehabilitation and Improvement Financing Program (RRIF), are playing a particularly important role in supporting innovative finance for projects across the country.

Some state and local governments have entered into public–private partnerships (P3s) to finance, construct, and operate transportation infrastructure. P3s can provide an alternative source of financing that can accelerate projects and save taxpayers money when used appropriately under the right circumstances. However, because P3s typically take more resources to evaluate and procure than conventional projects, and private financing costs are often higher than the costs of public financing, P3s are only appropriate for complex, high-risk projects. Consequently, P3 investments account for only a small portion of overall transportation investments. Between 2007 and 2013, $22.7 billion of public and private funds were invested in P3s, about 2% of overall capital investment in the nation's highways during that same period [27].

6.4 Improving Efficiency

Transportation agencies across all modes have adopted new ways to use limited resource more efficiently using data to inform decisions and increase accountability. Over the past 30 years, public agencies have developed increasingly sophisticated measures to guide asset management and safety decisions. However, few state agencies have found effective ways to accurately measure how transportation investments can affect outcomes in critical goal areas such as economic development and environmental sustainability.

Competitive, multimodal, discretionary federal transportation programs that incentivize performance have also emerged in recent years. The Transportation Investment Generating Economic Recovery, or TIGER Program, is a competitive grant program that was initiated as part of ARRA in 2009. TIGER grants fund planning and capital projects across different surface transportation modes.

6.5 Key Policy Options

These trends point to a number of possible policy responses:

- Ensure adequate revenues to address critical needs, through existing taxes, new excise taxes, user fees, tolls, congestion pricing, VMT fees, or other funding mechanisms
- Reduce spending to match revenues
- Prioritize investments based on performance outcomes
- Improve coordination of investment between states, MPOs, and local governments.

7 Conclusion

Beyond Traffic analyzes the latest data and trends shaping transportation in order to objectively frame critical policy choices that need to be made. Our goal is to understand how to prepare our policies and our institutions to lead us to the best possible future. *Beyond Traffic* provides Americans with a common basis of fact for a larger national discussion about the future of transportation.

Beyond Traffic elaborates a basic set of principles to guide us in making sound transportation policy decisions. These are common sense, nonpartisan principles that can guide public agency transportation decisions at all levels:

1. Recognize the perilous forces that threaten our transportation system and address those forces honestly, transparently, and in a fact-based manner framed by data and analysis.
2. Develop new mechanisms to adapt to changing circumstances and advancing technologies with speed and flexibility.
3. Reevaluate and simplify the roles of various levels of government and engage the private sector to foster collaborative solutions and partnerships to achieve common goals.
4. Assure adequate resources to preserve, sustain, and build transportation assets and support options for funding and/or financing new investments in twenty-first century assets.
5. Advance balanced and sustainable economic growth without exacerbating income inequality or social division.
6. Support technological innovation, while ensuring the preeminence of safety, security, and privacy.

The challenges that face our transportation system are daunting, but we cannot give into fatalism. We can make choices based on fact and guided by principle. By analyzing the long-term trends affecting our transportation system and describing the implications of those trends, *Beyond Traffic* provides a framework by which Americans can debate those choices and make the decisions we need to shape a better future.

References

1. U.S. Census Bureau, 2014 National Population Projections: Summary Table 1. www.census.gov/population/projections/data/national/2014/summarytables.html (2015). Accessed 20 June 2016
2. Mather, M., Pollard, K., Jacobsen, L.A.: First results from the 2010 census. www.prb.org/pdf11/reports-on-america-2010-census.pdf (2011). Accessed 20 June 2016
3. U.S. Census Bureau, Population Estimates. www.census.gov/popest/data/historical/index.html (2015). Accessed 20 June 2016
4. FHWA, Forecast of Vehicle Miles Traveled, May 2015. www.fhwa.dot.gov/policyinformation/tables/vmt/vmt_forecast_sum.pdf. Accessed 20 June 2016

5. FHWA, Traffic Volume Trends. www.fhwa.dot.gov/policyinformation/travel_monitoring/tvt.cfm (2015). Accessed 20 June 2016
6. FHWA, Summary of Travel Trends: 2009 National Household Travel Survey. http://nhts.ornl.gov/2009/pub/stt.pdf (2011). Accessed 20 June 2016
7. U.S. Census Bureau, Means of Transportation to Work by Selected Characteristics, 2010–2014. http://factfinder.census.gov/faces/tableservices/jsf/pages/productview.xhtml?pid=ACS_14_5YR_S0802&prodType=table (2014). Accessed 20 June 2016
8. FHWA, The Next Generation of Travel: Research, Analysis and Scenario Development. www.fhwa.dot.gov/policy/otps/nextgen_finalreport.cfm (2013). Accessed June 20, 2016
9. Fry, R.: More Millennials Living With Family Despite Improved Job Market, July 2015
10. U.S. Census Bureau: Quarterly Retail E-Commerce Sales, 3rd Quarter 2015, November 2015. www.census.gov/retail/mrts/www/data/pdf/ec_current.pdf. Accessed 20 June 2016
11. Jones, J.M.: In U.S. Telecommuting for Work Climbs to 37%, 19 Aug 2015. www.gallup.com/poll/184649/telecommuting-work-climbs.aspx. Accessed 20 June 216
12. Shaheen, S., Cohen, A.: Innovative Mobility Carsharing Outlook: Summer 2015, July 2015. http://innovativemobility.org/wp-content/uploads/2015/07/Summer-2015-Carsharing-Outlook_Final-1.pdf. Accessed 20 June 2016
13. BTS, National Transportation Statistics. www.rita.dot.gov/bts/sites/rita.dot.gov.bts/files/publications/national_transportation_statistics/index.html (2015). Accessed June 20, 2016
14. FHWA, Freight Facts and Figures, January 2013. http://ops.fhwa.dot.gov/freight/freight_analysis/nat_freight_stats/docs/13factsfigures/pdfs/fff2013_highres.pdf. Accessed 20 June 2016
15. U.S. Census Bureau, U.S. International Trade in Goods and Services, 4 Nov 2014. www.census.gov/foreign-trade/Press-Release/2014pr/09/ft900.pdf. Accessed 20 June 2016
16. American Association of Port Authorities, Western Hemisphere Port TEU Container Volumes, 1980–2013. www.aapa-ports.org/Industry/content.cfm?ItemNumber=900 (2013). Accessed 20 June 2016
17. EIA, Annual Energy Outlook 2014. www.eia.gov/forecasts/aeo/pdf/0383(2014).pdf (2014). Accessed 20 June 2016
18. Begos, K., Fahey, J.: Deadly Side Effect to Fracking Boom, 5 May 2014. http://bigstory.ap.org/article/ap-impact-deadly-side-effect-fracking-boom-0. Accessed 20 June 2016
19. EPA, Inventory of U.S. Greenhouse Gas Emissions and Sinks: 1990–2012, April 2014. http://www.epa.gov/climatechange/Downloads/ghgemissions/US-GHG-Inventory-2014-Main-Text.pdf. Accessed 20 June 2016
20. EIA: Total Oil Supply Data. www.eia.gov/cfapps/ipdbproject/IEDIndex3.cfm?tid=5&pid=53&aid=1 (2015). Accessed 20 June 2016
21. TRB: Potential Impacts of Climate Change on U.S. Transportation (2008)
22. National Climate Assessment and Development Advisory Committee, The Third National Climate Assessment (2014)
23. DOE: U.S. HEV Sales by Model. www.afdc.energy.gov/data (2015). Accessed 20 June 2016
24. FHWA and FTA: Status of the Nation's Highways, Bridges and Transit: Conditions and Performance. www.fhwa.dot.gov/policy/2013cpr/pdfs/cp2013.pdf (2013). Accessed 20 June 2016
25. FHWA: National Bridge Inventory: Deficient Bridges by State and Highway System. www.fhwa.dot.gov/bridge/nbi/no10/defbr14.cfm (2014). Accessed 20 June 2016
26. The White House: An Economic Analysis of Transportation Infrastructure Investment, July 2014
27. FHWA: Highway Statistics 2013. www.fhwa.dot.gov/policyinformation/statistics/2013 (2013). Accessed 20 June 2016
28. Texas A&M Transportation Institute and INRIX, 2015 Urban Mobility Scorecard, August 2015. http://d2dtl5nnlpfr0r.cloudfront.net/tti.tamu.edu/documents/mobility-scorecard-2015-wappx.pdf. Accessed 20 June 2016

Creating an Innovative Mobility Ecosystem for Urban Planning Areas

Dewan Masud Karim

Abstract Urban planning, infrastructure design, and mobility policy are up against a tough system-level challenge: the rapid adaptation of shared mobility. The new mobility is destabilizing the current auto-oriented transportation paradigm, and gradually moving toward a new mobility ecosystem. In order to capture the potential and create shared infrastructure, an innovative mobility planning model based on a scientific approach was developed to identify context-sensitive area solutions and the scaling of the proposed ecosystem for short- and long-term horizons. The aim of this model is to build capacities and competencies, enable municipal authority and system planners to quantify the scale and cost, and accurately model the potential impact and benefits of various innovative mobility strategies.

Keywords Innovative mobility ecosystem · City planning · Connected multimodal · Collaborative implementation

1 Introduction: A Mobility Megatrend

Rapid adaptation of sustainable mobility, particularly smart technology based on shared/on-demand service, is changing the current auto-oriented paradigm. Shifts in lifestyle, an engaging planning culture, demographic changes, and the rise of the concept of "Mobility-as-a-Service" [1] are paving the way for a new mobility ecosystem in urban multimodal planning while replacing the demand for traditional oversized, expensive, and complex physical infrastructure. These changes started to appear in public sector policies acknowledging shared mobility and smart

D.M. Karim (✉)
City of Toronto, City Planning, 5100 Yonge Street,
Toronto, ON M2N 5V7, Canada
e-mail: dkarim@toronto.ca

G. Meyer and S. Shaheen (eds.), *Disrupting Mobility*,
Lecture Notes in Mobility, DOI 10.1007/978-3-319-51602-8_2

technologies. However, innovative mobility planning in the public sector is struggling to adapt to the rapid advancement of the shared economy. With fragmented urban mobility management and lack of system wide assessment, a gap in collaboration in the planning process is developing as planners work to supply an appropriate level of vehicle infrastructure while adding a multimodal facility as an "extra feature." An imbalance is being created between the growing demand for sustainable mobility and the oversupply of vehicle assets. Absence of a natural balance is holding back change. Seamless multimodal mobility that adapts from "delivering transportation" to "delivering solutions" could unleash the full potential of the emerging innovative mobility model [2]. Recognizing the unprecedented growth of shared mobility systems in the last decade [3], a new innovative mobility master planning process is envisioned in this model. The process incorporates a quantitative mobility analysis and infrastructure assessment process, and addresses people-oriented needs while redesigning scarce public spaces and the mobility delivery system.

The rise of smart and connected mobility [4] along with prevailing social and economic changes have created dramatic structural and societal consequences to the current mobility model, culminating in direct conflicts and strains between public sector regulations and service providers. Around the world, city governments are increasingly facing pressure to change their current approach to public engagement and policymaking. Financial uncertainty in the public sector [3], declining road reinvestments [5], and potential annihilation of the traditional mobility industry are a new reality. Reallocation of public space [6] and parking [7] for efficient modes are facing steep challenges from traditional inefficient system entities. Intensifying remaining developable urban lands [8] and replacing surface parking spaces with infill developments [9] have emerged as leading city building strategies in the last decade to protect the last remaining green spaces. The breadth and depth of these changes herald the reshaping of public policy and the governance structure, introducing shared goals with service providers to form an approach toward greater efforts in global sustainability while striving for meaningful prosperity [10] and maintaining safe space within a city's operating boundary [11].

Despite equity and wage concerns and the potential threat to public transit and active transport, shared mobility brings social and economic benefits in addition to environmental gains. This includes providing access to those who cannot afford to own or operate a vehicle, providing opportunities for extra income using excess capacity, offering more choices and connections to public transit services [12], and reducing parking demand and decreasing traffic congestion intensity to free up land, allocating up to 20–30% of land for new shared mobility services [13, 14]. If innovative mobility policies and implementation strategies/incentives are developed around low-carbon mobility as a "core service," new technologies will bring sustainable benefits to the community, environment, and economic progress.

Despite growing attention to innovative mobility and continued progress in disruptive technology, there is a surprising dearth of literature or research on a quantitative mobility planning approach, practical public policymaking and, more specifically, the scalable impact on traditional mobility planning, management, and governance. Planning assessment models generally exclude shared mobility services. The aim of an innovative mobility planning process is to reinvent multimodal mobility assessment with innovative options and smart technologies. Recognizing the impact of shared mobility options on city policies and planning practices, the model reinforces the low-carbon option while addressing environmental and health benefits, and social equity in mobility planning for all users. Bringing shared mobility research findings into implementation tools and repurposing land and reallocating space to sustainable and shared mobility facilities, the innovative approach ensures new shared and connected technologies do not become just another platform of exclusion, and avoid the disproportionate burden vehicle throughput placed on unban quality of life.

2 Developing a Framework for the Mobility Ecosystem Model

Driven by the prospects of disruptive innovation, future mobility planning will improve the quality of life of its residents by forging a positive relationship between technology, business and the environment [15]. The new mobility model will assess, measure, and integrate every possible element of the mobility ecosystem. To achieve this objective, the development of this model incorporates several pioneer and recent concepts in mobility system and city building approaches. The conceptual mobility ecosystem framework was built on the premise that physical space constraints and economic and resource constraints will increasingly set the "safe operating limits" of a city's carrying capacity, i.e., in recognition of the basic philosophy of "planetary boundary" [11, 16]. A Dutch model of Spatial Planning and Design [17] is reviewed and modified to reflect the complex layers of the ecosystem, the interrelation between modes of travel, and the fundamentals of the natural environment [18]. In order to capture the transformative power of new transportation technologies and social trends, the SMART model was envisioned to transform the automotive industry's business toward sustainable transportation approach at multi-scale and dynamic coupled systems [19]. Finally, demand management strategy has become the focus of recent policy discussion [20]. Three fundamental strategies to reduce emissions from the transportation sector, collectively known as the Avoid-Shift-Improve approach [21], are gradually being commonly accepted due to the prospect of the impossibility of future road improvements satisfying unlimited traffic growth. Combining automotive and new mobility service integrators into a multimodal model, a set of basic principles associated with the limitations of mobility infrastructure and service was integrated into the mobility ecosystem model (Fig. 1).

Identifying future aspects of the symbiotic relationship between six fundamental interactive elements in a mobility ecosystem, the proposed innovative mobility planning model (see Fig. 2) envisions a novel urban morphology, shapes a new

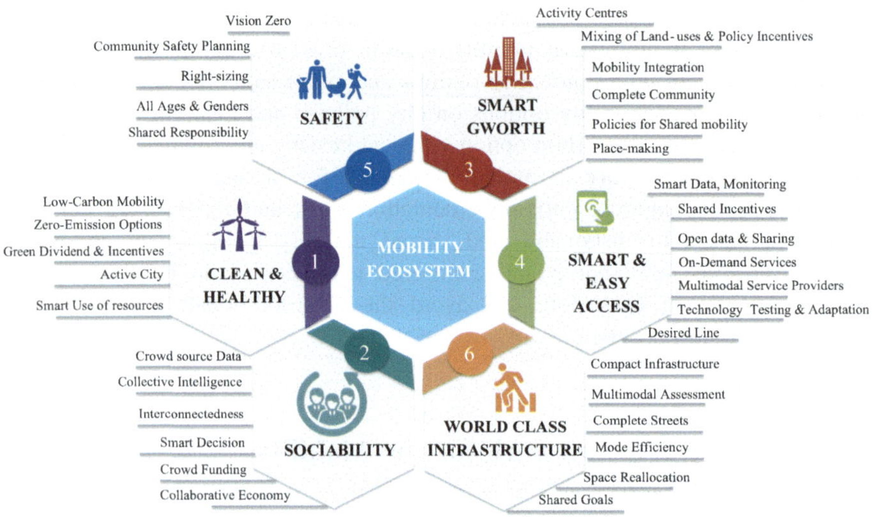

Fig. 1 Basic concepts of mobility ecosystem principles (adapted from Ohta [20])

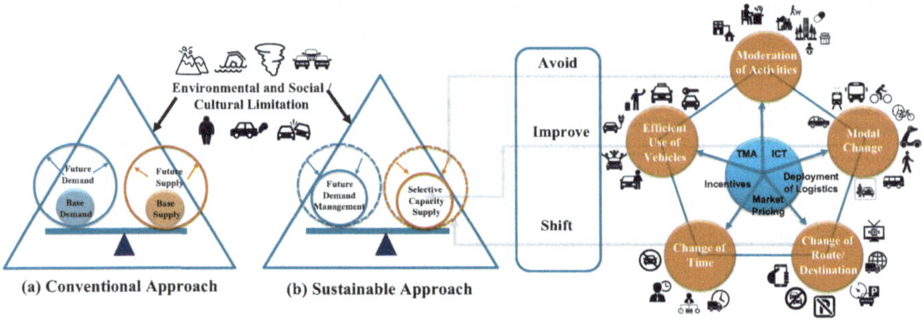

Fig. 2 Conceptual framework of innovative mobility ecosystem for urban planning areas

experience of urban space, and turns unexplored assets into an ecosystem of vibrant, sustainable innovation. The six fundamentals elements (and associated planning policies) of the mobility ecosystem are: (1) healthy environment, low-carbon footprint and clean energy; (2) smart growth principles linking the right mobility mix; (3) sociability and livability by enhancing social capital and innovation; (4) smart and easy access to all types of mobility services; (5) sustainable safety by reducing crash risk and severity; and (6) world class infrastructure that prioritizes sustainable and shared uses.

Through a rigorous literature review and background analysis, the following sections describe the development of the key principles, and the factors that limit each fundamental element and guide the implementation of smart technologies in the quantitative process of mobility planning.

2.1 Clean and Healthy Cities: The Challenges of Urban Mobility and Smart City Building

The demand for physical space for new human development generates driving alone commuting which comes with unused excess capacity and unsustainable use of limited natural resources. In spite of immense technological development and progress, our economies and societies still fundamentally depend on ecosystems to provide us with a hospitable climate, clean water, food, fibers and numerous other goods and services. Two planetary processes, fossil fuel emissions by private vehicles and auto-oriented sprawling land use, are gradually pushing the safe thresholds of "planetary boundaries" [11]. By 2050, urban mobility systems will use 17.3% of the planet's bio capacities, five times more than they did in 1990 [2]. Following this global trend, the transportation sector in the City of Toronto has grown exponentially to become the largest source of green-house-gas (GHG) emissions (41%, excluding rail, plane and boat) [22]. Linking mobility patterns and greenhouse gas emissions, a Greater Toronto Area study [23] concluded that most emissions are caused by "extreme commuters," people who work in the old City of Toronto, but live in the outer suburbs and commute by private vehicle. Thus, unlike last century's city planning, the focus of this new mobility model is to create a low-carbon "urban ecosystem" [17] by mixing land-use with appropriate density, addressing the depletion of natural and financial resources, and continuing to manage sustainable growth within "planetary boundaries" that will shift mobility patterns to achieve the target of GHG emissions.

2.2 Sociability: Changing Socioeconomic Structure and Travel Patterns

The emergence of a new social order and collaborative consumption is driving our society in exciting new directions for future mobility, and reshaping almost every aspect of society. The rise in the importance of "sociability" (instead of efficiency) and citizens' environmental preferences appear to be important drivers in the pursuit of specific emission measures and the adaption of climate plans [24].

Firstly, a new collaborative economy is disrupting the traditional ownership-based mobility paradigm. Highlighting the influence of a new model of organic economic growth and ecological necessity [10], evidence is emerging that beyond a certain point, growth does not increase human well-being and that the ultimate solution lies with new sustainable mobility investment policies (such as the rate of return on investment should be lower, around 1.5%) [25] to achieve "Our Common Future" [26]. New collaborations are emerging between political platforms and economists. These economic parameters and social changes are reflected in the proposed model in the estimation of life cycle assessment [27, 28], environmental benefits [29] and social impact to quantify urban livability.

Secondly, demographic changes have profound impact on urban mobility. For instance, roughly 60% people live and work within Toronto downtown. Single women have become an important share (roughly 30%) in the real estate market and most of them walk, cycle or use transit to daily destinations. A global trend of stagnated vehicle growth [31] is reflected in changing travel patterns. In Toronto, for instance, only half of people use a vehicle and over one-third use transit. Cycling (current mode share 2.2%) has emerged as the fastest growing of transportation modes (annual growth 7.5%) followed by walking (1.5%) [30]. As a result, vehicle mode share has been falling 0.5% annually while transit share has been increasing at same rate. These findings lay the foundation of a future modal share pattern that reduces the number and length of vehicle trips opening the door for appropriate density and diversity of land-use in urban centers and corridors.

Thirdly, facing a changing mobility landscape and affording people more choices, the automotive industry (Original Equipment Manufacturers, OEM) is forced to rethink the diversity in their business models and is gradually moving toward multimodal urban mobility solutions [32]. OEMs are introducing car sharing and ebike with a major focus on a "shared transit" system. Technology companies and new players are entering the market as service integrators. Reflecting the changes in the mobility industry, the proposed mobility model develops the capacity of all the possible elements using industry parameters to capture the appropriate level of local and technological context and determine an expected level of shared mobility usage.

Finally, the power of collaboration and sharing through digital technologies is helping to transform consumption pattern, design goods to last longer while reducing production, and move toward distributed, connected communities that will be control general people through peer trust [33]. Unprecedented global urbanization is recreating the city as economic center, giving rise to increasing online and immediate delivery services that replace the need for trips and lead to an increased number of shorter trips. These socioeconomic variables and consumption patterns are reflected in the proposed model as key indicators for estimating the scale of new mobility demand.

2.3 Smart Growth Principles Linking the Right Mobility Mix

Smart growth is a set of principles that promote more compact and mixed development, and create sustainable mobility. Smart growth reduces urban sprawl, parking demand and vehicle pollution, and maximizes the effectiveness of investment. Smart growth is often confused with 'density' and bad development, causing angst with local communities and local government. The reality is that when done well, with 'appropriate' density and mix, development based on smart growth principles can result in several economic, environmental and social benefits [34].

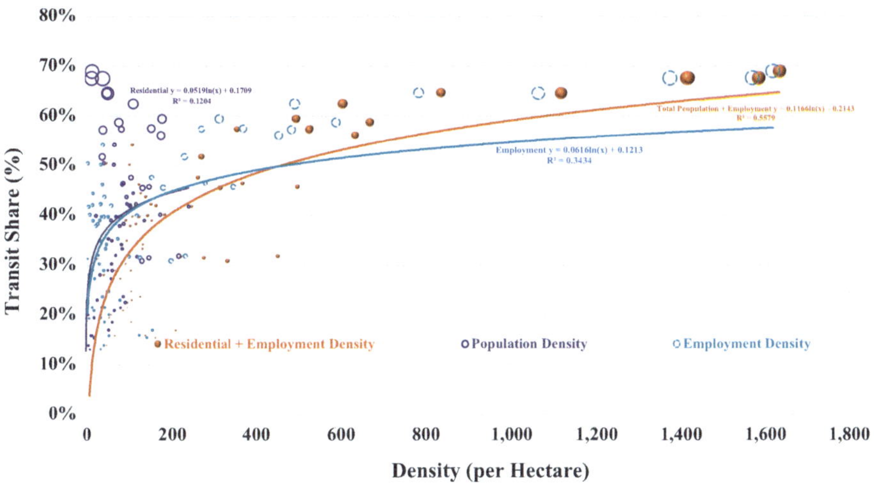

Fig. 3 Integration and limiting boundary of land use and transit relation [30]

The appropriate density varies with the area and overall context. In-depth analysis using Toronto data (within 500 m of subway stations) reveals benefits of density diminish beyond a "density sweet spot" [30]. While the minimum subway density threshold is usually 100 population and employment per hectare, the optimum transit share (i.e., 40–50% transit) is achieved when density is around 200–450. The downtown core, with a density exceeding 450, leads to a marginal increase in transit mode share (Fig. 3). While employment is the key ingredient of maximum transit usage, the appropriate share of diverse land-use (25–40%) is critical to providing access to daily needs. The reason behind low-performing subway stations (around 58%) are vehicle focus retail or employment usage, poor physical and digital connectivity or lack of real-time information, and pointed density around rapid transit stations [35]. Optimum limits of density also determine emission outcomes [36]. Total on-road CO_2 increases rapidly with population density below 1650 persons per square kilometer while per capita emissions decline as density rises (1650–3500 persons per square kilometer) and emissions begin to rise again as density exceeds 4000 persons per square kilometer. These boundaries set the limits of mode split, appropriate density, and the extent of diversity of land-use that maximizes self-contained trips.

The supply of parking, an intersection between mobility and land-use, entirely depends on minimum parking requirements that fail to account for complex relationships between parking supply and demand. Minimum parking requirements in cities are a likely cause of increased driving among residents and employees and higher cost of housing [37]. To the contrary, underground parking remains half empty whereas on-street parking is close to capacity in Toronto's major urban centers. This indicates a shift in land-use and changing demographics that prefer easily accessible parking spaces. Innovative mobility, particularly on-demand/shared systems, has profound implications for a city's parking requirements and may enable inhabitants to live without a car. Although municipalities update their parking requirement to reflect high-density uses, the impact of sustainable and shared mobility and market-based pricing on parking demand is largely unknown. This model aims to quantify the demand for space and parking of sustainable and technology users.

2.4 *Smart and Easy Access*

Unlike vehicle usage patterns, the effective use of sustainable and shared/on-demand modes depends on multiple layers of accessibility features. Firstly, pedestrian movement and social activities dictated by a '400 m rule' of pedestrian shed [38] and optimal street patterns with ideal connectivity [39] influence access time and shorten the distance to mobility service locations. Secondly, easy access from neighbourhoods through street networks to transit stops/station [40] and other service locations [41] optimize social benefits of mobility schemes. Thirdly, digital technologies with real-time information on trains, buses and on-demand/shared service availability can shrink the "reliability buffer" [42], the extra time a traveler builds into a trip to account for possible delays, and significantly reduce the "time window" improving quality of service. Research indicates that providing people with access to real-time transit information results in 15% less time spent waiting at bus stops [43], increases average daily ridership by 2% [44], and results in $5 million per year in additional fare revenue [45] with total potential savings up to $60B [15]. If transit wait time was eliminated using technology, the urban mobility score would be doubled [46]. The proposed model applies acceptable physical access distance, connectivity and access measures, and quality of service standards to redesign street network, minimize distances and optimize connections to sustainable and shared mobility service locations.

2.5 *Safety in the Planning Process*

Traffic safety plays a central role in increasing active transportation and connecting shared mobility modes to conventional public transit. However, traffic fatalities are traditionally framed as individual and mechanical failures rather than systematic flaws in mobility planning, urban and street design [47]. More recently though, a planning focused safety approach has emerged. In the 1980s, a Dutch safety model commonly known as the "sustainable safety traffic system" developed several quantitative targets to reduce the number and severity of collisions through better-integrated community and street planning [48]. Scandinavian and East Asian nations advanced the Dutch concept, treating collisions as a preventable disease.

Beneath these fundamental safety principles, evidence points to two root causes of traffic safety problems: longer driving distances per driver are a strong predictor of crashes [49]; and the combination of wider streets/intersections with wide lanes [50, 51] and unwarranted/unused right-turn lanes with an island [52] lead to higher number of crashes when higher proportion of seniors and greater number of sustainable and shared mobility users. Oversized infrastructure with higher design speeds tends to reduce interaction between street users and which ultimately increase collision risks [53]. Policies that work toward the systematic reduction of vehicle traffic while increasing pedestrian and cycling usage [54] and redistributing space and rescaling urban infrastructure [55] have emerged as important safety solutions. Recent "Complete Street" design approaches improve overall safety and create an opportunity for multimodal mobility [56]. The proposed model uses a scientific approach based on safety performance functions [57] to investigate the root cause of current safety issues. The model identifies several safety boundaries by using forecasts of multimodal trips and corresponding reduction of vehicle traffic due to shared mobility services, i.e., a combination of "sustainable safety" [48], nonlinear risk behavior [54] and "community safety planning" [58] concepts.

2.6 *World Class Infrastructure*

Best practices of sustainable mobility policies are currently shifting from the concept of "predict and provide" to "optimality and sustainability." Creating streets as places for trip destinations flows from the "Link" and "Place" concept [59, 60], and identifies context-sensitive land-use [61]. A detailed planning practice that unifies the role of different professionals and provides guidance in developing a comprehensive two-dimensional street classification has recently been developed such as 30 by 30 street downsizing strategy (30 kmph speed and 30 m right-of-way) that aligns with compact and dense city living ideas [60]. The world class infrastructure ideas in this model were developed from scientific evidence and creating street and intersections at all levels for safe human interaction. Traffic engineering solutions have kept adding lanes to reduce vehicle delays, but, limitations to capacity have to be recognized. Expanding intersections above a certain size has proven to be an expensive, ineffective and short-lived solution to traffic congestion problems [62]. Secondly, too many lanes lead to increased traffic volume, and increased distances traveled, leading to an increase in collision frequency [49, 57]. It is clear that road widening carries the seeds of a future decline in a city's livability. Thirdly, the system faces economic, political and environmental challenges including the question of an scale and size for transportation infrastructure [55]. Ignoring these challenges could lead to system failure if the system breaks down due to the implications of events such as an aging population, extreme weather due to climate change, or infrastructure that is unused due to social and technological changes [63]. In the model, the issues discussed here form the basis for the maximum size of infrastructure, while assessing the future demand for new mobility systems, a shared mobility modes demand that is traditionally ignored. This approach prevents frequent system breakdown such as excessive delays, crash-prone clusters environmental degradation and the funding trap of maintaining oversized "complete street" infrastructure.

3 Formulation of a Mobility Ecosystem in the City Planning Process

With the uncertainty surrounding new mobility systems and their impact on sustainable and shared mobility, the proposed planning model aims to answer two questions which is generally not considered in the traditional mobility planning model: (1) what quantitative process in mobility planning can take into account the optimum size of infrastructure or services while maximizing social, environmental and economic well-being of inhabitants?; and (2) what policies can create a mobility ecosystem that keeps "sustainable mobility as core" service and provides incentives to integrate innovative mobility options through the rethinking of land-use strategies and the reallocation of public space or assets toward space and time efficient modes? To establish a link between two objectives, this section formulates the path to quantification and integration of all ecosystem elements into the city's mobility planning process while identifying public policies to achieve shared goals.

3.1 Process and Resources for Mobility Ecosystem Planning Model

Truly smart mobility planning only emerges if inhabitants participate through a transparent process that includes, for example, networking capabilities that link inhabitants to government policy making, smart open crowdsource data, and an appropriate mobility assessment and implementation process. The resulting smart system offers a sociable and more efficient system without imposing order from city planners or traffic engineers. Figure 4 shows the layers of mobility ecosystem planning model and development process. The model first lays out an overall path of transformation to a future ecosystem that maximizes the social, environmental and economic well-being of users. The development of basic principles including limiting boundaries or constraints and interdependency between six fundamentals elements is performed. The second stage establishes the link between the policy variables of land use and mobility options while formulating the multimodal

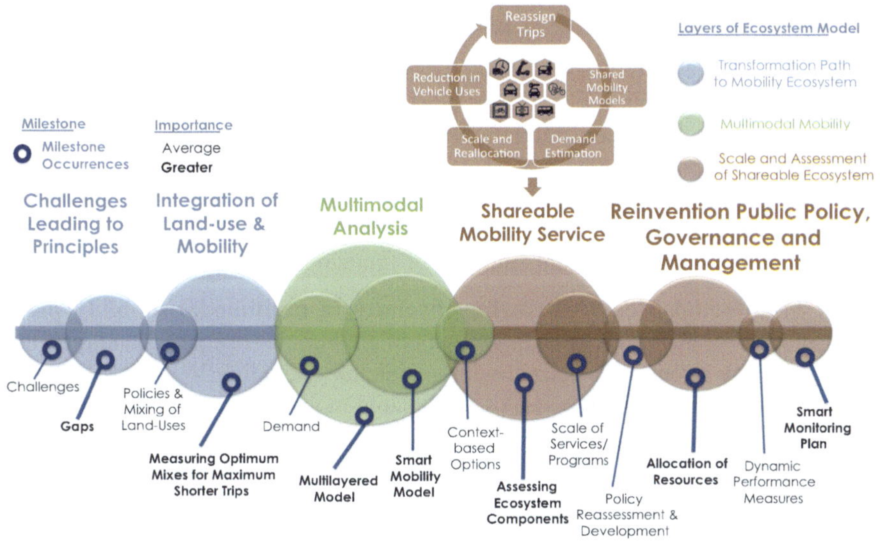

Fig. 4 Development of layered framework of mobility ecosystem planning model

demand forecasting and infrastructure needs assessment process. Finally, the demand and supply scale of shared mobility is developed with innovative policy and implementation strategies that minimize or eliminate negative impacts on quality of life.

3.2 Development of Quantitative Model for Mobility Ecosystem Planning

Transformation Path to Mobility Ecosystem that Maximizes Human Well-being According to the concept of "Systems approach to Sustainable Mobility" [19, 64], exponential expansion of new systems is not sustainable. If sustainability constraints (*E*) are introduced into system dynamics through economic fundamentals (*F*) and mobility policies (*P*), a new innovative mobility ecosystem (IM) would be produced within a practical timeframe (*t*) while identifying initial conditions of mobility (EM) and satisfying a set of necessary conditions for economic, environmental, and social effects to maximize the well-being of the community. The well-being of the community is summarized in a genuine progress indicator, GPI, which represents the quality of life of city residents. The path to a desired level of mobility ecosystem (MES) is

$$\mathrm{MES}_{t+1} = \mathrm{EM}_t + \Delta\mathrm{IM}_t(F, P)$$
$$\Delta\mathrm{GPI} = \mathrm{MES}_{t+1} \pm \sum E_{i(t+1)} \geq 0$$

Aggregate sustainability constraints will be positive and identified through investigation of the limiting boundaries of each of the six fundamental ecosystem elements (FE) and investigation of the net benefits achieved through the progress of sustainable, shared mobility using policy incentives and strategies.

$E = \int$ (FEi) where i represents number of fundamental ecosystem elements.

Initial Multimodal Model—Bringing diversity in multimodal mobility planning: Smart Growth policies pertaining to built environment variables and mobility accessibility are strongly associated with vehicle use including Vehicle Kilometres Travelled (VKT) and determinants of sustainable mobility (such as Smart Growth Index (SGI) Model [65, 66]). VKT is strongly correlated to measures of accessibility to destinations and street network design variables. The policy variable (P) is a function of six city building fundamentals, denoted here as 6 Ds. The 6Ds are: 'density'—residents plus employees divided by land area; 'diversity'—the jobs–population ratio; 'design'—a combination of sidewalk completeness, route directness, and street network density; 'destination'—regional accessibility; 'distance'—the distance to the nearest transit or ecomobility stations or stops; and 'digital access'—information and telecommunication technologies [3] for sharing/on-demand services).

$P = \int$ (Density, Diversity, Design, Destinations, Distance to Transit, Digital Access).

These key policy variables enable city residents to take shorter trips and minimize the burden on peak hour travel. Shorter and Internal trips (I) are generally less than 5 km, an ideal distance for a combination of walking and cycling or innovative options such as bike share, micro-mobility, or shared mobility options. Maximizing internal trips is a principal indicator of complete communities and a function of diversity of land use, density, and physical and information access to the nearest sustainable mobility services [66].

Internal Trips (I) = $\int$ (Density, Diversity, Mode Share and Access)

The remaining external trips (E), trips that are relatively longer, could be completed by public or shared/on-demand transit, rideshare, carshare, and carpooling options. Every trip from any land use starts as a person trip (T_p) and a combination of internal and external trips of traditional mobility modes (m): vehicle (V), transit (T), bicycle (B), and walk (W) where VO is vehicle occupancy [67]

$$T_p = \text{Internal Trips}\,(I) + \text{External trips}\,(E) = (T_v * \mathrm{VO}) + T_T + T_W + T_B$$

Person trips are usually derived from trip rates (T_m) for each mode with intensity [gross floor area (GFA)] and unit measurements. A is the unit of GFA used for comparing land-use types (l)

$$T_p = T_m * \mathrm{GFA}_l / A$$

Finally, trips for each mode (MT_m) are generally estimated using area modal share (MS) of all types, original sustainable modes (m), and major destinations (dir) within or outside the city

$$\mathrm{MT}_m = T_p * \mathrm{MS}_l^{\mathrm{dir}}$$

Final Mobility Ecosystem Model—Scale and Assessment of New Shareable Ecosystem Elements: However, traditional multimodal models generally ignore the simple reality that travel behavior could have been different if smart technologies, real-time information, and easy access to multiple shared/on-demand mobility options were available to individual users. Adoption rates (AR) for innovative and shared mobility options (n) determine the nature and scale of new demand created by new technologies. Adoption rates are a key factor affecting vehicle–owners and drive alone trips (a_d), and nondrivers or persons without vehicle access (a_O). Nondrivers and persons without vehicle access are particularly important in estimating the use of innovative options available in their area. While total innovative mobility trips (IMT) will increase with increasing values of a_d and a_O, vehicle traffic will reduce under an $a_d >> a_O$ scenario (a positive adoption rate) and increase under $a_d << a_O$ [14]. Hence, innovative mobility trips for different contexts (geographic location, i) and level or technology (t) can be estimated.

$$\mathrm{IMT}_n^i = \mathrm{MT}_m * \mathrm{AR} = \mathrm{MT}_m * (a_d + a_o)_t$$

City building policy incentives with appropriate density, easy access to alternative options for different demographic groups and socioeconomic activities, and connecting technology to transit and other sustainable modes of network (P_{IMT}) increases positive adoption rates.

AR ($a_d >> a_O$) = f (demographic variables, socioeconomic variables, density, level of smart technology available) = f (P_{IMT}).

Supply constraints on innovative and shared mobility services determine the service parameter (SP) of each system. Therefore, the number of adjusted innovative mobility trips for a certain area can be derived.

$$\mathrm{IMT}_n^i\,(\text{adjusted}) = \mathrm{SP}_n * f\,(P_{\mathrm{IMT}})$$

Adjusted total innovative mobility trips produce the ultimate mode share of the final mobility system model.

4 Results and Discussion on Outcome of Mobility Ecosystem Model

This section describes all the elements of the proposed mobility ecosystem model, starting with the basic layers of the model for each fundamental element, followed by a brief analysis of the results and impact on current public policies and strategy aiming to implement a new mobility management process. As described above, interdependent modules for each fundamental element are measured against limiting boundaries or constraints to maintain the sustainable carrying capacity of a city.

4.1 Sociability: Reinventing Multimodal Mobility with Social Innovation

Identifying the social and demographic lifestyle changes and how they will transform the three key mobility planning factors (adoption rates of innovation mobility options, corresponding emerging travel patterns, and configuration and service parameters of the new mobility system [12, 68–70]) is the most difficult part of the proposed model. Public policies and economic fundamentals will determine the scale and levels of these planning variables. Real time and open or crowdsource data, therefore, is a vital part of identifying the trends that turn into mobility needs and developing quantitative process and evaluation models for each mobility option. The proposed model of a future mobility ecosystem must be able to support people and the choices they desire with minimal constraints. In order to achieve this, three levels of adoption rates were assumed: Level 1 is the basic shared systems currently available, Level 2 takes into account available connected and real-time technology, and Level 3 recognizes how people's values are shifting under economic realities and imminent environmental pressures. Among the shared vehicles systems, three levels were assumed: basic sharing technology, connected and electrical products, and fully autonomous technologies.

However, quantifying adoption rates of different emerging modes or services is a perilous task that may lead to overestimation of technology capacity and the timing of full market penetration. To avoid this pitfall, available mobility options and emerging technologies and the response to them were tested iteratively, particularly in a set of pilot project partnerships. These tests identified the challenges and lessons and learning process required to develop fair policies that balance public safety and well-being and create an opportunity for service providers. Through this trial and testing process, a new mobility ecosystem emerges for each planning area (see Fig. 5 for Toronto examples) with an improved understanding of context and local mobility cultures and demographics.

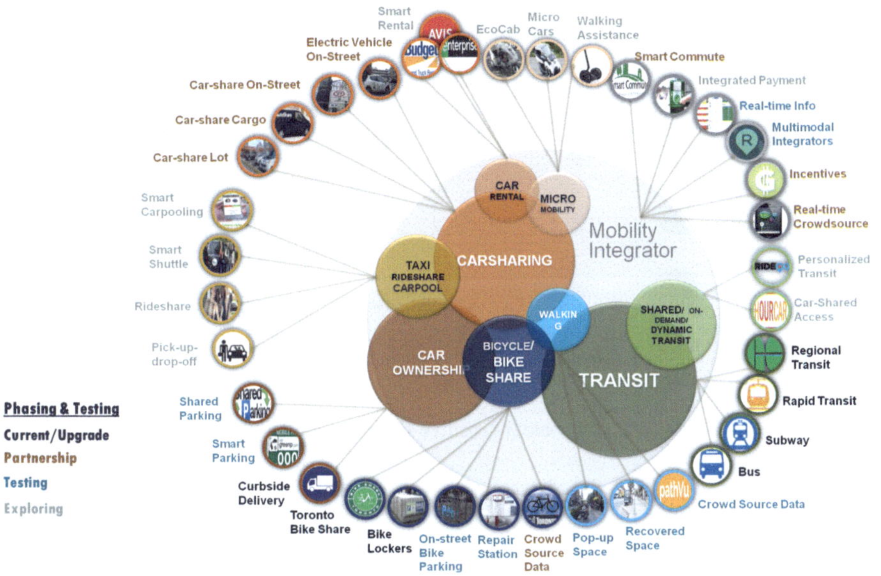

Fig. 5 Emerging social framework and adaptation and testing of mobility ecosystem

4.2 *Smart Growth: Integration of Land-use and Mobility*

A complete mobility ecosystem depends on diversity of mobility options, and variation in smart land-use policies. To establish the link between smart growth polices and the mobility ecosystem elements described in Sect. 2.3, the model integrates five layers of policy variables (Fig. 6). Firstly, appropriate share and right mix of land-use policies were tested against shared internal trips [30] by several activity centers in the city. Secondly, through iteration and testing [71] in the second stage, internal trips in planning areas were estimated against the optimum share of nonresidential and intensity of diverse land-use. Thirdly, modal shares of all fundamental modes were produced for all land-use and directions of travel within or outside the city. Context-sensitive and reliable targets were adopted using limiting boundaries of density and sustainable mode share (Fig. 4) and a citywide internal trips scale (Figs. 6). Fourth, person trips, multimodal trips, and parking space demand [72] for each mode of mobility were estimated to realize the scale and number connections for each area. Existing multimodal trips, mode share of transit station users, and trips generated by other developments immediately next to planning areas were added to the total future multimodal trips. Finally, multimodal trips were reassigned into the "shareable mobility service" mode to generate the scale of demand for all available or potential future shared mobility systems within the planning areas (Fig. 7). The impact of shared mobility including all demand management measures on parking was reassigned in similar fashion. Depending on

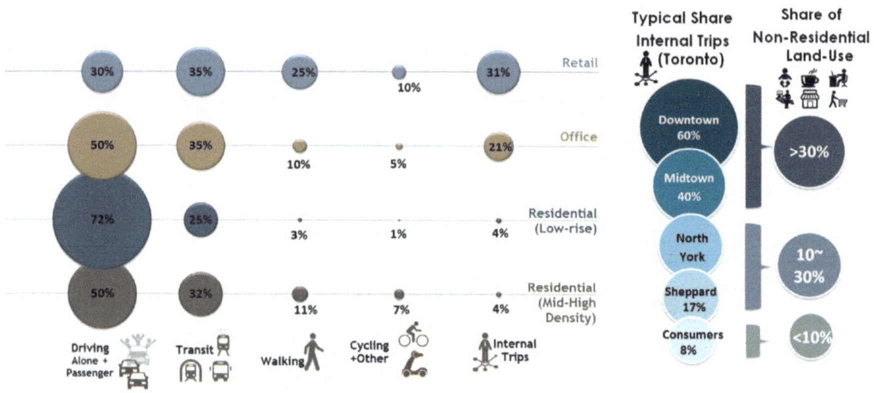

Fig. 6 Diversity of land use and context-sensitive nature of human travel pattern

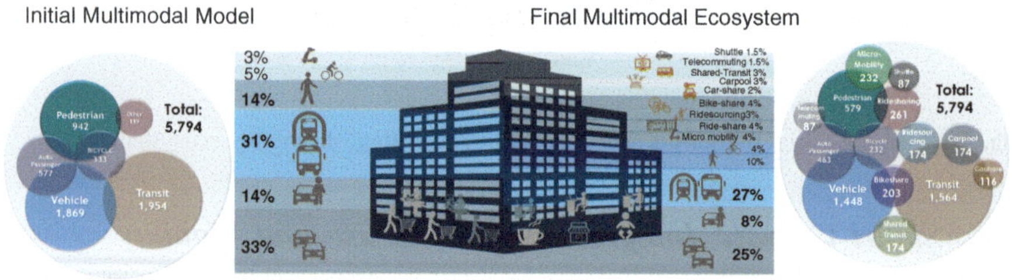

Fig. 7 Example of multimodal innovation mobility model on land-use intensification

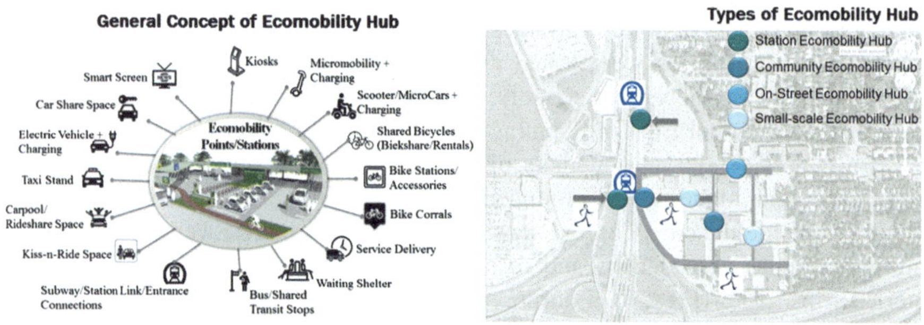

Fig. 8 *Example* of Smart and Easy Access: Ecomobility concept—a multimodal one-stop points (*Centre Image Source* Sophia von Berg, Multi-mobility, Institut für Verkehrsmanagement, 2016)

area context, the model estimates a reduction of 15–26% in the number of vehicle trips and a reduction of 20–40% in parking demand Relevant polices and incentives relevant to area specific zoning by-laws were developed to encourage optimum share and diversity of land-use.

4.3 Smart and Easy Access for All Mobility Users

Quantitative assessment generates the total demand for each mobility modes (both traditional and new) in the ecosystem and the facilities required for each planning area. Acceptable walk sheds to different modes or service station locations is applied to make sure people can access modes easily and walk safely (walk shed varies with mobility options: Bike parking/walkway 100 m, Bus/ped crossing 200 m, Bikeshare 300–400 m, Car-share 530 m, rapid transit 800–900 m). Common measures of multimodal area wide level-of-service [73], the connectivity index (for active modes, the acceptable range of the index is 1.6–1.9) and the pedestrian directness index (the acceptable range is 1.5 or less) are used to ensure that acceptable levels of the physical network infrastructure are in place for accessing mobility service locations. Ecomobility station ideas and short walking distance to neighborhood promenades or hubs where all mobility services are available were developed to ensure integration for easier and smarter access to existing transit or future mode infrastructure. Quality of service in terms of waiting time and service frequency was identified for each service mode to make sure reliability and convenience services are maintained. This enables the development of capacity of all modes or services to match future total mobility demand for planning areas. Finally, connected technologies and real time display or smart screen requirement policies inform area residents or visitors about available service, service status, location or service disruption. Connected technologies also ensure users can pay, book, and locate services. Using nine shared mobility sub-models, the scale of demand, location ecomobility hubs and distribution of shared service were plotted on a base mobility network in order to determine how existing/future public space and connecting private space need to be redesigned and how to reallocate space accordingly (see Fig. 8).

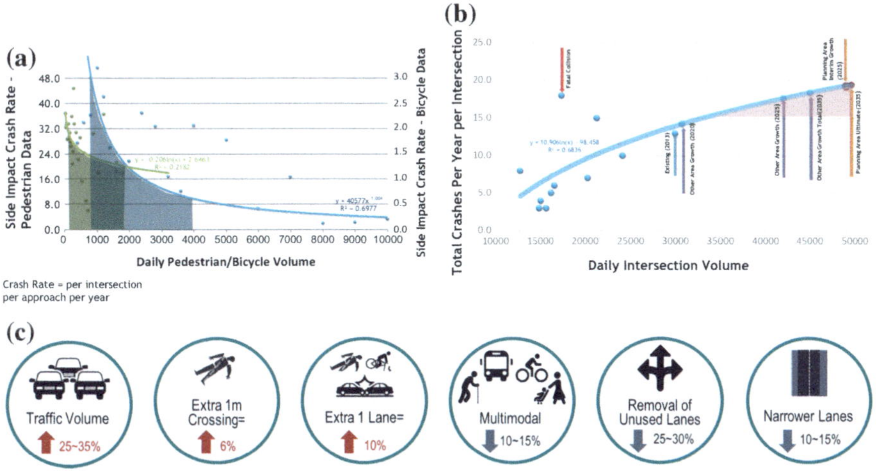

Fig. 9 Summary of sustainable approach to safety in community planning model

4.4 Safety First Approach

Introduction of safety into a proposed mobility ecosystem establishes interconnection between public policy around the safety and overall benefits of sustainable and shared mobility systems. Using local area crash data, initial and effective boundaries (see shaded area in Fig. 9a) of pedestrian and bicycle usage are established where the "safety in numbers" effect becomes strongest, reducing all types of crashes, and where the slope of crash rate decline stabilizes. Since pedestrians, cyclists and vulnerable citizens experience higher crash rates before the initial boundary is reached (roughly 1000 pedestrians and 200 cyclists per intersection in peak hour or 250 pedestrians and 50 cyclists in peak hour of any street segment), it is absolutely critical to implement safe infrastructures and definitive safety policies as fast as possible to shorten the path to achieve initial boundary conditions. The final boundary is drawn where additional land-use intensification encourages more pedestrians and bicycles gradually diminish. A second layer of safety policies is supported by the vehicle traffic reduction strategy, i.e., polices that reduce the number of vehicles in favor of higher transit and shared vehicle usage. Comparing similar proxy sites, an expected crash level is established using a safety performance function approach. However, the expected crash level may be higher in less safe cities (see shaded area in Fig. 9b) and it may be unacceptable to continue the current trend. A community safety planning approach [58] incorporates social and demographic variables and combines crowdsource hazard data and public input and statistical analysis to generate net safety benefits by reducing the number of crashes. The final layer of safety analysis investigates detailed traffic and geometric conditions and identifies major causes of higher crash rates (Fig. 9c). Net safety benefits are estimated to be a 20–40% reduction in the crash rate. These findings indicate right-sizing streets and intersections and reuse unused vehicle spaces are critical to address safety issues in the mobility planning process.

4.5 Recommended World Class Infrastructure

Quantification of the scale of demand and supply of sustainable and shared mobility programs and infrastructure from the model provides an excellent opportunity to redesign and reallocate public spaces to complement the area's mobility needs. From a political and human psychology perspective, it is difficult task to retrofit existing infrastructure. It is relatively easy for new neighbourhoods if mobility stakeholders understand and are able to visualize their mobility challenges. In order to achieve the objective of quantifying infrastructure needs, the infrastructure of existing streets, public spaces and parking was reviewed (Fig. 10). The results showed that enlarging an intersection by adding lanes reduced capacity by at least 25–30% compared to normal intersections comprising a five-lane cross section.

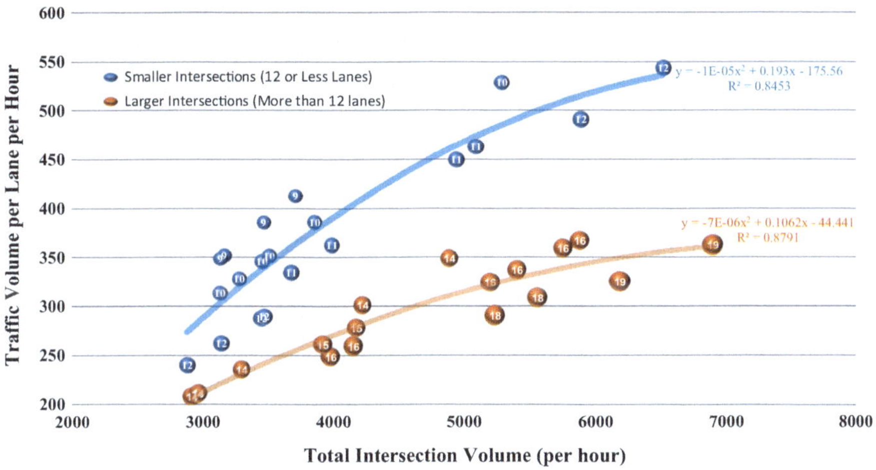

Fig. 10 Limiting boundary of physical capacity expansion of mobility infrastructure

The "wider is safer" approach without any scientific basis resulted in 21–25% of pavement dedicated to vehicles being unused. A capacity review of right-turn lanes also reveals that most of these lanes are not warranted as they are used less than 10% of the time during peak hours (only 3% in 24 h). A review of existing speed reveals that more than half of drivers disregard posted speed limits, with 15% drivers traveling more than 65–70 km/h, roughly the average speed of highways. These results have lead to specific policies that change traffic engineering practice and set limits on unnecessary infrastructure expansion.

Several strong and direct policies can be developed with the help of quantified future shared mobility demand, and a comprehensive review of existing space, street space, and parking area. First, redesign existing curb space or lanes toward shared and sustainable mobility uses. Second, reallocate unused right-turn lanes to create space for short and easy access to shared mobility services. Thirdly, reallocate corner spaces and reduce capacity of local streets to create parking laybys for priority users and shared mobility services. Fourthly, reuse recovered corner space for publicly accessible bikeshare, placemaking, and enhanced streetscape. Fifth, develop partnerships with private property owners to create ecomobility stations and maintain/operate services that provide access to tenants and visitors while sharing unused parking spaces. This is achievable through connected technologies and the release of idle capacity. Finally, multimodal level of service and risk indices were applied to quantify the service improvements by downsizing intersections and streets, and introducing frequent safe crossing locations (Fig. 11). Early results obtained from sites with world class infrastructure indicate that better street design did not slow down regular vehicles, but did slow down speeding vehicles. Livable street designs and reclaimed places invite people to interact with people, express themselves and play—a sign of a healthy and livable city.

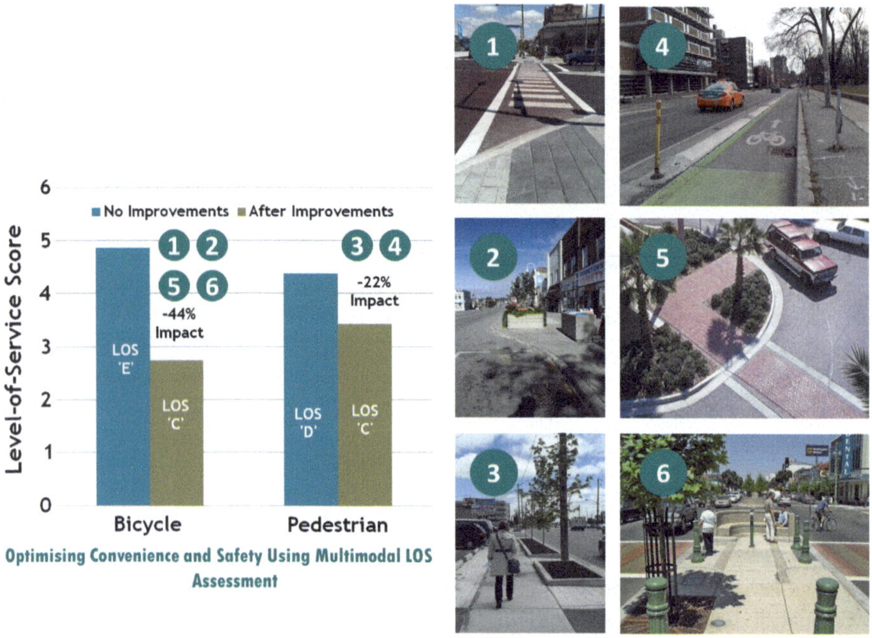

Fig. 11 Example of world class infrastructure planning and assessment

4.6 *Smart Use of Energy, Environment, and Healthy Planning*

The new quantitative science of cities is becoming possible because of the increasing availability of information, particularly the availability of key performance indicators from quantitative mobility ecosystem models. Combining model outcomes, novel measures of human and social activity, and scientific tools or standards developed by leading organizations, the model estimates net environmental footprint reduction [74], energy consumed by passenger vehicles [75], reduction of private vehicles uses per household [29], and the health impact [27] and economic benefits of pedestrian and cycling policies and infrastructure [28]. Compared to low-density land use, the proposed mobility ecosystem for suburban centers along with the mixing of land-use is expected to reduce energy resources and pollution from vehicles by roughly half (Fig. 12). Urban growth centers or downtown areas combined with sustainable and shared modes could reduce up to 55% of energy, vehicle usage and the carbon footprint. An additional 15 and 5% of health benefits can be achieved through policies encouraging sustainable and

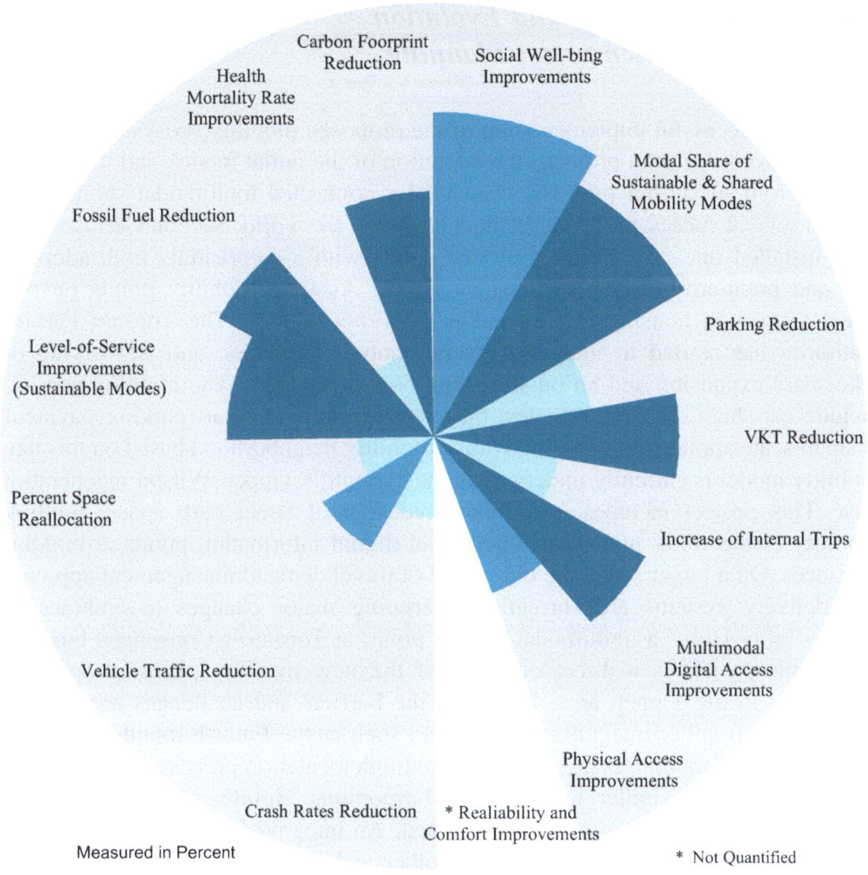

Fig. 12 Estimation of social, economic and environmental improvements within the city's carrying capacity and boundaries

walking and cycling infrastructure, respectively. Figure 12 summarizes the aggregate sustainability benefits of the complete urban mobility planning ecosystem.

5 Moving Forward: Implementation Process and Partners

The complex structure of the mobility ecosystem requires a departure from traditional isolated governance and the resulting fragmentation of the mobility delivery system. Private stakeholders, community partners and third-party service providers will all play a critical role in implementing this new model of mobility planning in conjunction with public sector.

5.1 *Incremental Steps and Evolution of Mobility Ecosystem Planning*

Key to the successful implementation of the proposed mobility ecosystem are a set of initial prototype pilot projects, the evolution of the initial model, and the lessons learned from successful projects. Demand for connected multimodal services has created several successful business models around the world. Several German cities have installed one-stop mobility service points with a surprisingly high adoption rate and popularity among the residents [76]. Austrian mobility points provide direct service to housing estates and neighborhoods [77]. The Toronto Parking Authority has started to integrate several mobility services, and is working on bike-share expansion and an on-street app booking system. The mobility services include car-share, electric vehicles, bicycle parking, and smart parking payment. A small scale application of an innovative mobility neighborhood based on this new mobility model is currently underway in the Toronto's Tippett-Wilson regeneration area. This project includes a complete redesign of street curb space, multiple mobility connections, and smart screen and digital information points at building entrances. On a larger scale, the old model of travel demand management approach and delivery systems are currently undergoing major changes to embrace an "Ecomobility Hub," a multimodal service point, at Toronto's Consumers business park. This project is a direct outcome of the new mobility planning approach introduced in the Tippett area. However, the barriers and challenges are endless. The lack of mobility integrators in Toronto, such as the Finnish monthly mobility package (MaaS model), is holding back the implementation process. A multimodal service by OEMs similar to Toyota's Harmonious Mobility or Ford's shared-transit-based system is currently being tested. An integrated payment system, such as the system available from Moveel or other technology companies, will fill the large void in the current fragmented user interface. The introduction and adaptation of these new applications through appropriate regulations, policies and hard and/or soft infrastructure in the proposed ecosystem will likely eliminate current mobility gaps.

5.2 *Critical Changes and Stakeholder Function*

The following major changes will be dominant forces in the new mobility ecosystem where different stakeholders play different roles, multidisciplinary public–private innovation become common practice, and private and public leaders develop a shared vision:

- **Mobility-as-a-service package**: The private sector may take the lead in bringing innovative products to general users while the public sector cooperates to facilitate the creation of a platform that regulates and promotes multimodal options, allowing these options to thrive.

- **OEMs multimodal business**: Similar to the model envisioned in SMART [19], automotive manufacturers and technology companies develop end-to-end mobility products and deliver services through the public sector with proper regulation and security/safety assurance.
- **Ecomobility points**: Public parking authorities and multimodal business providers cooperate with private developments to install a network of ecomobility points or stations that provide one-stop service points that create seamless link between all modes.
- **Public policy development**: The public sector revamps official plans to recognize new private sector mobility products/services and their new hard and soft infrastructure requirements. The public sector also reduces or eliminates unnecessary infrastructure or services that promote unsustainable use of vehicles.
- **Redesign of streets and curb side management**: The public sector initiates the process of street space allocation while local business improvement or community organizations maintain certain portion of streets or facilities along building frontages.
- **Redesign of building frontage**: The real estate and commercial sectors redesign building access points to follow public accessibility policies promoted and regulated by the public sector.
- **Digital access points**: Supported by the public sector's demand management policies, private sector communications companies develop and install innovative and integrated realtime information systems with smart screen display.
- **Rethinking public space**: The public sector works with the retail and commercial sectors to create or reallocate public spaces as part of the move toward a new and shared commercial/public economic model.
- **Shared and smart parking**: The public sector creates demand based parking policies. The real estate and development sectors partner with smart technology companies to introduce the infrastructure required for the demand based parking policies.

6 Summary and Conclusions: Vision for People-Oriented Mobility Ecosystem

Today, the dynamics of mobility technologies and options include environmental and health issues, and the need to establish a sustainable society. In reality, easy access, safer, reliable and comfortable multimodal systems change daily travel patterns, particularly work trips. If shared mobility is gradually implemented through future growth, connected mobility systems will significantly alter travel patterns. Since these mobility services are shared, coordinated area approaches are critical to securing and implementing sustainable and shared mobility services.

Given that transit infrastructure is the backbone of the Canadian mobility landscape, the study recommends the following process for the adaptation of innovative options and the integration and transformation of traditional sustainable modes: (1) if done properly, while increasing society and natural well-being, innovation and shared/on-demand technologies have a greater impact on parking supply and reduction of single occupant vehicle uses, and thus, reduce crashes, environmental pollution and low-carbon footprint; (2) instead of an adversarial reaction to new systems, test new options and technologies, and integrate into mobility planning processes once a system becomes a mature and viable; (3) integrate existing and emerging mobility and smart growth options in planning processes through collaboration between different levels of public agencies, mobility integrators and knowledge institutions; (4) using a quantitative planning model, estimate the scale and impact of innovative mobility options and evaluate and monitor progress using smart and crowdsourcing data; (5) create implementation tools and policies from scientific evidence through best practices of technological adaptation, and encourage policies and incentives to reduce inefficient use of vehicles and discourage negative impacts of technology that may become a threat to sustainable mobility modes; and (6) develop public policies that change the process of infrastructure planning, and make it easier to redesign public spaces, repurpose lands, and create ecomobility hubs and community interaction places through connected technologies and real time access to mobility service locations or programs.

Echoing an ancient Peruvian proverb, the marriage between technology and future mobility planning without improving social well-being will be worthless. Believing that mobility planning practitioners should support a more efficient and modern scientific innovations mobility system, the study recommends natural adaptation of the emerging mobility paradigm through the reinvention of people-oriented public policies—with shared incentives and goals between collaborative governance structures and mobility integrators—that improve quality of life of residents and improve genuine progress indicators. A steady and organic adaptation process of innovative technologies will enable cities to replicate nature's model of intricacy and sophistication into a new mobility ecosystem that rebuilds human social capital through peer trust.

Disclaimer The views expressed in this article are those of the author and do not necessarily reflect the views of the City of Toronto or other cities where the "Mobility Ecosystem" framework was applied.

References

1. Heikkilä, S.: Mobility as a service—a proposal for action for the public administration: case Helsinki. Master's Thesis, Aalto University (2014)
2. Little, A.: The future of urban mobility 2.0: towards networked, multimodal cities of 2050. International Association of Public Transport (UITP) (2015)

3. Cohen, B., Kietzmann, J.: Ride On! Mobility Business Models for the Sharing Economy, vol. 27, no. 3, pp. 279–296. SAGE Publications (2014)
4. Schwab, K.: The Fourth Industrial Revolution: What It Means, How to Respond, World Economic Forum, Global Agenda, Jan 14, 2016
5. Ausubel, J.H.: The Evolution of Transport, The Industrial Physicist, American Institute of Physics, vol. April/May, 20–24, 2001
6. ITDP and EMBARQ: The Life and Death of Urban Highways, Mar 13, 2012
7. ITDP: Europe's Parking U-Turn: From Accommodation to Regulation, Jan 11, 2011
8. Government of Ontario: Places to Grow Act, Ministry of Municipal Affairs and Housing (2005)
9. Listokin, D., Voicu, I., Dolphin, W., Camp, M., Jay, D., Leavey, M., Sherry, J.: Infill development standards and policy guide. Center for Urban Policy Research, Rutgers University for New Jersey, Department of Community Affairs, April 2007
10. Jackson, T.: Prosperity without Growth: Economics for a Finite Planet. Routledge, June 2011
11. Steffen, W., Richardson, K., Rockström, J., Cornell, S., Fetzer, I., Bennett, E., Biggs, R., Carpenter, S., Vries, W., Wit, C., Folke, C., Gerten, D., Heinke, J., Mace, G., Persson, L., Ramanathan, V., Reyers, B., Sörlin, S.: Planetary boundaries: guiding human development on a changing planet. Science **15**, 1–10 (2015)
12. Shared Use Mobility: Reference Guide. Shared-use Mobility Center (SUMC), Chicago (2015)
13. Doctoroff, D.: Panel Discussion, Disrupting Mobility Summit. Google Sidewalk Lab, Cambridge, USA (2015)
14. Alexander, L.P., González, M.C.: Assessing the impact of real-time ridesharing on urban traffic using mobile phone data. 4th International Workshop on Urban Computing, Sydney, Australia, Aug 10, 2015
15. Manyika, J., Chui, M., Bisson, P., Woetzel, J., Dobbs, R., Bughin, J., Aharon, D.: Unlocking the Potential of the Internet of Things. McKinsey Global Institute, June 2015
16. Berg, P.: The Finite Planet: How Resource Scarcity Will Affect Our Environment, Economy and Energy Supply, Create Space Independent Publishing Platform, Sept 16, 2011
17. Priemus, H.: The network approach: Dutch spatial planning between substratum and infrastructure networks. Eur. Plan. Stud. **15**(5), 667–686 (2007)
18. Forman, R.: Urban Ecology: Science of Cities. Cambridge University Press (2014)
19. Zielinski, S.: New Mobility: The Next Generation of Sustainable Urban Transportation, The Bridge—Linking Engineering and Society, vol. 36, no. 4. National Academy of Engineering, Winter 2006
20. Ohta, K.: TDM measures toward sustainable mobility. IATSS Res. **22**(1) (1998)
21. Strompen, F., Litman, T., Bongardt, D.: Reducing carbon emissions through transport demand management strategies: a review of international examples. Final report, GIZ China, Transport Demand Management in Beijing, 2012
22. City of Toronto: For a Healthy, Equitable, Prosperous Toronto (2015)
23. Weichenthal, S., Ryswyk, K., Goldstein, A., Shekarrizfard, M., Hatzopoulou, M.: Characterizing the spatial distribution of ambient ultrafine particles in Toronto, Canada: a land-use regression model. Environ. Pollut. **47**(PT A), 1–8 (2015)
24. Millard-Ball, A.: Do city climate plans reduce emissions? J. Urban Econ. **71**(3), 289–311 (2011). Elsevier
25. Weitman, M.: A review of the stern review on the economics of climate change. J. Econ. Lit. **XLV**, 703–724 (2007)
26. World Commission on Environment and Development: Our Common Future. Oxford University Press (1987)
27. WHO: Health Impact Assessment: Promoting Health Across All Sectors of Activity. World Health Organization (2012)
28. Anderson, J., Weiland, C., Muench, S.: Green Roads Manual, Version 1.5. University of Washington (2011)

29. EPA: Inventory of U.S. Greenhouse Gas Emissions and Sinks: 1990–2011, Chapter 3 (Energy), Tables 3-12, 3-13, and 3-14, U.S. Environmental Protection Agency, Washington, DC. U.S., EPA #430-R-13-001 (2013)
30. University of Toronto: Transportation Tomorrow Survey (TTS) (1985–2011)
31. Millard-Ball, A., Schipper, L.: Are we reaching peak travel? Trends in passenger transport in eight industrialized countries. Transp. Rev. **31**(3), 357–378 (2010)
32. Sustainable Urban Infrastructure: Vienna Edition—Role Model for Complete Mobility, Siemens Mobility, 2015
33. Botsman, B.: What's mine is yours: the rise of collaborative consumption. Harper Bus. (2010)
34. Litman, T.: Can smart growth policies conserve energy and reduce emissions? Cent. Real Estate Q. J. (2011)
35. Gordon, D., Janzen, M.: Suburban nation? Estimating the size of Canada's suburban population. J. Archit. Planning Res. **30**(3), 197–220 (2013)
36. Gately, C., Hutyra, L., Wing, I.: Cities, traffic, and CO_2: a multidecadal assessment of trends, drivers, and scaling relationships. PNAS **112**(16), 4999–5004 (2015)
37. McCahill, C., Garrick, N., Atkinson-Palombo, C., Polinski, A.: Effects of parking provision on automobile use in cities: inferring causality. Transp. Res. Board (2015)
38. Mehaffy, M., Porta, S., Rofe, Y., Salingaros, N.: Urban nuclei and the geometry of streets: the emergent neighbourhoods model. Urban Des. Int. **15**(1), 22–46 (2010)
39. Barthelemy, M., Flamini, A.: Modeling urban streets patterns. Phys. Rev. Lett. **100**, 138702 (2008)
40. Al Mamun, M., Lownes, N.: A composite index of public transit accessibility. J. Public Transp. **14**(2) (2011)
41. City of London: Measuring Public Transport Accessibility Levels: PTALs—Summary, Transport for London, April 2010
42. Santi, P., Resta, G., Szell, M., Sobolevsky, S., Strogatz, S., Ratti, C.: Quantifying the benefits of vehicle pooling with shareability networks. PNAS **111**(37), 13290–13294 (2014)
43. Watkins, K., Ferris, B., Borning, A., Rutherford, G., Layton, D.: Where is my bus? Impact of mobile real-time information on the perceived and actual wait time of transit riders. Transp. Res. Part A Policy Pract. **45**(8), 839–848 (2011)
44. Tang, L., Thakuriah, P.: Ridership effects of real-time bus information system: a case study in the city of Chicago. Transp. Res. Part C **22**, 146–161 (2012)
45. Brakewood, C., Macfarlane, G., Watkins, K.: The impact of real-time information on bus ridership in New York City. Transp. Res. Part C Emerg. Technol. **53**, 59–75 (2015)
46. Caywood, M., Cochran, A., Schade, M.: Urban Mobility Score: Quantifying Multimodal Transportation Access. Disrupting Mobility Summit, Cambridge MIT (2015)
47. Dumbaugh, E., Rae, R.: Revisiting the relationship between community design and traffic safety. J. Am. Plann. Assoc. **75**(3), 309–329 (2009)
48. Van Schagen, I., Janssen, T.: Managing road transport risks: sustainable safety in the Netherlands, risk management in transport. IATSS Res. **24**(2), 18–27 (2000)
49. Luoma, J., Sivak, M.: Why is road safety in the U.S. not on par with Sweden, the U.K., and the Netherlands? Lessons to be learned. Report no. UMTRI -2013-1, University of Michigan Transportation Institute, January 2013
50. Welle, B., Liu, Q., Li, W., Adriazolasteil, C., King, R., Sarmiento, C., Obelheiro, M.: Cities Safer by Design: Guidance and Examples to Promote Traffic Safety through Urban and Street Design, Version 1.0, World Resources Institute (2015)
51. Karim, D.: Narrower Lanes, Safer Streets, Canadian Institute of Transportation Engineers, Annual Conference, Regina, Saskachewan (2015)
52. Fitzpatrick, K., Schneider, I.V., William, H.: Turn Speeds and Crashes within Right-turn Lane, Report 0-4365-4, Texas Transportation Institute (2005)
53. Marshall, W.E., Garrick, N.W.: Evidence on why bike-friendly cities are safer for all road users. Environ. Pract. **13**(1) (2011)
54. Elvik, R.: The non-linearity of risk and the promotion of environmentally sustainable transport. Accid. Anal. Prev. **41**, 849–855 (2009)

55. Robertson, L.: Transforming our Cities to Foster Responsive, Affordable Mobility: Lessons from Detroit and Berlin, UN High Level Dialogue on Sustainable Cities and Transport, Berlin (2013)
56. Transport Canada: Complete Streets: Making Canada's Roads Safer for All (2009)
57. Hauer, E.: A Case for Evidence-Based Road-Safety Delivery, AAA Foundation for Traffic Safety (2007)
58. Lovegrove, G., Sayed, T.: Using Macro-Level Collision Prediction Models in Road Safety Planning Applications, Transportation Research Record No 1950, August 2006, pp. 73–82 (2006)
59. City of London: The London Plan: Review of Official Plan (2015)
60. Jones, P., Boujenko, N., Marshall, S.: Link and Place: A Guide to Street Planning and Design. Landor Press, London (2007)
61. Andrés, D., Chellman, C., Hall, R., Swift, P.: Smart Code Module, Center for Transect Studies. Duany Plater-Zyberk & Co., Version 2.0 (2009)
62. Musci, K., Khan, A.M.: Effectiveness of additional lanes at signalized intersections. ITE J. 26–30 (2003)
63. Dizikes, P.: New approaches to urban infrastructure. Conference at the Center for Advanced Urbanism, Plan 88: Article, MIT News Office (2014)
64. Yevdokimov, Y., Mao, H.: A systems approach to measuring sustainability of transportation. Proceedings of the International Conference on Transportation Systems Planning and Operation, pp. 519–528. Chennai, India, Allied Publishers Pvt. Ltd., 18–20 February, 2004
65. Criterion Planners/Engineers Inc.: Smart Growth Index: A Sketch Tool for Community Planning, Version 2.0, U.S. Environmental Protection Agency (2002)
66. Bochner, B., Hooper, K., Sperry, B., Dunphy, R.: Enhancing Internal Trip Capture Estimation for Mixed-Use Developments, NCHRP Report 684, Transportation Research Board, Washington D.C. (2011)
67. Kenchappagoudra, M.: Estimation of Person and Multimodal Trips Using Baselines Site Trip Generation Data, Transoft Solutions Inc. (2015)
68. IDTP: The Bikeshare Planning Guide, Institute for Transportation and Development Policy, New York (2014)
69. Rayle, L., Shaheen, S., Chan, N., Dai, D., Cervero, R.: App-based, On-demand Ride Services: Comparing Taxi and Ridesourcing Trips and User Characteristics in San Francisco, University of California Transportation Center, UCTC-FR-2014-08 (2014)
70. Shaheen, S., Martin, E.: Unravelling the modal impacts of bikesharing. Access **47** (2015)
71. Ewing, R.: Traffic Generated by MXD: New Prediction Methods Ahead, Planning: The Magazine of the American Planning Association (2011)
72. Arrington, G.B., Cervero, R.: Effects of TOD on Housing, Parking, and Travel, Transit Cooperative Research Progra, TCRP Report 128, Transportation Research Board (2008)
73. Steiner, R., Bond, A.: Future Directions for Multimodal Areawide Level-of-Service Handbook Research and Development, Florida Department of Transportation (2004)
74. Smith, J., Heath, L., Nichols, M.: U.S. Forest Carbon Calculation Tool User's Guide: Forestland Carbon Stocks and Net Annual Stock Change. General Technical Report NRS-13 revised, U.S. Department of Agriculture Forest Service, Northern Research Station (2010)
75. IPCC: IPCC Guidelines for National Greenhouse Gas Inventories. Intergovernmental Panel on Climate Change, Geneva, Switzerland (2006)
76. Berg, S.B.: Multimodal mobility concepts: development opportunities for public services in public transport with special consideration of sustainable mobility objectives. Unpublished thesis, Cologne (2013)
77. MO.Point., Wo Mobilität zu Hause ist., MO.Point Mobilitätsservices GmbH (i.G.) (2016)

Part II
Sharing Economy and Multimodal Mobility

How Disruptive Can Shared Mobility Be? A Scenario-Based Evaluation of Shared Mobility Systems Implemented at Large Scale

Francesco Ciari and Henrik Becker

Abstract This paper reports on ongoing work on getting a deeper insight into possible integrations of different shared vehicle systems. It introduces an original methodology in three stages, which helps dealing with the complexity of the problem. Using a simulation tool, different scenarios are assessed. The paper presents preliminary results obtained by simulating two extreme-case scenarios with large-scale car-sharing and bike-sharing schemes. The results suggest that shared mobility, if supplied at large scale and in the right mix, could indeed serve a large share of current travel demand without substantial losses in terms of generalized costs.

Keywords Shared mobility · Integration · Large-scale · Car-sharing · Bike-sharing · Ride-sharing · Agent-based simulation · MATSim

1 Introduction

Shared mobility is often mentioned for its potential to disrupt the current transportation system and to help creating a more sustainable one. This idea is supported by the incessant growth of shared mobility systems worldwide within the last decade and a relatively large literature assessing its benefits. However, the current market

F. Ciari (✉)
Swiss Federal Institute of Technology (ETH) Zurich, Institute for Transport Planning and Systems (IVT), Stefano-Franscini-Platz 5, F 33.2, 8093 Zurich, Switzerland
e-mail: ciari@ivt.baug.ethz.ch

H. Becker
Swiss Federal Institute of Technology (ETH) Zurich, Institute for Transport Planning and Systems (IVT), Stefano-Franscini-Platz 5, F 34.1, 8093 Zurich, Switzerland
e-mail: henrik.becker@ivt.baug.ethz.ch

G. Meyer and S. Shaheen (eds.), *Disrupting Mobility*,
Lecture Notes in Mobility, DOI 10.1007/978-3-319-51602-8_3

share of such modes is still generally low and consequently, their actual (positive) impact on the transportation system is not large. Things might change soon, if such systems grew further. Yet, given the complex nature of a transportation system, it is hard to predict how they will interact and what outcome can be expected at the urban scale. For example, in order to accurately predict the impact of shared modes on urban transportation, it will be crucial to understand if and under which circumstances they are complements or competitors. The work presented in this paper is part of an ongoing research project funded by the Swiss National Science Foundation within the "Energy Strategy 2050" scheme aiming at developing new technologies in order to substantially reduce energy consumption in Switzerland by 2050. The project specifically looks at the large-scale implementation and integration of different instances of shared mobility in order to estimate potential energy savings. To achieve this goal, an innovative methodology is used, which comprises three stages. In the first stage, several hypothetical scenarios, representing different combinations and availability levels of shared mobility options, will be generated and evaluated in terms of cost and benefits using the agent-based simulation MATSim [1] (www.matsim.org). To give more substance to the simulation part, the second stage focuses on the acceptability of the different scenarios generated in the previous part. This is done through specifically designed surveys, administered to a sample of the population of the study area. The results of the survey are used to estimate behavioral models, which are then implemented in the simulation. This improves the realism of the simulation and provides more solid insight on the combined use of various shared mobility concepts.

This paper provides two main contributions. First it describes in depth the methodology of the whole project in its three stages and explains its innovative aspects. Second, it provides preliminary results of the first stage dealing in particular with some extreme shared mobility scenarios and their impact on the transportation system.

2 Background

In recent years, shared mobility has been a focus for various disciplines. All present variations of car-sharing, bike-sharing, and ride-sharing have been investigated by researchers from around the world. Yet, the paths to today's success of these modes have been long and bumpy. The systems were ideated, and the first implementations were attempted, in between the late 1940s (car-sharing and formal carpooling) and the early 1960s (bike-sharing). For different reasons these attempts neither lasted long nor inspired immediate followers. Car-sharing was ideated to share a resource, the car, which in 1947, when the Sefage program was started in Zurich [2], was useful but expensive and not yet considered as a "must have" object for every household. The fast motorization of the following decades was not the ideal context for the success of this idea. Carpooling has a similar history: the American government promoted it during and immediately after WWII in order to limit oil consumption but the policy went largely forgotten once the political and economic situation had changed. It is not a coincidence that during the oil price shock of the '70s, carpooling was actively promoted again and then again forgotten when prices went back to normal levels and until the diffusion of HOV lanes in the US gave momentum to the idea again. The first attempt with bike-sharing—namely the White Bikes Program in Amsterdam in 1960—failed mainly because of vandalism. This attempt has made clear, that without the possibility to have a tighter control on bikes and users, bike-sharing has only limited chances for success. Other attempts have been deployed over the years, both car-sharing and bike-sharing, but none of those programs was successful enough to spread the "virus" of shared vehicles systems. Only recent advances in information technology have been able to unlock the potential of collaborative mobility solutions and have led to the creation of new ones like free-floating car-sharing, peer-to-peer car-sharing or dynamic ride-sharing.

The amount of research in the field has largely followed the fashion of the various ideas, and in the case of carpooling has followed its ups and downs too. The overwhelming majority of scientific literature on car-sharing and bike-sharing has been written in the last 20 years [3, 4] whilst carpooling was quite popular among transportation scientist in the '70s [5, 6] and in the '90s [7, 8]. In contrast, the recent appearance of app-based dynamic ride-sharing services has just started to attract the attention of researchers [9, 10].

The research on car-sharing has produced agreement on several issues. For instance, it is widely accepted that the most suitable markets are dense urban areas with good public transport supply [11, 12] or that the prototype user is relatively young, affluent, and well educated [13, 14]. In the case of carpooling, in contrast, there is still disagreement, for example, on the effects of HOV lanes [15, 16] or on motivations to participate to carpooling [17–20]. Literature on bike-sharing is much less abundant and is mostly concerned with the optimal location of bike-share stations as well as relocation processes, necessary to compensate temporally and spatially imbalances in demand (for example [21, 22]). The research was complemented by identifying user types [23, 24] and usage patterns [25]. Due to their only recent surge in market share, scientific literature about dynamic ride-sharing schemes is very limited and often deals with legal discussions (e.g., [10]). However, the market potential and current use pattern for such systems have already been investigated [9, 26].

The impact of these collaborative modes on the transport system has also been investigated by several researchers. The works focusing on station-based car-sharing were able to confirm several positive impacts like less vehicle travel and lower emissions [27] reducing the need for parking [28, 29] by reducing private vehicle holdings [30]. Similar studies for free-floating or one-way car-sharing have found that the service may at least partly compete public transportation resulting in a still unclear net impact [31, 32]. The impact of bike-sharing on the transportation system and travel behavior has also been recently addressed by researchers [33, 34] finding that it can be an effective measure to shift suburban residents' mode choice towards public transportation, although savings in vehicle kilometers traveled may be more than offset by relocations. When determining the effect of carpooling, it has to be differentiated between inter-household carpooling, which has been found to reduce vehicle miles traveled [35, 36], and intra-household carpooling, which might incur substitution and trip induction effects [37]. Given the very limited selection of literature on dynamic ride-sharing, its effects on travel behavior are still to be studied.

This short excursus shows that the scientific literature about shared mobility modes has grown in scope and number in recent years. There is already a large corpus of literature that deals with many different aspects of these mobility options. A large part of the research on collaborative mobility is of descriptive nature, but quantitative methods, rather rarely adopted in the early works, are now increasingly popular. Nevertheless, there are still some evident research gaps. First, the explicit modeling of demand for this kind of modes has not yet been thoroughly investigated despite being crucial to forecast how different levels and types of supply would impact the demand. Second, they are very often considered as "stand-alone" systems ignoring the whole complexity of the interactions with other (shared) modes. Therefore, it is not yet possible to estimate, how large-scale, integrated systems of collaborative mobility will impact the transportation system. However, this can be particularly relevant in the future, since their growth rates and some societal changes suggest that these modes could gain much larger shares of travel.

3 Methodology

How is it possible to assess the large-scale use of shared mobility modes? Experimenting with the implementation of shared mobility solutions at large scales cannot be undertaken in real life, because of the high financial investments necessary. Even if some pilot projects could be organized, it is not possible to deploy such systems at large scale, and try several combinations, just to get an insight on them. Thus, a methodology is needed, which allows to evaluate and compare different future scenarios in a cost-efficient way. The software MATSim has already been used for the simulation of car-sharing [38–40] and has all the necessary functionalities. However, the nature of the research questions implies other challenges which need to be addressed. The simulated scenarios represent hypothetical future situations, in which shared modes would have a much larger market share than today. The actual constellation of services will depend on the development path of such services and on modifications of people's preferences in the area. These two items are interdependent. It was therefore necessary to ideate a methodology to circumvent this potential chicken-egg problem. The resulting approach is described in the following subsections. Each of the next three subsections describes one particular stage of this methodology, whilst the fourth briefly describes the simulation framework MATSim.

3.1 Prescreening of Possible Shared Mobility Scenarios

In order to account for future modal preferences, a stated preference survey may be an obvious tool. However, choice situations should reflect the hypothesized future scenarios in order to guarantee consistence. Given the large amount of possible scenarios, this can be challenging. Therefore, in a first step, a prescreening of possible solutions is performed. To this end, MATSim simulations are used to understand, how the new modes would substitute the existing patterns of modal use and especially of private vehicles and what would be the best way to combine them according to preselected criteria. Given the overall goal of the project, the reduction of energy consumption will play a key role in defining the "best" scenarios.

3.2 Surveys and Models

To give feedback to the simulation, the second stage aims to get insight on the acceptability of the scenarios generated at the previous stage. This will be done through a specifically designed survey. Assuming different scenarios (that is, different levels of supply and prices for the innovative modes considered) the respondent will be asked if and how their mobility behavior (mode choice, location choice) would change. The scenarios described in the questionnaires will be based on a small set of the previously simulated scenarios. In particular, it will include the scenarios (combinations of modes), which turned out to be more impactful in terms of energy consumption reduction. Using this procedure—allows asking the participants more precise questions, describing tangible scenarios instead of generically asking about single mobility options. This also helps to envision possible policies—which would also inspire some of the questions of the survey—therefore providing a better understanding on the feasibility of the best scenarios. The results of the survey will be used to estimate behavioral models, in the form of discrete choice models, which are then implemented in the simulation.

3.3 Final Assessment of Shared Mobility Scenarios

At this point, the new simulation runs will entail more sophisticated behavioral models, because mode choice will be based on stated preferences obtained at the previous stage. Other attributes and behaviors (mobility tool ownership, activity chains) will be varied to take into account possible long-term impacts. This second series of simulations will give a final answer on what can be achieved in terms of reduction of energy consumption and other externalities with a large-scale, combined use of various innovative mobility concepts given the appropriate fleets, especially the share of electric vehicles. An additional aspect to be explored is the use of autonomous vehicles. Although they do not yet belong to our daily life, some scientists think that they may soon become an important factor in transportation. The implications are far too broad and complex to be investigated thoroughly in the context of this project. However, if autonomous vehicles will be used for car-sharing and ride-sharing the scope of these systems could dramatically change, and thus, some scenarios will be based on this assumption.

3.4 The Simulation Tool

The software MATSim [1] (www.matsim.org) has already been used for the simulation of shared mobility in several studies (e.g., [39, 41]). The simulation is based on a synthetic population of agents representing census data of the study area. The population acts autonomously in a virtual world, which reflects the supply side (road network, land use, available transport services and activity opportunities). Each agent acts according to an individual, predefined plan which contains a chain of activities which are to be performed during the simulation day [42]. As a general rule, performing activities gives a positive utility, whilst travel gives negative utility. One virtual day is simulated iteratively. From iteration to iteration, a predefined set of agents is allowed to change some of their daily decisions in order to search for a plan with a higher utility. The set of choice dimensions can be varied according to the exact purpose of the study, but standard dimensions are: trip starting time, duration of activities, location of secondary activities, mode of transport and route. The simulation follows a coevolutionary iterative process. At the end of the simulation, the plan that each agent has in use is a plausible approximation of the real world behavior of an individual with similar characteristics. Since the simulation represents individual travelers, it is possible to build scenarios making assumptions at the individual level rather than at the systemic level. Therefore, the model is much more intuitive, as it is based on simple observable behavioral rules.

4 Scenarios and Preliminary Results

This section deals with scenario generation and their prescreening (preliminary assessment) with the agent-based simulation MATSim.

4.1 Scenario Generation

The very first step in this process is the generation and evaluation of some "extreme scenarios" which will provide insight on the possible impact of an extremely wide diffusion of a particular shared mode. This step is necessary because it allows to understand, which kind of trips can realistically be made with which mode, what kind of potential overlap in supply exists, and what kind of cost/benefits can be expected. Ultimately, it will provide fundamental knowledge to create integrated scenarios (scenarios where several instances of shared mobility are all implemented at large scale).

4.1.1 Car-Sharing

It is safe to say that, despite impressive growth in the last decade, no car-sharing scheme has achieved a substantial market penetration yet. Even in Switzerland, which is the only country with a seamless, nationwide car-sharing system, only 2.5% of the license holders are car-sharing members [43]. Nevertheless, membership numbers are expected to grow further, with the proliferation of other mobility-as-a-service schemes and most importantly with the possible entry of self-driving vehicles into car-sharing operations. Therefore, the first extreme scenario assumes that all private car trips are substituted by car-sharing trips. In other words, a large-scale free-floating car-sharing system is put in place. A further assumption is that travelers will only accept a vehicle, which is at most 5 min away.

4.1.2 Bike-Sharing

Despite Switzerland being the cradle of car-sharing [44], and having some interesting traits in term of "sharing culture" (for example in condominiums, it is quite common to share the washing machine among all apartments), bike-sharing is not very diffused. A possible reason is the hilly topography of most larger cities. The rapid diffusion of e-bikes and their use in bike-sharing schemes is expected to overcome the burden of elevation and some schemes are about to launch their services. Therefore, in this scenario, a large e-bike-sharing system is assumed and all trips between 750 m and 10 km from the base scenario are now made by this mode.

4.2 Simulations: Preliminary Results

The car-sharing scenario was run several times in order to determine the optimal number of vehicles required to offer the desired level of service. As shown in

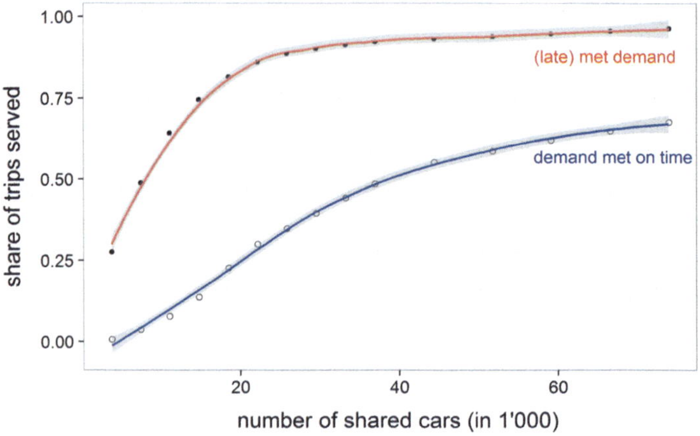

Fig. 1 Fleet size necessary to serve current travel demand

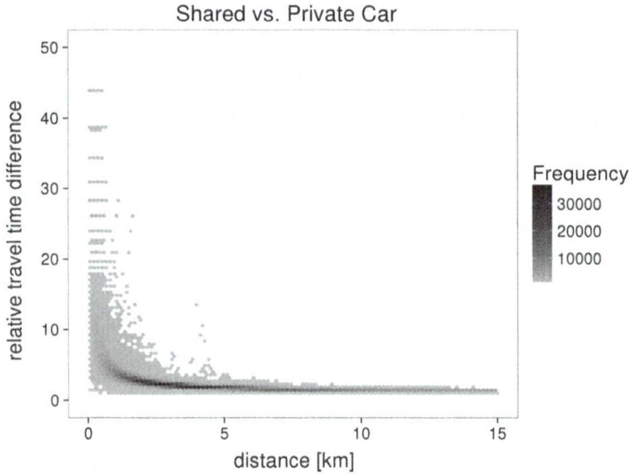

Fig. 2 Travel time difference between shared and private car depending on traveled distance

Fig. 1, around 60,000 vehicles would be sufficient to meet almost 100% of the demand (of which over 60% would be served within 5 min).

Compared to the base scenario, this constitutes a reduction of about 180,000 cars. Consequently, all private cars could be replaced by such a car-sharing program. At this stage, car-sharing is regarded equivalent to private cars in terms of utility, except for the access time at the beginning of the trip. Therefore it is intuitive, that longer car trips are more likely to be substituted by car-sharing, because the access time becomes less important in the evaluation (generalized costs) of the whole trip (Fig. 2).

Therefore, it is necessary to check if e-bikes would be a good complement for this car-sharing system and are be able to capture the rest of the demand. Figure 3

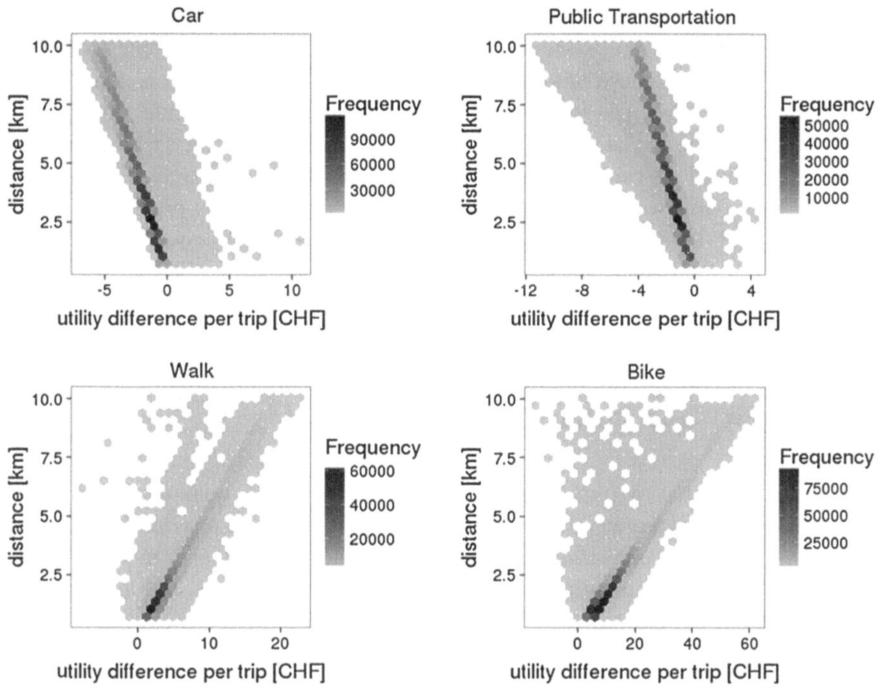

Fig. 3 Utility difference between e-bikes and travel with other modes for a predefined set of trips

shows, how the utility of using e-bikes compares to the utility of using car, public transit, bicycle, or walk for the same trip.

It turns out, that e-bikes are more convenient than walk and regular bikes for short trips and become less attractive as the distance grows. E-bikes are less convenient than car and public transport and the difference grows for longer distances. However, the difference is not very large for short trips especially with respect to public transportation. This basically means that there is potential for e-bikes to complement the car-sharing system and capture the demand for shorter trips.

4.3 Discussion

The results presented above show that shared e-bikes and car-sharing could be usefully combined in order to capture a large part of current travel demand, in particular car travel. It seems however, that for medium distances (5–10 km), it could be necessary to integrate an additional option for this range as e-bikes are not very competitive against private cars any more, and car-sharing in the suggested form is not yet competitive due to the relatively high effect of the access time in this distance range. Ride-sharing could be this additional option, as it would also have a certain, probably similar, access time, but may be cheaper. This requires the exploration of further single-mode extreme scenarios and also of some combined scenarios with two of these modes or even all three. The main point is, that finding an equilibrium between a large-scale car-sharing and a large-scale ride-sharing scheme will not be trivial. It has been shown that a car-sharing system with the selected specifications can substantially reduce the size of a city's car fleet and that it would be possible to totally avoid private car ownership whilst providing a good level of service. However, if ride-sharing would be based on private cars, a large enough fleet of them should still be available. If this would be rather a shared taxi scheme, one would need to find another equilibrium.

5 Summary and Future Work

This paper describes a three-stage project with the main goal to find optimal combinations of collaborative mobility solutions, which would provide a substantial reduction of energy consumption without reducing individuals' mobility. This allows to get precious insights on how collaborative mobility solutions could be combined. Additionally, this also helps to understand, which policies could help to achieve a more sustainable, less energy intensive, transportation system. Local governments in many countries have supported the diffusion of collaborative mobility solutions—one of the most prominent examples are bike-sharing systems—although there is only limited evidence on how they impact the transportation system as a whole, especially if they are scaled up and combined. This research will provide such local governments with more awareness on how to invest their limited resources. For shared mobility operators, this research will provide additional insights into the potential of the single solutions and possible combinations. This helps understanding which growth strategies are the most appropriate. If a given threshold, in terms of diffusion/public's patronage, is surpassed, such modes might cannibalize each other's customer base. Although this research, will not specifically study possible competition among different operators of the shared mobility sector, the results will help operators to navigate the market. To get such insight on possible future scenarios, in which shared mobility systems would be implemented at large scale, the agent-based simulation MATSim is used. The preliminary results suggest that e-bikes and car-sharing could serve a large part of the current demand. However, it should not be forgotten that the simulation as used at this stage does not entail a high level of detail. In fact, some assumptions are rather coarse (for example car-sharing having the same utility of private cars). Nevertheless, it should be stressed, that this series of simulations is intended to explore the solution space and produce a meaningful basis for the generation of stated preference exercises, which are the core of the next stage of the project. The data collected through this survey will be used to obtain discrete choice models, which will be implemented in the simulation. In the final stage of the project, it will be possible to run new simulations with fully functional representations of car-sharing, bike-sharing, and ride-sharing in MATSim. This will provide a plausible insight on how shared mobility modes could be integrated at large scale, capturing a large part of the current travel demand whilst reducing transport-related energy consumption.

Whilst minimizing energy consumption is the main focus of this research, in the future, different dimensions could be included. For example, one could extend the scope to life-cycle energy consumption which would also include embodied energy, or generalize the analyses to include broader environmental and social benefits. Finally, it is worth mentioning that this research covers scenarios, in which shared mobility systems are already available at large scales. The transition from the current system, which is mainly based on vehicle ownership, towards one in which shared mobility is prevalent, is beyond the scope of this research. Although this is without doubt a worthwhile research topic (for example how pricing, availability, comfort, safety, and other features will affect the large-scale implementation of shared mobility), the exercise presented here has a fundamentally different approach. Using insights on desirable future scenarios, it will be possible to understand how such scenarios could be realized (i.e., through certain regulatory framework or policy making). In a phase of profound and quick changes, we consider it more important to first determine desirable future scenarios. Subsequently, policies and implementation strategies which could help realizing such scenarios can be deduced from the insights into those scenarios, and indeed, this strain of research will most likely be an important part of the future work.

Acknowledgements This research project is part of the National Research Programme "Managing Energy Consumption" (NRP 71) of the Swiss National Science Foundation (SNSF). Further information on the National Research Programme can be found at www.nrp71.ch.

References

1. Horni, A., Nagel, K., Axhausen, K.W. (eds.): The multi-agent transport simulation MATSim. Ubiquity Press, London (2016)
2. Harms, S., Truffer, B.: The emergence of a nationwide carsharing co-operative in Switzerland, report for the EAWAG (Eidgenössische Anstalt für Wasserversorgung, Abwasserreinigung und Gewässerschutz), Dübendorf (1998)
3. Millard-Ball, A., Murray, G., ter Schure, J., Fox, C., Burkhardt, J.: Car sharing: where and how it succeeds. TCRP Report 108. Transportation Research Board of the National Academies, Washington, DC (2005)
4. Shaheen, S.A., Guzman, S., Zhang, H.: Bikesharing in Europe, the Americas, and Asia: past, present, and future. Transp. Res. Rec. **2143**, 159–167 (2010)
5. Kendall, D.C.: Carpooling: status and potential. U.S Department of Transportation Report DOT-TSC-OST-75-23 (1975)
6. Ben-Akiva, M., Atherton, T.J.: Methodology for short-range travel demand predictions: analysis of carpooling incentives. J. Transp. Econ. Policy **11**(3), 224–261 (1977)
7. Giuliano, G., Levine, D.W., Teal, R.F.: Impact of high occupancy vehicle lanes on car-pooling behavior. Transportation **17**(2), 159–177 (1990)
8. Brownstone, D., Golob, T.F.: The effectiveness of ridesharing incentives, discrete-choice models of commuting in Southern California. Reg. Sci. Urban Econ. **22**(1), 5–24 (1992)

9. Rayle, L., Shaheen, S.A., Chan, N., Dai, D., Cervero, R.: App-based on-demand ride-services: comparing taxi and ridesourcing trips and user characteristics in San Francisco. University of California Transportation Center, Berkeley (2014)
10. Rassman, C.L.: Regulating rideshare without stifling innovation: examining the drivers, the insurance "gap" and why Pennsylvania should get on board. J. Technol. Law Policy **15**, 81–100 (2014)
11. Stillwater, T., Mokhtarian, P.L., Shaheen, S.A.: Carsharing and the built environment, geographic information system-based study of one U.S. operator. Transp. Res. Rec. **2110**, 27–34 (2008)
12. Grasset, V., Morency C.: Carsharing: analyzing the interaction between neighborhood features and market share. Paper presented at the 89th Annual Meeting of the Transportation Research Board, Washington, DC (2010)
13. Burkhardt, J., Millard-Ball, A.: Who's attracted to car-sharing? Transp. Res. Rec. **1986**: 98–105
14. Kopp, J., Gerike, R., Axhausen, K.W.: Do sharing people behave differently? An empirical evaluation of the distinctive mobility patterns of free-floating car-sharing members. Transportation **42**, 449–469 (2015)
15. Giuliano, G., Levine, D.W., Teal, R.F.: Impact of high occupancy vehicle lanes on carpooling behavior. Transportation **17**(2), 159–177 (1990)
16. Parkany, E.: Can high-occupancy/toll lanes encourage carpooling? Case study of car-pooling behavior on the 91 express lanes. Transp. Res. Rec. **1682**, 46–54 (1998)
17. Buliung, R.N., Soltys, K., Bui, R., Habel, C., Lanyon, R.: Catching a ride on the information super-highway: toward an understanding of internet-based carpool formation and use. Transportation **37**(6), 849–873 (2010)
18. Ciari, F., Axhausen, K.W.: Choosing carpooling or carsharing as a mode: Swiss stated choice experiments. Paper presented at the 91st Annual Meeting of the Transportation Research Board, Washington, DC (2012)
19. DeLoach, S.B., Tiemann, T.K.: Not driving alone? American commuting in the twenty-first century. Transportation **39**(3), 521–537 (2012)
20. Vanoutrive, T., Van De Vijvera, E., Van Malderenc, L., Jourquinc, B., Thomasd, I., Verhetsela, A., Witlox, F.: What determines carpooling to workplaces in Belgium: location, organisation, or promotion? J. Transp. Geogr. **22**, 77–86 (2012)
21. Singla, A., Santoni, M., Bartók, G., Mukerji, P., Meenen, M., Krause, A.: Incentivizing users for balancing bike sharing systems. Paper presented at the 29th AAAI Conference on Artificial Intelligence (2015)
22. Lin, J.R., Yang, T.H., Chang, Y.C.: A hub location inventory model for bicycle sharing system design: formulation and solution. Comput. Ind. Eng. **65**(1), 77–86 (2013)
23. Fishman, E., Washington, S., Haworth, N., Watson, A.: Factors influencing bike share membership: an analysis of Melbourne and Brisbane. Transp. Res. Part A **71**, 17–30 (2014)
24. Vogel, M., Hamon, R., Lozenguez, G., Merchez, L., Abry, P., Barnier, J., Robardet, C.: From bicycle sharing system movements to users: a typology of Vélo'v cyclists in Lyon based on large-scale behavioural dataset. J. Transp. Geogr. **41**, 280–291 (2014)
25. Tran, T.D., Ovtracht, N., d'Arcier, B.F.: Modeling bike sharing system using built environment factors. Procedia CIRP **30**, 293–298 (2015)
26. Zhang, D., He, T., Liu, Y., Lin, S., Stankovic, J.A.: A carpooling recommendation system for taxicab services. IEEE Trans. Eng. Topics Comput. **2**(3), 254–266 (2014)
27. Martin, E., Shaheen, S.A.: The impact of carsharing on public transit and non-motorized travel: an exploration of North American carsharing survey data. Energies **4**(11), 2094–2114 (2011)
28. Millard-Ball, A., Murray, G., ter Schure, J.: Carsharing as parking management strategy. Paper presented at the 85th Annual Meeting of Transportation Research Board, Washington, DC (2006)

29. Shaheen, S.A., Rodier, C., Murray, G., Cohen, A., Martin, E.: Carsharing and public parking policies: assessing benefits, costs, and best practices in North America. Report CA-MTI-10-2612, Mineta Transportation Institute (2010)
30. Martin, E.W., Shaheen, S.A., Lidicker, J.: Impact of carsharing on household vehicle holdings, results from North American shared-use vehicle survey. Transp. Res. Rec. **2143**, 150–158 (2010)
31. Suiker, S., van den Elshout, J.: Wirkungsmessung Einführung car2go in Amsterdam. Paper presented at Nationaler Verkehrswissenschaftskongress, Amsterdam (2013)
32. Le Vine, S., Lee-Gosselin, M., Sivakumar, A., Polak, J.: A new approach to predict the market and impacts of round-trip and point-to-point carsharing systems: case study of London. Transp. Res. Part D **32**, 218–229 (2014)
33. Fishman, E., Washington, S., Haworth, N.: Bike share's impact on car use: evidence from the United States, Great Britain, and Australia. Paper presented at the 93rd Annual Meeting of the Transportation Research Board, Washington, DC (2014)
34. Martin, E.W., Shaheen, S.A.: Evaluating public transit modal shift dynamics in response to bikesharing: a tale of two U.S. cities. J. Transp. Geogr. **41**, 315–324 (2014)
35. Minett, P., Pearce, J.: Estimating the energy consumption impact of casual carpooling. Energies **4**(1), 126–139 (2011)
36. Shewmake, S.: Can carpooling clear the road and clean the air? Evidence from the literature on the impact of HOV lanes on VMT and air pollution. J. Plann. Lit. **27**(1), 363–374 (2014)
37. Morency, C.: The ambivalence of ridesharing. Transportation **34**(2), 239–253 (2007)
38. Ciari, F., Bock, B., Balmer, M.: Modeling station-based and free-floating carsharing demand: test case study for Berlin. Transp. Res. Rec. **2416**, 37–47 (2014)
39. Ciari, F., Balac, M., Balmer, M.: Modeling the effect of different pricing schemes on free-floating carsharing travel demand: a test case study for Zurich, Switzerland. Transportation **42**(3), 413–433 (2015)
40. Ciari, F., Balac, M., Axhausen, K.W.: Modeling carsharing with the agent-based simulation MATSim: state of the art, applications and future developments. Accepted for publication in Transp. Res. Rec. (2016)
41. Dubernet, T., Axhausen, K.W.: A multiagent simulation framework for evaluating bike redistribution systems in bike sharing schemes. Arbeitsberichte Verkehrs- und Raumplanung, IVT, ETH Zurich, Zurich (2010)
42. Balmer, M.: Travel demand modeling for multi-agent transport simulations: algorithms and systems. Dissertation, ETH Zürich, Zürich (2007)
43. Swiss Federal Statistical Office (BFS): Mobilität in der Schweiz – Ergebnisse des Mikrozensus Mobilität und Verkehr 2010, Neuchâtel 2012
44. Shaheen, S., Sperling, D., Wagner, C.: Carsharing in Europe and North America: past, present, and future. Transp. Q. **52**(3), 35–52 (1998)

Transit Systems and the Impacts of Shared Mobility

Joe Iacobucci, Kirk Hovenkotter and Jacob Anbinder

Abstract Mobility is fundamental to quality of life and economic prosperity in cities. Public agencies have an interest in better understanding how the emergence and evolution of shared mobility services is changing the transportation playing field. Transit Center commissioned Sam Schwartz Consulting, in partnership with the Shared Use Mobility Center, to perform a national study with the goal to better understand the evolving relationship between fixed route transit systems and emerging forms of shared mobility such as ridesourcing, bikeshare, carshare, and microtransit. This work is aimed at informing public transit providers, transportation agencies, public officials, and the public, all of whom play a role in ensuring that cities maintain effective, efficient, and equitable system.

Keywords Shared mobility · Public policy · Transit · Governance · Public–private partnerships

1 Introduction

In the last ten years, many new transportation options have emerged in cities across the country. From New York to Portland and dozens of cities in between, travelers can make use of bicycles and cars that are shared among users rather than owned by a single person. In other places, the public can take advantage of new taxi or ridesourcing companies such as Uber and Lyft. Early estimations show that the number of people using these shared-use modes of transportation has grown

J. Iacobucci (✉)
Sam Schwartz Consulting, 505 N LaSalle St., Chicago, IL 60654, USA
e-mail: jiacobucci@samschwartz.com

K. Hovenkotter · J. Anbinder
TransitCenter, One Whitehall Street, 17th Floor, New York, NY 10004, USA
e-mail: khovenkotter@transitcenter.org

J. Anbinder
e-mail: janbinder@transitcenter.org

G. Meyer and S. Shaheen (eds.), *Disrupting Mobility*,
Lecture Notes in Mobility, DOI 10.1007/978-3-319-51602-8_4

rapidly. At the same time, transit ridership continues to grow nationally, yet similar to the new transportation options, uptake is uneven across regions. Given how quickly the new services have emerged, the U.S. has few significant national or local policy initiatives to integrate the new transportation options into existing public transportation. In some parts of the country, politicians and local leaders have welcomed shared-use mobility options with excitement. Elsewhere, a hesitant approach has taken hold—one that is wary of their potential impact on the public transportation that governments have provided for the last 60 years.

As more of the public becomes familiar with shared-use mobility and as more providers enter the market, the urban transportation landscape will dramatically change. The shared aspect of the new services, combined with the growth of transit demand generally, suggests a massive potential to decrease car ownership and promote car-free and car-light lifestyles in cities across the country. A policy framework is necessary for governments to integrate shared-use mobility into their existing transportation networks to realize this potential and to ensure that new services support the long-term transportation vision of the city. The framework was the result of findings built from in-person interviews with 29 government agencies, other government organizations, and private-sector stakeholders in Seattle, Los Angeles, San Francisco, Boston, Chicago, and Washington D.C. More than 80 interviews elicited key findings about the current state of practice and the shortcomings of current policy

This paper covers a portion of the full report. The expanded report will provide additional analysis, examples of best practices, and more in-depth findings on the barriers and opportunities related to this issue.

2 Study Methodology

At the time of the study (2015) only two regions had proposed rule-making on a local jurisdiction level. Both Boston and Seattle had agreements for data-sharing from the ridesourcing companies, but this data was limited policy and planning decisions required much more data than provided by these ridesourcing companies.

For this reason, much of our methodology was based on interviews with transportation officials across the country. The interview questions were open-ended and provided an opportunity for transportation officials to identify (1) what was occurring, (2) how their jurisdictions were currently responding, (3) barriers to jurisdiction response, (4) what responses would be ideal, and (5) how they felt about the future.

The full interview questionnaire is listed below.

Sam Schwartz Engineering (SSE), in partnership with the Shared Use Mobility Center(SUMC), is performing a national study identifying the relationship between fixed route transit systems and emerging forms of shared mobility (TNC's, car share, bikeshare, shuttles, etc.*). The study will identify the impacts—and response to the impacts—of shared mobility that may indirectly benefit fixed route transit.*

A major component of this project is interviews with six pilot cities containing robust transit and shared-use options (Seattle, Chicago, Boston, Washington D.C., Los Angeles, and San Francisco). Our goal is to create a compilation of policy and planning initiatives that transit agencies, cities, and MPO's should adopt to maximize mobility in their respective service areas.

In addition, TCRP has funded SUMC and SSE to perform a quantitative analysis of impacts on ridership, personal choices for people living in areas with robust transit and shared mobility resources. The sum of both of these projects will lead to a greater understanding of the changing dynamics of urban mobility.

Introduction

Purpose: To identify the relationship between fixed route transit systems and emerging forms of shared mobility and best practices for integration of shared mobility.

Process/Partners/Definitions
Shared use is defined as follows:

- *Car Share*
- *Point-to-Point Car Share*
- *Bike Share*
- *TNC's (Uber/Lyft/Sidecar)*
- *Private Shuttles*

Interview Questions

1. *Please describe any current and/or planned policies/partnerships your agency has with shared mobility.*

Examples:

- *Provision of parking for carsharing (i.e., park-and-ride)*
- *Marketing/advertising support or subsidy for SM operators*
- *Use of property/facilities for bikeshare operations*
- *Ownership of bikeshare system*
- *Seed funding for carshare program*
- *Capital and/or operating support for bikesharing*
- *Joint marketing of first-last mile trips w/companies*
- *Outsourcing of vanpooling/carpooling trips to SM companies*

2. *What is influencing these policies/partnerships?*

Impacts on Transit Systems

3. *To date, have you observed an impact on transit ridership from shared-use companies?*

4. *Has your agency observed patterns in these effects (i.e., peak travel periods, late night trips, certain geographies)?*
5. *How do you anticipate the growth of these companies will impact transit ridership in the coming years?*
6. *What role (if any) do you see shared mobility companies playing in serving paratransit trips. Are there policies in place that encourage/discourage shared mobility companies to provide this service.*
7. *How has the growth of shared-use mobility influenced your agency's IT strategy, Facilities, Real Estate:*

Technology

8. *Please describe any technology solutions your agency is pursuing (internal and/or customer-facing) that take advantage of innovations in on-demand transportation.*
9. *What type of interaction has been experienced with Network Aggregation Companies?*

Future of Mobility

10. *What would you describe as a future dystopia of shared mobility?*
11. *What would you describe as a future utopia of shared mobility?*

Other Thoughts/Questions
Next Steps

3 Interview Results

Private companies' increasingly innovative shared-use mobility options pose a challenge to the public sector, which finds it must support, tax, and regulate private providers in a consistent manner in order to protect the public interest. Moreover, many interviewees grappled with what these new options portend for scheduled, fixed route transit.

To their credit, some city governments and transit agencies have already begun to offer shared-use services with the aim of providing mobility more efficiently and to greater numbers of people using existing transportation infrastructure. It became apparent that there are multiple opportunities to provide better and more equitable transit service to the public through improved relationships between transit and shared-use mobility providers. The findings from interviews elaborated below are a foundation for constructing more comprehensive policies regarding shared-use mobility that optimize the public good.

Over eighty interviews with urban transportation stakeholders yielded these key findings:

- By improving relationships between traditional transit and emergent shared-use mobility providers, there is substantial opportunity to provide better and more equitable transit service to the public.
- Employees of transit agencies, city departments of transportation, and metropolitan planning organizations recognize the tremendous potential of shared-use options, but are skeptical that they can be embraced in a way that will not result in shared-use "cannibalizing" transit rather than complementing it.
- Some cities have begun to integrate shared-use mobility options into their transportation policies. The extent of this integration varies significantly from city to city and from agency to agency. There is a widely shared sense that most U.S. transit agencies are not nimble enough to take advantage of the technological advancements that are essential to collaborating with new shared-use providers.
- Transit agencies and city governments define their separate missions narrowly as providers and maintainers of discrete transportation infrastructure services and networks. Overcoming this inflexibility is a significant challenge to creating the flexible and collaborative policy-making, planning, and regulatory approaches required in today's increasingly complex transportation environment.
- Public sector transit agencies and transportation departments have vastly different approaches to organized labor compared to shared-use companies. Questions of wages, work rules, and the legal relationship between workers and employers have yet to be addressed, and no U.S. city has yet to offer a comprehensive solution.

4 Recommended Policy Framework

As noted in the presentation at the Disrupting Mobility Conference, the role of government in providing and managing mobility is quickly evolving. Where once operating mass transit was the single goal of transit agencies, now they are changing include an enhanced role of integrating shared-use mobility.

Transit agencies and city governments must reclaim a leadership role to take advantage of shared-use mobility options that would further the public interest. The emergence of shared-use services has been the most rapid change to transportation in recent history, and too often, cities and traditional transit providers have been primarily reactive, creating patchwork solutions calibrated to the provider's terms. Cities and agencies must now act swiftly to reclaim the agenda.

Policy makers must seize the opportunity to integrate shared-use mobility options into traditional transit networks in ways that benefit government, the private sector, and the public. In most parts of the country, major impediments still exist. This stated, cities are already taking the initial steps to integrate shared mobility into the larger transportation network. Integration with carshare and open data platforms prove that transit agencies can adapt in the right circumstances. In addition, a common theme expressed from our interviews is the intention of integrating shared mobility in the future.

This report's recommendations comprise three distinct but related areas: understand the new choices in the urban transportation marketplace, coordinate transit planning and governance to incorporate those new choices in an efficient and equitable manner, and learn by implementing projects that benefit the public and maximize mobility.

4.1 Better Understand the Twenty-First Century Transportation Marketplace

From our interviews, a recurring theme was the lack of knowledge some city officials and transit agencies have about the state of transportation use. It starts with lack of knowledge about shared-use mobility, both within their own regions and in peer cities. Private providers rarely fall under existing regulations and are not compelled to share their data with government. Though a few examples of such data-sharing exist, most shared-use companies view their data as a trade secret. Where cities do collect provider data, the data is often limited. Further, more facts pertaining to the shared-use industry are needed: how their businesses are structured, who uses them, and how. Knowing about existing regulatory policies in other cities and countries would also be helpful. Finally, the interviews revealed substantial knowledge gaps regarding current users of *public* transportation. Many transit providers and regulators do not have comprehensive information about who rides mass transit in their cities and why they do so.

Collect, Demand, and Publish More Data

As former New York City Mayor Michael Bloomberg famously said, "In God We Trust, all others bring data." Effective shared-use policy simply cannot be created without data; not just from shared-use providers, but also from cities, transit agencies, transit users, and would-be transit users.

By routinely conducting surveys on traveler behavior–not just from those who use transit but also from those who choose *not* to use transit–cities and transit agencies can learn how people take advantage of shared-use options and how those options currently, or could in the future, complement the existing transit network. Agencies that already conduct rider surveys will gain even deeper insights by expanding their surveys to include questions that capture the entire journey, including use of shared options. Purchasing data from independent mobility data companies, such as StreetLight and Teralytics, will complement city and agency-collected data on rider behavior.

By making data-sharing a prerequisite for private shared-use providers' ability to operate within their borders, city and state governments will gain a better understanding of the impact of shared-use options on the transit system. Authorizing legislation or regulation should require origins, destinations, duration, distance, a map of the trip, the price paid by the traveler, and the customer's rating of the driver be provided to the regulating agencies or held in a third-party repository for data. Transit agencies that do not have regulatory authority should require data-sharing in any future contractual arrangements between transit agencies and providers.

Finally, if cities publicize standardized data about transit and shared mobility use, policymakers and observers will be able to compare trends across modes and cities. This, in turn, will create more opportunities for technology developers, and will help shared-use mobility providers better plan their services.

Use learning networks to establish and share best practices and to create predictable operating environments

Just as a lack of data-sharing has stymied progress on shared-use policy, so too has the lack of knowledge of precedent and best practices among cities.

Create learning networks among cities and transit agencies to increase knowledge and share model examples of how cities and transit agencies are regulating, collecting data from, contracting with, and implementing other policies regarding shared-use mobility providers.

Ensure predictable operating environments in which private operators can do business by establishing standards of best practice within a single city, and between cities. Create consistent standards for data, develop common regulatory frameworks, and build a body of evidence to inform future planning efforts.

The venue for inter-city public–private collaboration on shared-use best practices, standards, and policies could be housed within an existing organization, such as the National Association of City Transportation Officials, the National Association of Regional Councils, or perhaps even the League of Cities or another local government association.

4.2 Integrate City and Transit Agency Planning, Regulation and Other Actions that Affect Transportation Markets and Systems

Cities can actively incorporate shared-use options into their long-range transportation vision. Coordination between city planners and transit agencies is needed for the successful integration of shared-use mobility and public transit. Most cities and regions implement transportation policy in a manner that is too siloed to adequately address shared-use options. This is hardly a new phenomenon, but the rise of private shared-use providers has brought renewed attention to the intergovernmental tensions and gaps that stymie coordinated policy-making in many

American cities. If the problem is not addressed, the proliferation of private shared-use options will only exacerbate them.

Establish a coordinator of public- and private-sector shared-use mobility options within the transportation system
Cities and transit agencies rarely play a proactive role in coordinating transportation options across modes and sectors. Instead, cities tend to maintain physical infrastructure, such as streets and sidewalks, while transit agencies provide specific, discrete services, such as bus or train service that they operate themselves. Municipalities also regulate taxi and for-hire vehicle providers, but often as part of an entirely different city department. It is rare that taxi regulators, city planners, and transit planners work for the same agencies, nor would it be likely that their paths would cross given typical departmental structures. This distributed oversight of transportation can be particularly challenging for companies looking to operate a new transportation service that crosses traditional modal boundaries.

One solution is to create a Director of Mobility position or mobility office tasked with coordinating the relationship between transit agencies, cities, and shared-use mobility providers. This position would be the point of contact for private shared-use providers who want to do business with the city and coordinate with public sector transit providers to improve travelers' experiences across modes. The institutional location of this position will necessarily vary by region, but it will be most effective if it holds regional authority over these stakeholders.

As governments and private shared-use providers take their first steps toward coordination, there is potential to evolve the position or office into a regional coordinating body, whose sole role is to coordinate across jurisdictions for fare system integration, contractual requirements. The ideal institutional setting in the United States might be within a metropolitan planning organization, but such intergovernmental regional organizations that are common in European countries are established independent of federal and state government funding and policies. Regional transit authorities with the ability to contract out can also provide the necessary institutional setting.

Simplify payment options for riders
Simplifying the user experience of transit is essential to increasing its appeal. Cities and agencies need to design a common fare structure and medium that works across transit and shared-use modes, addresses unbanked populations, and encourages would-be riders.

Rather than relying on proprietary fare media and payment systems, instead work toward the adoption of a universal payment network that allows for multiple payment options. Account for the payment methods that people already use, such as credit and debit accounts, while also incorporating smartphone payment technologies such as Android Pay and Apple Pay.

The implementation of new payment systems need not be top-down. Cities could issue RFPs inviting third-party developers to create the best multimodal fare payment system in exchange for a small transaction fee. Require shared-use providers to accept payment via the universal payment system through specifications in contracts with city governments and transit agencies.

Optimize service to increase the efficiency of the existing transportation network

The greatest promise of shared-use mobility providers is in their potential to encourage city dwellers to adopt car-free or car-light lifestyles. With the exception of bikeshare, however, an individual's sole reliance on shared-use options without transit is more expensive than car ownership. A behavior shift away from car ownership will only occur if a frequent, reliable, easy-to-use transit network exists to complement carshare, bikeshare, and ridesourcing services.

Redesigning transit service networks not only with travelers' current mobility needs in mind, but with consideration for how new shared-use options can fit into an ideal transit network, would actually reinforce more sustainable travel behavior. Transit agencies could focus on improving service on lines with high ridership and high demand. They can partner with shared-use providers to funnel riders to the core parts of the network from areas with less demand for transit. They can communicate this network to riders with branding and marketing through schedules, maps, and smartphone apps.

No network redesign will make transit fit the mobility needs of every traveler all the time, but if a network redesign makes transit a more viable option for a larger number of trips, travelers will grasp the money- and time-saving potential of not owning a car and instead use shared-use modes to "fill the gaps" in the transit network. In the long run, city governments and transit agencies may wish to consider actively integrating shared-use options into their networks via contractual arrangements to replace routes with low ridership and high per-rider operating costs or to add capacity to overcrowded routes in high ridership corridors.

Use more comprehensive performance measures

Broadening the metrics by which cities evaluate and measure a transportation system will highlight the complementary nature of shared-use services. Simply measuring ridership or level of service no longer effectively captures the benefits that an integrated, multimodal system can provide. These are adequate performance metrics for how well an agency is doing, but they do not answer the question of how the overall suite of services performs for the public.

For decades, transit agencies have used a relatively small number of metrics to determine institutional performance: ridership, vehicle-hours, vehicle-miles, vehicles operated in maximum service, the "peak-to-base" ratio, the farebox recovery ratio, and money expended per passenger-mile are some of the most common.

If agencies and city governments are to take advantage of shared-use options to maximize mobility for citizens, a broader set of performance metrics is needed. These could include passenger trips instead of vehicle trips, total trip time across modes, reliability, and accessibility, among others.

4.3 *Launch Immediate Policy Initiatives and Pilot Programs to Find Synergies and Uncover Needed Institutional and Policy Reforms*

Cities and transit agencies cannot wait for the perfect data set or the ideal plan to implement projects that take advantage of shared-use innovations. The material benefits of immediate action are substantial, and there are steps cities and transit agencies can take to integrate shared-use options into their transportation policies right away. Through pilot projects, proof-of-concept demonstrations, technological investments, and regulatory actions, the public sector can take an iterative approach to integrating shared-use options before embarking on more fundamental institutional changes.

Open contracts to shared-use mobility providers to improve existing services and control costs
Across the country, transit agencies already pay private providers substantial sums to provide vanpool programs, paratransit, and guaranteed-ride-home programs. When these contracts expire and are rebid, enumerate specific performance goals and invite new shared-use providers (especially ridesource companies) to submit bids. Their technological solutions developed through significant consumer testing could introduce much-needed competition into what may be a stagnant procurement market, saving money in the long run.

Make development and street design support shared-use and transit
A future in which multimodal connections between transit and shared-use are seamless is a future in which physical connections between transit networks and the places people travel are also seamless.

Carshare, ridesource, and bikeshare systems rely on a certain level of dense urban development to function. Ridesource companies do serve sprawling suburbs but they usually charge more for the service, as the lower density of potential riders reduces the incentive for drivers to operate in the area. As cities and transit agencies cooperate on building transit-friendly streets, investments that integrate shared-use options with traditional transit would create the infrastructure to enable car-free and car-light lifestyles. Major transportation hubs such as train stations, airports, and bus terminals are ideal places to start. Cities and transit agencies can quickly demonstrate the ease of multimodal connections by moving bikeshare stations onto transit agency property, reserving nearby on- or off-street parking spaces for carshare vehicles, and clearly marking the presence of shared-use options through legible wayfinding.

Several pioneering cities have begun to break down the institutional barriers between transportation and land-use planning by building street infrastructure that prioritizes transit. Some of the fixes are quite minor: exclusive bus lanes, signal priority at important intersections, queue-jump lanes that allow buses to move ahead of parallel traffic at an intersection, and bulb-outs that reduce the need for buses to move out of traffic to collect passengers. These innovations are becoming more common, but are not nearly as widespread as they need to be.

Improving the way privately owned land adjacent to transit is used in American cities through land-use regulations is another underused strategy in transportation planning. Cities can reform zoning codes to promote development near transit and eliminate rules that require developers to construct unnecessary parking. Reward developers for accommodating shared-use infrastructure, such as bikeshare docks and parking spaces reserved for carshare vehicles, in new commercial and residential developments as a means of reducing costly publicly subsidized private car ownership. Just as environmental advocates have called for carbon-neutral construction, cities could require new developments to be VMT-neutral through mitigation strategies such as limiting parking and free transit passes for tenants.

Ensure smartphone apps provide multimodal directions and use real-time data

Smartphone apps have become central to urban travelers' decision-making. With few exceptions, however, apps remain siloed by mode—a savvy traveler probably has on his or her smartphone separate apps for Uber, Lyft, carshare, bikeshare, and transit. Even Google Maps does not integrate all available modes into a particular set of directions.

Trip-planning apps serve the public best when they render the transportation system as a panoply of options, rather than a single dominant service. If an agency or city has commissioned apps that include trip planning, issue RFPs calling for seamless multimodal directions that are agnostic as to the provider–public or private–offering the service, allowing riders to make direct comparisons between shared-use providers.

In particular, emphasizing the interoperability of bikeshare, carshare, ridesource, and buses or trains is key to encouraging the car-light lifestyles that shared-use innovations make possible.

In particular, transit agencies must ensure the data used in third-party transit apps are accurate and reflect reality by improving the quality of the data that developers use. To that end, we encourage transit agencies to update their fleets to provide high-quality automatic vehicle location data, and to make such data available in a standardized format for app developers.

In the future, multimodal apps could not only provide directions and compare options, but would actively arrange connections between transit and shared-use services. For example, on a journey that involves a bus trip followed by a last-mile Uber connection, such an app could automatically "hail" an Uber car as the traveler's bus approaches the stop. If multiple app users are aboard the same bus, their last-mile trips could be combined into one UberPOOL ride.

Level the playing field through consistent fees and reasonable regulations

Cities should ensure the public health and safety by regulating all shared-use options in a consistent way, whether they are bikeshare companies, rideshare providers, taxis, or so-called "Transportation Network Companies." Ridesourcing companies such as Uber and Lyft have been successful at providing the same services as taxis, offering more choices for residents by operating outside of the strict regulatory environment in which taxis exist. Examine current taxi and for-hire vehicle regulatory regimes and determine the purpose–and intended beneficiary–of existing regulations.

Regulate to prioritize the public interest, consumer and worker safety and protection foremost. As long as these protections are in place, regulations should not stand in the way of market competition and entrepreneurial innovation. Such a regulatory stance creates an atmosphere in which providers can introduce new services to consumers who may or may not choose to use them.

Regulations that must be applied consistently include pricing transparency, minimum licensing and insurance requirements, rules regarding driver background checks, data reporting requirements, accessibility for people with disabilities, and vehicle inspection requirements. Rather than maintaining tight barriers to entry on conventional taxi services while allowing rapid proliferation of app-enabled for-hire vehicles, these types of services are essentially the same for consumers and should be treated accordingly.

Establish reasonable fees on for-hire vehicles that ensure cities' streets are being used in the most efficient way possible and articulate a transportation vision. Fund efficient uses of streets with these fees that incentivize transit, bikeshare, high-occupancy vehicles, biking, and walking.

5 Conclusion

Paired with the growing demand for transit, new shared-use transportation options can transform our cities into walkable, healthy, vibrant places where car ownership is no longer a necessity. Policymakers must seize the opportunity created by the rapid emergence of shared-use services by elaborating a comprehensive twenty-first century vision for their cities' transportation networks.

By better understanding the current transportation marketplace, coordinating policy across agencies and departments, and putting those policies into practice in legislation and on city streets, cities can proactively ensure that the benefits of shared-use mobility and transit are widely and equitably shared. If cities and agencies do not take the reins, they risk becoming irrelevant to the future of urban transportation–an outcome that citizens cannot afford.

Cities and agencies must now act swiftly to reclaim the agenda for the public good. In the interviews that inform this report, it became apparent that while large challenges exist on issues of data, bureaucracy, and implementation, the desire on the part of public servants at cities and transit agencies to address these issues is even larger. We hope this framework addresses their questions, hopes, and fears, and lights a path for the future.

Shared Mobility in Asian Megacities: The Rise of the Apps

Katja Schechtner and Melinda Hanson

Abstract In August 2015, the Philippines became the first country in Asia to legalize app-based shared mobility services by defining a framework for "Transport Network Companies" (TNCs). With the country's long history of shared transport, the underlying concept was already culturally ingrained. However, given that only around 31% of Filipinos have bank accounts, with an estimated 4% access to credit card, and smart phone penetration around 21%, the current market for TNC services is limited to a small segment of the population, compared to the overall shared transport market. While it remains unclear whether TNC services will add to congestion by helping to spur some suppressed demand trips, or ultimately reduce car ownership by providing an alternative shared model, the quick uptake and growth of the TNC services also show that they are improving the overall mobility of certain population segments. However, the services remain out of reach for the majority of the population and add to mobility inequality. Both issues traffic congestion and inequality of access reflect the discussion in the developed countries, but are magnified by the extreme growth rates of cities like Manila, Jakarta, and Bangkok, where the future of those services will likely be shaped and decided by daily practice far ahead of the West.

Keywords Shared mobility · Informal transport · Megacity · Transport network services · Ridesharing · Asia · Manila · Uber · Grab

K. Schechtner (✉)
MIT Media Lab, Massachusetts Institute of Technology,
77 Massachusetts Ave, Cambridge, MA 02139, USA
e-mail: katjas@mit.edu

K. Schechtner
Institute of Urban Planning, Vienna University of Technology,
Karlsplatz 13, 1040 Vienna, Austria

M. Hanson
Urban Project Collective, 892 Bergen St #6A, Brooklyn, NY 11238, USA
e-mail: melinda@urbanprojectcollective.com

G. Meyer and S. Shaheen (eds.), *Disrupting Mobility*,
Lecture Notes in Mobility, DOI 10.1007/978-3-319-51602-8_5

1 Introduction

Cities around the world are growing rapidly and are struggling to meet transport demand. Nowhere is this truer than in Asia, currently home to 14 of the world's 25 megacities. In addition to existing megacities, rural to urban migration is creating new cities at an unprecedented rate. With this urban growth come added problems around congestion and associated increases in air pollution: currently the ADB estimates that 5000 people die from air pollution-related illnesses every year in the Philippines. As urbanization and car ownership increases, this figure stands to increase.

Though car ownership rates in most Asian megacities remains under 20%, cities including Jakarta, Beijing, and Manila have some of the worst traffic congestion in the world. With inefficient, underperforming, and sometimes even nonexistent public transit options, private operators have stepped into provide shared transport services. Privately operated shared modes have long been a staple of the Metro Manila transport system, with the majority of residents relying on a combination of buses, light rail, jeepneys, and AUVs. Jeepneys alone are estimated to account for more than 40% of daily trips in Manila [1].

In August 2015, the Philippines became the first country in Asia to legalize app-based shared mobility services. With the country's long history of shared transport, the concept was already deeply culturally ingrained. While this clears the way for companies such as Uber, Lyft, and GrabTaxi to grow in Philippine cities, a number of operational and regulatory questions remain.

This paper explores the experience with app-based shared mobility services in Metro Manila to date. We discuss the existing shared transport network and explore the regulatory and enforcement structures currently used to manage the networks. Next, we explore the process and policy structure that led to the new policy, as well as public reception to date. Finally, we discuss the role of app-based shared mobility in the context of two major issues urban transport faces in developing megacities: traffic congestion and inequality of accessibility.

2 Shared Mobility in Manila

2.1 Current Shared Transport Operations

As mentioned above, Manila's transport system includes a variety of modes. Public transit options include the city's Metro Rapid Transit (MRT) and Light Rapid Transit (LRT) modes, as well as city buses. Fares are distance based, priced depending on number of stations, and range from 13 to 30 Philippine pesos (PHP) per trip (about $0.28 to $0.65). Riders are picked up and dropped off at designated locations along an assigned route. Bus services are concentrated along major roads, while the LRT and MRT span a total of around 50 km (31 miles) across the city [2].

Asian Utility Vehicles (AUVs) are another common mode in Manila. AUVs are sports utility vehicles that can hold around 8–10 passengers. Similar to city buses, AUV routes are concentrated on major roads in Metro Manila [2]. AUVs take passengers between assigned pick up and drop off locations, without offering stops along the way. Users pay approximately $0.04 (PHP 2) per kilometer traveled.

The iconic Jeepneys have been an integral part of the Philippine transport system since the 1950s. The vehicles are custom built using parts repurposed from US military vehicles and tend to be elaborately decorated with glassless windows for ventilation. Each Jeepney carries around 16–20 passengers, who board through the open back door and sit shoulder to shoulder on benches. Similar to a taxi, jeepneys are flagged down by riders standing by the side of the road. Routes across the city are dispersed, and frequencies are extremely high on some roads, with volumes approaching 500 vehicles per hour in a single direction [2]. While there are some individual owners, most jeepneys are rented by individual drivers from large fleet owners. A driver and a fare collector oversee operations. In busy corridors, competition can be fierce, and so fare collectors also work to drum up business. Jeepneys cost $0.15 (7 PHP) for the first four kilometers and charge an additional $0.03 (1.5 PHP) for each additional kilometer. For most Filipinos, jeepneys are still the cheapest way to get around, costing about $0.22 (10 PHP) for the average ride.

A number of taxi services are also readily available throughout Manila. Taxis make up roughly 35% of the 1.9 million cars in Metro Manila [3]. As with most cities, drivers lease cars from fleet owners. Services are metered, costing $0.87 (40 PHP) for pick up and an additional $0.26 (12 PHP) per kilometer. Taxis can be hailed by passengers street side, or by using GrabTaxi, an app that has been in use in Manila since 2013 and allows riders to hail a taxi via smartphone. Drivers rent their cars from car owners on a daily basis for a rate of $17 to $30 (800–1400 PHP) per day. Drivers keep earnings above this rent minus fuel costs. Finally, human powered and motorized tricycles provide first–last mile transport, and serving the narrow alleys in poorer neighborhoods.

While all of these shared modes exist, overall mobility in the city is extremely low. For road-based modes, traffic congestion is a plague, limiting average speeds to fewer than 10 km per hour during peak commuting times. The LRT and MRT suffer from regular shutdowns and overcrowding—by some estimates, passenger density during peak commuting hours is four times recommended best practice. Further, jeepney routes are hard to decipher, and the various modes are not well linked. It is common for commutes in the city to take upward of 2 h, and for passengers to transfer modes two to three times each direction.

2.2 *Governing Shared Transport*

The modes listed above are regulated by a patchwork of government agencies. Metro Manila includes 16 cities and one municipality, all with their own mayors and government bodies, which has historically made transport planning and development a challenge. The Department of Transport and Communications (DoTC), a federal agency, oversees planning, development, and maintenance of the country's transport and communications systems and infrastructure. The Land Transportation Franchising and Regulatory Board (LTFRB), a division of DoTC, is responsible for administering, enforcing, and monitoring compliance of policies and regulations of public land transportation services.

Competition and differing views and priorities among the various cities in Metro Manila has made the construction of public transit infrastructure a challenge. Each city government faces a multitude of lobbies for the different transport modes, ranging from very vocal local jeepney and taxi advocates to a few very influential and politically well connected bus operators, who are well versed in pitching the different agencies against each other in order to resist any change. While the Metro Manila Development Authority exists to help coordinate efforts across the city agencies, they have been largely ineffective.

3 App-Based Mobility in Manila: The Uber and Grab Experience

As discussed above, Manila has a long history of transport entrepreneurship, with private operators stepping into provide services that the government is unable to offer. Manila's residents were quick to adopt Grab and Uber, and, in May 2015, the Philippines became the first country in the world to adopt nationwide regulation to govern Transport Network Companies (TNCs). A "TNC" is defined as "an organization, whether a corporation, partnership, sole proprietor, or other form that provides prearranged transportation services for compensation using an online-enabled application or platform technology to connect passengers with drivers using their personal vehicles" [4].

3.1 Launch and Early Operations

The app-based service GrabTaxi, a subsidiary of Malaysia's MyTeksi, launched in Manila in 2013. The service allowed for users to hail local taxis through smart phones for a small fee. Drivers and vehicles on the service are screened, and, in its early days, the company promoted the safety benefits and convenience of using the app-based hailing service. In 2015, GrabTaxi changed its name to Grab and began offering a number of additional services—including Uber competitor GrabCar (launched in 2014) and GrabBike for shared motorcycle rides (launched in 2015)—in addition to the original GrabTaxi service. Uber began operating in Manila in February 2014.

Uptake of both services in the country was rapid, and tensions began to mount in mid 2014, with taxi and AUV protests. Technically, at the time, both services were illegal as they did not have proper verification to operate as public transport vehicles.

In late 2014, the Metro Manila government began to threaten to take action against these services. In October 2014, an Uber driver was fined $4000 (PHP 200,000) because Uber did not have a franchise to operate as a public vehicle [5]. User outrage surrounding the incident encouraged the national government to begin taking steps to legalize shared app-based mobility services.

3.2 The Transport Network Vehicle Service Classification

Early debates for the new TNC regulations concentrated on the fluctuating fare structure of the app-based shared mobility services. The LTFRB has legal mandate to regulate public transport fares, and pushed for regulation on TNC fares. A February 2015 article in the Philippine Inquirer outlined the stance of Uber and the LTFRB, with an Uber representative noting that no other city regulates the fare structure and that drivers make more and passengers pay less than traditional taxi services, and so, government agencies in Manila should stand by and allow them to "do what they do well." As quoted in the same article, LTFRB Chair Winston Ginez responded, "We know the economics of this [business] and we will not be unreasonable. But we want to have the final say. We do not want prices to be abused" [6].

After months of intense lobbying for regulations that legalize the service without fare oversight on the part of Uber and Grab and, on the other side, lobbying from taxi operators pushing for tight restrictions, a new law was passed. On May 8, 2015, the Department of Transportation and Communications (DOTC) Issued Department Order No. 2015–2011, thus making the Philippines the first country to legalize TNCs nationwide. The new regulation requires Transport Network Companies (TNCs) to apply for accreditation, an application process which requires submitting a business permit, certificate of registration issued by the Bureau of Internal Revenue (BIR), a detailed business model proposal with proposed fare rates and service charges, and an explanation of the complaint mechanism against vehicle owners and drivers. Drivers must be certified as screened by the TNC, and register as franchisees with the government. Further, the accreditation requires that ride requests and payment be made through the app—drivers are not allowed to pick up passengers on the streets, and riders are not allowed to use cash.

In the announcement, DOTC Secretary Joseph Emilio Abaya called for broader innovation in the transport sector, stating, "these [regulations] will motivate current PUV operators to modernize, upgrade, and innovate as well" [7]. TNC representatives commended the decision, with Uber's Senior Vice President saying "this first-of-its-kind order is a shining example of how collaboration between government and industry can advance urban mobility, create new economic opportunity and put rider safety first" [8]. The new rules providing a legal framework for country-wide operation was seen as a big step in the shared, app-based transport movement, and this news was reiterated up by several international publications, including Wired [9].

As in many places, the local taxi association quickly organized to protest the new rules, claiming that the requirements on traditional taxi drivers compared to Uber and Grab drivers are disproportionate and unfair. As reported in a June 2015 Rappler article, Philippine National Taxi Operators Association President Jesus Manuel Bong Suntay said: "Regular taxi operators went through tedious process and inspection just to get franchise for our units. Foreign companies, like Uber, are given leeway, while local operators had difficult time just to get license" [10].

While protest from the taxi industry continued, in July 2015 Grab complied with all government requirements, thus becoming the first legally operating TNC in the Philippines. In early August, the government announced its plans to begin apprehending and fining all unregistered operators starting Friday, August 21, 2015. Uber filed its paperwork on August 19, thus becoming the second licensed TNC in the country [11].

3.3 Latest Developments (As of Spring 2016)

The number of TNC drivers and riders continues to grow rapidly. As of November, 2015, there were a total of 4465 accredited TNVS units to date, with around 3500 applications pending approval with the LTFRB [12]. The TNCs have continued pursing creative marketing campaigns to attract new customers. For example in 2015, Uber offered free "uberCHOPPER" rides for a day.

Groups opposed to TNCs have continued to protest. On December 2015, Quezon City Regional Trial Court granted a petition in favor of the Stop and Go Coalition's claim that app-based transport services were causing "grave and irreparable injury and damages to the association's officers and members because of their claim that they suffer less or low incomes" [13]. The president of the coalition claimed that income of taxi and UV express services had declined by up to half as a result of the new vehicle categories, and a Quezon City judge suspended the processing of new applications for app-based transport services Uber and GrabCar for 20 days [13]. Misleading reporting of the incident was common; with several Web sites claiming the injunction had suspended all operations.

Even as the regulatory battles continue, TNCs keep innovating and adding new services. In February 2016, Manila became the third city in the world to adopt UberHOP, a ridesharing option available during rush hour that allows commuters to join in a shared ride vehicle for a flat fee. The company estimates the new service will reduce user costs by more than 70% as compared to an UberX ride [14]. And Grab recently announced the launch of GrabBike, which allows for users to opt for a ride on a motorcycle. However, the service was quickly halted by the LTFRB in

February, 2016, with no indication on when or if it may be rolled into the TNC regulation [15].

3.4 Uptake and Response

Given that only about 4% of Filipinos used a credit card in 2011, and the smart phone penetration was around 30% in 2015, the market for TNCs is currently limited compared to the overall shared transport market. However, in a city of 18 million people, even this small segment is a sizeable business opportunity, especially as it was recently named one of the fastest growing smartphone markets, and is projected to increase 20% year-over-year [16]. This upper-class user group is demanding and able to afford the added comfort that TNCs provide, which was reflected in the user interviews we conducted in 2015/2016.

More than 90% of our interviewees said that they increasingly prefer Uber and Grab because of the added safety and convenience of these services, and also noted the relief that comes from not having to negotiate with taxi drivers. This sentiment was echoed in online forums. From interviews with taxi drivers in the Manila area, we learned that it was not uncommon for taxi drivers to lease taxis for 24-h periods, and drive the duration without sleeping. Several interviewees noted lack of sleep as a safety concern, along with stories of robberies perpetrated by taxi drivers in the Metro Manila area.

Safety issues aside, it is typical in Manila for taxi drivers to resist using the meter, and instead rely on a prenegotiated rate. During peak commute hours, some drivers charge a very high rate. Even during nonpeak hours, drivers will sometimes attempt to negotiate higher rates citing traffic congestion, which is omnipresent in Manila. The stable fare structure offered by app-based shared mobility services, as well as the perceived extra vetting of the drivers, interviewees noted, made using the service less of a hassle then negotiating with drivers.

As the TNC companies are still in early days, and fighting regulatory battles, both Uber and Grab have set up their payment structures to be particularly generous toward drivers and inexpensive for riders. Thus, according to Straits Times, an industry source said that around 30–40% of drivers who signed up for GrabCar and Uber fleets are former taxi drivers [17]. Users report the services are reasonable as compared driving or taking taxis.

While the current cost structure has been effective at attracting drivers and users, it may not be sustainable. As noted in a 2014 Wired article, "As the incentives Uber has put into place to spur growth are being phased out, drivers' salaries are apparently taking a hit so that these fleet owners can break even" [9]. In cities such as New York, San Francisco, and London, where Uber previously took a similar approach, drivers are actively protesting recent fare cuts. While drivers in the Philippines according to our interviews appear to be happy with the current setup, similar pushback from drivers is to be expected if Uber employs the same strategy in the future.

At the same time opposition from traditional taxi drivers and Jeepney operators continues. For example, in Kuala Lumpur, Malaysia, the home base of Grab, taxi drivers have begun conducting "citizens arrests" of GrabCar drivers. While it has not yet resported to this level of action in the Philippines, they remain actively opposed to the TNCs.

Similarly, municipal administrators face aggressive marketing tactics by the TNCs, who build ad hoc customer pressure through social media campaigns to force quick regulations in their favor, which led one DoTC employee to declare that they felt like they were working for Uber, who was taking up much of the capacity of the government employees for a service that, serves only a small, rather rich share of Manila's population.

3.5 Differences Between TNCs

Local experts and government officials have noted the difference in data sharing and approach toward government partnerships between the different services. At the 2016 APEC meetings in the Philippines, one World Bank official noted that while Grab has been open to share some of its origin-destination data, which can be useful for planning better overall transport infrastructures and services, Uber has been less cooperative.

While Grab and Uber are begrudging allies vis-a-vis regulatory bodies, they work hard to attract new and loyal customers for their respective services. Analyzing online discussion groups and user interviews both said Uber and Grab are very similar in regard to safety considerations and availability in the Metro Manila area. Surge pricing—where pricing for a trip changes depending on user demand—is a common critique of Uber, and is less common and also organized differently in the Grab application, which may apply the "Rush Hour Rate" on peak hours for and fare increase capped at 1.5 or twice the standard rate, while Uber has been known to quadruple their fares during special occasions [18, 19]. Users also noted that they prefer to know upfront what they are going to pay, regardless of traffic, a service that Grab provides but noted that average fares may therefore be higher on Grab.

4 Challenges and Opportunities for Shared App-Based Transport in Developing Megacities

While the regulatory issues have moved comparatively fast, operating TNCs in developing Asian Megacities presents its own unique challenges. As noted above, middlemen have emerged to reduce benefits to drivers in Metro Manila. Further, the income inequality in the city limits the service to middle and high-income users. Traffic congestion remains a major problem and estimated arrival times and travel times are highly unreliable in Metro Manila. Finally, a major complaint is the inaccurate digital representation of addresses: Roads and addresses are poorly marked, both in the real and digital realm, making it challenging to communicate one's location and destination to the driver.

4.1 *Social and Employment Impact Considerations*

It appears that social benefits to riders are more limited in Metro Manila than in other parts of the world. While TNCs offer a convenient and, for some, affordable way to get around Metro Manila on demand, thus supporting overall mobility of these users, given current credit card ($\sim$4%) and smart phone penetration ($\sim$30%), few of Manila's residents are currently able to access the services. Further, as shown in Table 1, TNC services cost around ten times as much as Jeepneys or AUVs, and are similar in price to Taxis. Finally, given traffic congestion, benefits from travel time savings that may be realized in some cities provide less benefit to users here. While further cost analysis is needed, our initial review suggests that though TNCs may help to fill in gaps of the existing transport system in cities in North America and Europe, these benefits are currently limited in developing Asian megacities.

Benefits that may accrue to drivers from job creation are also limited as compared to more developed markets, largely because drivers in Manila do not own their own vehicles. As in other markets, the question around quality of employment, including rights and benefit allocation (health and liability insurance, living wages, etc.), remain on the forefront of discussions. However, many of our interview partners—from government, industry, and the public alike—noted that those concerns are rather "quaint" in developing countries, as there are generally lower or nonexisting laws to protect workers, especially in the somewhat "informal" shared mobility services that are currently providing the major amount of transport services in Metro Manila.

4.2 *Impact on Congestion*

Some argue that the existence of TNCs stands to reduce overall congestion in cities by making individual car purchases unnecessary. As the TNCs target a market that

Table 1 Cost of shared transport by mode in Metro Manila

Mode	Pickup cost	Cost per km	Total cost 10 km
Jeepney	\$0.15	\$0.03 (after first 4 km)	\$0.33
AUV	\$0.00	\$0.04	\$0.40
Taxi	\$0.87	\$0.26	\$3.47
TNC (private)	\$0.86	\$0.04/min + \$0.12/km	\$3.90

Assumes 45 min 10 km commute, based on Google maps travel time estimate during commute hour would potentially be able to afford a car, the service might allow them to forego car ownership thus reducing the number of cars on the street (especially for parking traffic and parking space) and potentially reducing traffic congestion and overall energy consumption and air pollution. There is an ongoing discussion if app-based shared mobility has the ability to support "leapfrogging" of private vehicle dependence, particularly among the urban professionals in Asia's megacities, who are more used to seeing mobility as a service rather than an individually owned product.

Currently, though, in a city like Manila, the addition of any new vehicle is a problem. Since middlemen are largely purchasing new cars to be used in TNC services, these are brand new cars added to the road that would not otherwise be there. While current estimates suggest that it is only around 15,000 new cars (as compared to $\sim$2 million cars in the city), the city's roads are already well over capacity. The car may offer another transportation option for a select few, but the city's congestion problem will only continue to worsen without significant investment is high volume mass transit.

4.3 Data Sharing Needs and Considerations

The data about urban mobility behavior collected via TNC apps could provide detailed insights into the origin-destination matrices of a specific user group in a city. While TNC's remain targeted at a relatively small group of urban travelers in Manila and other developing country cities, the information about routes, travel times, etc. would allow to draw a much clearer picture of transport and support the planning of new (public) infrastructures and services.

While Uber and other ride sharing platforms remain unwilling to share their full data sets, GrabTaxi, the World Bank, and the Philippine Department of Transportation and Communications (DOTC) in spring 2016 launched the OpenTraffic initiative in Manila and Cebu City that taps into Grab Taxi data sets containing speeds, flows, intersection delays, etc. via GPS data and open-source tools. For transport planners, the focus will lie on peak hour analysis along key corridors, the analysis of travel time prediction reliability, how vulnerable certain transport corridors are to difficult weather situations and if it will be possible to identify road incident black spots in order to prioritize areas with high road crash probability for improvement.

While Mara Warwick, World Bank's Country Director for the Philippines is hopeful that "*By leveraging advances in open-software and big data collaborations with companies like Grab, transport managers and city planners can have access to the most advanced congestion management analytical tools available.*" It is yet unclear if the data provided by Grab Taxi will be comprehensive enough to help understand and alleviate the urban transport challenges beyond the EDSA corridor and Makati and Bonifacio Global City, the two most affluent Metro Manila districts.

5 Conclusion/Final Thoughts

Services like Uber and Grab add another piece to Manila's overall transport system, which is made up of a variety of private and public transport modes. But while existing services like Jeepneys and AUVs serve a large sector of the population, TNC services are limited to a small but growing middle and upper-class market. The biggest issue for all transport modes in developing megacities remains the rise of private car transport and the resulting traffic congestion. While mass transit with dedicated right-of-way (e.g., Bus Rapid Transit, Metros) offer significant travel time savings benefits to riders and are priced to be accessible to the general population, TNC services currently offer none of those benefits.

As of yet, it is unclear whether TNC services will ultimately add to congestion by spurring suppressed demand trips, or reduce congestion by supporting shared services over individual car ownership. Limited only to those citizens that have the financial means to access app-based mobility, TNCs remain painfully out of reach for the majority of the population and add to the current mobility inequality in developing countries. Both issues traffic congestion and inequality of access reflect the discussion in the developed countries, but are magnified by the extreme growth rates of cities like Manila, Jakarta, and Bangkok, where the future of those services will likely be shaped and decided by daily practice far ahead of the West.

The Philippine transport authorities were the first in Asia to develop a legal framework to support integrating app-based services into the transport network, a framework that continues to be reshaped. The existing comfort with shared mobility among transport users, combined with the demand from TNCs to enter the Philippine market puts Manila officials in a unique position to be a global leader in integrating app-based mobility into the transport system in a way that broadens social benefits. To achieve greater social benefits, Manila's government could think about implementing a tax on TNCs and directing proceeds to mass transit projects. The government may also leverage TNCs to support technology development that improves existing AUV and jeepney services. While further research is needed to explore the potential implications and benefits of these ideas, more needs to be done to ensure TNC's contribute to overall mobility improvements.

References

1. National Economic Development Authority (NEDA): Roadmap for transport infrastructure development for metro manila and its surrounding areas (Region III & Region IV-A) (2014)
2. Institute for Transportation and Development Policy, EDSA BRT Conceptual Plan, 2015
3. Asian Development Bank, Philippines: Transport Sector assessment, strategy, and road map (2012)
4. Philippines Government: Land transportation franchising, and leasing board, rules and regulations to govern the accreditation of transport network companies (2015)
5. Rappler, No franchise: Uber falls in LTFRB sting operation, 23 Oct 2014. www.rappler.com/business/industries/215-tech-biz/72898-ltfrb-uber-sting-operation. Accessed 16 June 2016

6. Camus, M.R.: Uber technologies expanding in Philippines. Business Inquirer, 6 Feb 2015. http://business.inquirer.net/186170/uber-technologies-expanding-in-ph. Accessed 16 June 2016
7. Ineraksyon: Uber hails Philippines as government, in a global first, recognizes—and regulates—'ridesharing', 11 May 2015, www.interaksyon.com/business/110339/uber-hails-philippines-as-government-in-a-global-first-recognizes—and-regulates—ridesharing. Accessed 16 June 2016
8. Uber, Arriving Now: Progressive ride sharing regulations in Philippines, 10 May 2015. https://newsroom.uber.com/philippines/arriving-now-progressive-ride-sharing-regulations-in-the-philippines. Accessed 16 June 2016
9. Alba, D.: The Philippines just made Uber legal everywhere, 11 May 2015. www.wired.com/2015/05/uber-philippines. Accessed 16 June 2016
10. Paz, C.D.: PH taxi industry protests new rules, 4 June 2015. www.rappler.com/business/industries/208-infrastructure/95290-philippine-taxi-industry-dotc-uber. Accessed 16 June 2016
11. Paz, C.D.: LTFRB approves Uber as transport network company, 19 Aug 19 2015. www.rappler.com/business/industries/208-infrastructure/103061-ltfrb-approves-uber-accreditation. Accessed 16 June 2016
12. Paz, C.D.: No Uber, GrabCar applications for 20 days—court, 4 Dec 2015. www.rappler.com/business/industries/215-tech-biz/114902-grabcar-uber-application-tro-quezon-city. Accessed 16 June 2016
13. Viray, P.L.: Court suspends Uber, GrabCar operations, 4 Dec 2015. www.philstar.com/nation/2015/12/04/1528956/court-suspends-uber-grabcar-operations. Accessed 16 June 2016
14. Uber: UberHop launches in Manila, 9 Feb 2016. https://newsroom.uber.com/philippines/uberhopmnl. Accessed 16 June 2016
15. Adel, R.: LTFRB stops GrabBike operation, 4 Feb 2016. www.philstar.com/headlines/2016/02/04/1549665/ltfrb-stops-grabbike-service-operation. Accessed 16 June 2016
16. International Data Corporation: The Philippines is now the fastest growing smartphone market in ASEAN, 17 June 2016. https://www.idc.com/getdoc.jsp?containerId=prAP41533516. Accessed 30 Aug 2016
17. Cheng, G.: Should you quit your taxi job to join Uber or Grab Car, 25 Nov 2015. http://business.asiaone.com/career/news/should-you-quit-your-taxi-job-join-uber-or-grabcar. Accessed 16 June 2016
18. PassportChop.com: Uber/Grabcar vs. taxis—what you as a commuter should consider: safety, price, quality assurance, 23 Nov 2015. www.passportchop.com/featured/uber-grabcar-vs-taxis. Accessed 16 June 2016
19. eCompareMo.com: Uber vs. Grabcar: the benefits of ride sharing services in the metro, 28 Aug 2015. www.ecomparemo.com/info/uber-vs-grabcar-the-benefits-of-ride-sharing-services-in-the-metro. Accessed 16 June 2016

What Drives the Usage of Intelligent Traveler Information Systems?

Christopher Lisson, Margeret Hall, Wibke Michalk and Christof Weinhardt

Abstract Rising mobility demand and increasing complexity of transportation options put a higher pressure on transportation systems and are a challenge in urban areas. A solution requires changes on coordination and behavioral levels. Today's technology, e.g., omnipresent smartphones, comprises the capabilities to induce such change via supply and demand coordination through intelligent traveler information systems. To identify the driving forces behind the decision to use such systems on an individual level the UTAUT 2 is transferred to the context of mobility by enriching it with explanatory insights from transportation research. The results indicate that the driving forces are user-specific and depend on diverse influencing factors that exceed pure economic and socio-demographic dimensions.

Keywords Mobility services · ITIS · UTAUT · Mobility behavior · SEM-PLS · User-heterogeneity

---

C. Lisson (✉)
Karlsruhe Institute of Technology, Englerstr. 11, 76131 Karlsruhe, Germany
e-mail: christopher.lisson@kit.edu

M. Hall
University of Nebraska, 1110 S16th Street, Omaha, NE 68182, USA
e-mail: mahall@unomaha.edu

W. Michalk
BMW AG, Parkring 19, 85748 Garching, Germany
e-mail: wibke.michalk@bmw.de

C. Weinhardt
Karlsruhe Institute of Technology, Englerstr. 14, 76131 Karlsruhe, Germany
e-mail: christof.weinhardt@kit.edu

G. Meyer and S. Shaheen (eds.), *Disrupting Mobility*,
Lecture Notes in Mobility, DOI 10.1007/978-3-319-51602-8_6

1 Introduction

Considering life in urban areas two fundamental trends are challenging the transportation sector and thus transforming cities as a whole. The first is higher pressure on the transportation systems due to rising mobility demand induced by urbanization and limited space. The second is the increasing complexity of transportation options to satisfy this demand due to new mobility services.

In megacities like Beijing, the stress on the transportation system can lead to its collapse either through traffic congestion or air pollution [1, 2]. Building wider streets or increasing the capacities of conventional transport solutions alone cannot solve this problem—it requires changes on coordination and behavioral levels [3]. For example: Chin and Larson [4] have shown that potential mobility existing on the streets—in terms of cars driving around—can be exploited by developing autonomous and shared car fleets. In this regard the transition from single, private car to on-demand usage of transportation modes in flux can ensure the same level of overall mobility with just a fraction of the cars.

Today the omnipresence of smartphones and our experience in using apps facilitates the coordination capabilities that are required to establish such new ways of organizing mobility. Examples for upcoming transportation services that build upon technological innovations are shared mobility services like Uber[1] (RideSharing), and CitiBike[2] (BikeSharing). These transportation options add complexity to user's mobility decision-making process, especially since an overall service integration platform that orchestrates extant services along various means of transportation according to individual preferences has hardly been established until now [5]. Nevertheless, for Sussman [6] this vision of interrelated transport and information systems means the arrival of a new era in transportation. Hilty et al. [7] extend this view, claiming ICT stimulate the growth of public transportation and slow down the growth of private car traffic because of the ICT-induced time efficiency gains for the user of the public transportation. However, even though technically feasible and partly realized in intelligent traveler information systems (ITIS) services like Moovel,[3] it remains difficult to convince people to use such systems and even more to change their mobility habits permanently [8–10]. A better understanding of ITIS success drivers is a major contribution in developing these systems and therefore in establishing a behavioral change toward more sustainable mobility decisions [11]. This leads to the research question:

[1]www.uber.com.

[2]www.citibikenyc.com.

[3]www.moovel.com.

What factors influence the usage decision of intelligent traveler information systems on an individual's level?
This work addresses the question by combining the perspectives of technology acceptance, and transportation science. A review of extant literature is given in Sect. 2. As each domain addresses questions of service usage from different angles, a unifying context related model is proposed in Sect. 3, which is subsequently evaluated in Sect. 4 using an SEM-PLS approach based on the survey data. Section 5 draws implications upon the derived insights and sketches the future research agenda regarding this complex challenge supporting citizens to make sustainable mobility decisions.

2 Literature Review

2.1 Intelligent Traveler Information Systems

Transportation research distinguishes between intelligent transportation systems (ITS) and advanced traveler information systems (ATIS) as one of their integral components [12]. While ITS concentrate on the coordination of traffic in terms of optimizing the transport network as a whole, ATIS focus on helping customers with pre-, on-, and post-trip activities. With the omnipresence of smart connected devices, system interconnectedness over APIs, and the emergence of artificial intelligence technologies, transport and information systems become increasingly interrelated [74]. The advancement of AITS is not just intermodal by coordinating individual, public and third party private transportation, e.g., train and ride sharing. AITS can also intergrade mobility related services from different domains, e.g., parking or other location based services [13]. Denoted as intelligent traveler information systems (ITIS) these systems remember user-specific travel patterns and route choices from previous sessions. Thus, they are able to evaluate the accessibility of all integrated transportation options to provide personalized route guidance according to each individual's needs [12]. With enough travelers using such coordinating services a significant reduction in travel time, delay, fuel consumption, and emissions can be achieved by a more efficient distribution of travelers to the available resources of the transportation system in terms of routes and modes [7]. Thereby ITIS become an influencing and coordinating design instrument inducing behavioral change toward more sustainable mobility behavior on individuals as well as society levels [5]. Including intra and cross-domain services ITIS converge to a universally accessible network orchestration platform of service value networks. Thereby, they perform an automated on-demand composition of complex services from a steady, but open pool of complementary as well as substitutive standardized service modules [14, 15]. This implies that the value of ITIS results from the efficient and seamless integration of all its constituting parts, which range from pure information systems (IS)-related functionalities to the quality of the physical transportation [15].

2.2 User Heterogeneity and Mobility Decisions

Its composite character sets the investigation of ITIS' acceptance and usage at the intersection between IS research (using mobility related IS) and transportation science (making modal choices). Identifying the driving forces toward ITIS usage one must consider that mobility related decisions are not purely rational but contain emotional aspects as well [13, 16, 17]. This makes the set of motivational factors to use an ITIS as heterogeneous as the people who decide to use it. In the context of personal mobility, one can generally distinguish between economic needs and noneconomic needs [17–19]. Examples for economic needs are time and cost efficiency [20, 21], while mental effort, comfort, hedonistic motivation, and flexibility represent noneconomic needs [5, 19, 22]. According to Abu-Zeid [23] all activities in the mobility context are planned and undertaken to satisfy various human needs. This is also valid for the three hierarchical choices a person has to make in order to reach his destination [20]: The choice of trip or no trip, the choice of destination, and the choice of transport mode. Referring to Maslow [24], who defines multifaceted human needs as hierarchical, user-sided heterogeneity in the context of individual mobility can be interpreted as user-specific rankings over the importance of the diverse mobility needs driving each individuals' mobility decisions. A major reason for user-sided heterogeneity are peoples' individual characteristics. These can be of socio-demographic, psychological and behavioral nature [25–27]. Recent studies reveal that personality traits, past mobility behavior, and lifestyle are important elements to account for the user heterogeneity in personal mobility [28–30]. They further emphasize that clustering user-types according to individual characteristics, e.g., psychometric measures and mobility patterns, has higher predictive power than clustering based on pure socio-demographic characteristics [18, 25, 29, 31]. Established measures for personality characterization on the psychological side are the Technology Readiness Index 2.0 (TRI2.0) [32] and the Big-Five-Inventory [33], while modal mix, average travel distance, and the set of past destination choices proved to be good indicators [30, 34] for identifying mobility patterns.

2.3 Technology Acceptance in Transportation

Discussing current approaches to explaining choice of transportation modes, the underlying motivational driving forces, and thus peoples' mobility behavior, Anable [25] points out that the majority of work orientates toward the Theory of planed behavior (TPB) [35] and their decomposed extension (DTPB) [36].

The DTPB combines intention and innovation research, which more completely explores the dimensions of subjective norms, attitudes, and perceived behavioral control by decomposing them into specific belief dimensions. Its innovational aspect makes it interesting for the case of upcoming ITIS technologies. While both theories provide valuable insights concerning motivational aspects they neglect to account for habit, a factor that has been proved to be highly influential in the context of mobility decisions [9, 16, 37]. The UTAUT 2 model by Venkatesh et al. [38] aims at explaining the acceptance and usage of technologies and explains a higher percentage of variance compared to TPB and DTPB in the service domain. In addition to habit it also comprises the constructs performance expectancy, social influence, facilitating conditions, hedonic motivation, price value, and experience. It is thus capable to account for both economic as well as behavioral aspects, what makes it plausible for explaining ITIS acceptance and usage.

As its antecessor, UTAUT 2 distinguishes between the more acceptance-related intention to use a service and the actual service use in terms of frequency. In such a way it enables accounting for the fact the intention to use is just one out of a plethora of reasons with influence on the actual usage. This allows investigation of the phenomenon that even highly accepted services something are not frequently used even if desired—for example, due to a lack of service accessibility, occasions to use the service, or service quality. Especially service quality, measuring the extent to which the delivered service meets or exceeds the customer's expectations [39] is considered as a focal point in determining the behavioral intention of using a service [40–42].

3 Model Development

The development of the model explaining acceptance and usage of ITIS follows the seminal guidelines by MacKenzie et al. [43] and Gefen et al. [44]. In this study it is implemented in three steps: Conceptualization of an explanatory model based on extant literature; measurement validation and refinement of the theoretical constructs; and model evaluation using empirical data. The UTAUT 2 is selected as the theoretical baseline model. It comprises for habit, a construct that is underrepresented in existing mobility behavior models in transportation [45], it combines economic as well as noneconomic motivational aspects [18, 19], and has also demonstrated its explaining capabilities in mobile services [38]. In order to adapt UTAUT 2 to the context of personal mobility it is extended with attributes out of the transportation and service quality research that aim at explaining mobility behavior. Expert interviews are conducted according to Klein and Myers [46] to ensure the content validity of the adjusted model. In line with the principle of content-related saturation [47, 48], 25 persons are interviewed; seven from ITIS providers and 18 from the customer side. Experts from provider side were chosen due to their management responsibilities following the ITIS development process, while experts on customer side were split into three age classes (<20, 21–40, 40<)

and two location classes (rural, urban) as suggested by [30]. The sample contained a wide spectrum of usage experience: power users as well as people who have never used an ATIS or ITIS. The interview data analysis is conducted applying content analysis according to [49]. The following constructs constitute the initial context adjusted UTAUT 2 focusing on the usage of ITIS [50]:

- **Performance Expectancy** is defined as the degree to which an individual believes that using ITIS will help him or her to attain gains in his or her mobility performance. The constructs pertaining to performance expectancy are perceived usefulness [51, 52], extrinsic motivation [26, 53], relative advantage [54], outcome expectation [41, 45, 55], service fulfillment [13, 42], and travel mode performance [21, 56].
- **Effort Expectancy** is defined as the degree of ease associated with the use of the ITIS. The constructs pertaining to effort expectancy are perceived ease of use [52, 57], complexity [58], and ease of use [59], service convenience [41, 60], mental relaxation [19], process integration [13], flexibility, and comfort [5, 22].
- **Social Influence** is defined as the degree to which an individual perceives that their reference group believes that he or she should use ITIS. The constructs pertaining to social influence are represented as subjective norms [35, 52], social factors [55, 58], image [54], and symbolic motives [17].
- **Facilitating Conditions** are defined as the degree to which an individual believes that he has the capabilities to use the system or that the infrastructure exists to support use. The constructs pertaining to facilitating conditions are perceived behavioral control [36, 61], facilitating conditions [58], compatibility [54], and perceived control [42].
- **Hedonic Motivation** is defined as the fun or pleasure derived from using a technology and is based on the findings and proved to influence the intention to use technology [17, 19, 62, 63].
- **Time** is defined as the desire to save time using the ITIS and is one of the key drivers influencing the modal choices [21, 64].
- **Costs** is defined as desire to minimize the financial expenses to satisfy a certain mobility need and is one of the key drivers influencing modal choices [20].
- **Price Value** is defined as consumers' cognitive tradeoff between the perceived benefits of the applications and the monetary cost for using them [65]. Thus, it links to the quality of services to determine the perceived value of products or services [41, 42, 66].
- **Habit** has been defined as the extents to which people tend to perform behaviors automatically [37, 67]. Using previous experiences as guidelines for future behavioral performance [68] is widely acknowledged as an influential element of mobility decisions [18, 29].
- **Usage Intention** is defined as the intention of a person toward using a service or mode of transportation [38].
- **Usage** is defined in terms of the frequency a person actually uses a service or mode of transportation [38].

4 Model Evaluation

4.1 Data Acquisition

Data for statistical analysis is acquired via an online survey. The underlying questionnaire is developed following the guidelines by Malhotra et al. [69, 70], so that validated theoretical constructs remain in their original structure to the greatest possible extent while new items identified in the expert interviews have been generated following the recommendations by Groves et al. [69]. To ensure construct validity, the examination of the initial items involved a card-shuffling assessment with six business faculty members [71]. The suggestions collected from the card-shuffling panel improved the candidate items' readability and clarity; the feedback also indicated the removal of redundant items that tapped the same facet of a focal construct [42]. A cross-sectional design with a 5-Likert scale, anchored at 1 "strongly disagree" and 5 "strongly agree," is chosen for the study. The subjects have been recruited from a general student subject pool at a German research university addressing both highly experienced ATIS/ITIS using students as well as nonusers. Among the 1987 students contacted, 408 completed the online survey within one week. The respondents range between 20 and 29 years of age. These represent an innovation-affine target group that is used to intermodal mobility solutions. The gender composition seems slightly skewed with 68% male and 32% female students. After cleaning the data set 298 samples remained.

4.2 Measurement Model

To evaluate the measurement model an initial exploratory factor analysis (EFA) with oblique rotation was conducted [43, 72] using a 0.5 cut-off value for the factor loadings [73]. Items are dropped if their major loading or communality are smaller than 0.4, their cross-loading exceeds 0.4 or they logically lack to fit the content of a factor. Most items loaded on their respective dimensions, and overall loadings exceeded 0.7 on average [74, 75]. As a result, 27 items remain constituting nine influencing constructs as driving factors of ITIS acceptance. The usage model matches the proposed number of theoretical constructs. Assessing constructs reliability and validity, a subsequent confirmatory factor analysis with variance-based partial least square structural equation modeling (PLS-SEM) is conducted [76]. Thereby each constructs' convergent validity is examined in terms of its composite reliability (CR) [75], average variance extracted (AVE) [75], and Cronbach's α (Cr-α) [77]. In order to examine the discriminant validity the Fornell–Larcker criterion is applied, which compares the square root of the AVE values with the latent variable correlations [75, 76]. The results of all convergent and discriminative validity measures are summarized in Table 1.

Table 1 Convergent and discriminant validity measures

Construct	[COS]	[EE]	[FC]	[HAB]	[HM]	[UI]	[PE]	[PV]	[SI]	[TIM]	[USE]
Costs [COS] (CR = 0.929; AVE = 0.868; Cr-α = 0.911)	**0.932**										
Effort Expectancy [EE] (CR = 0.911; AVE = 0.720; Cr-α = 0.869)	−0.039	**0.848**									
Facilitating Conditions [FC] (CR = 0.940; AVE = 0.887; Cr-α = 0.883)	0.133	0.090	**0.942**								
Habit [HAB] (CR = 1.000; AVE = 1.000; Cr-α = 1.000)	0.026	0.290	0.066	**1.000**							
Hedonic Motivation [HM] (CR = 0.829; AVE = 0.618; Cr-α = 0.712)	0.341	0.111	0.030	0.149	**0.786**						
Usage Intension [UI] (CR = 0.907; AVE = 0.765; Cr-α = 0.846)	0.030	0.245	0.044	0.237	0.328	**0.874**					
Performance Expectancy [PE] (CR = 0.839; AVE = 0.515; Cr-α = 0.765)	0.068	0.559	0.111	0.359	0.266	0.292	**0.718**				
Price Value [PV] (CR = 0.930; AVE = 0.869; Cr-α = 0.850)	−0.08	0.240	0.094	0.234	0.068	0.250	0.193	**0.932**			
Social Influence [SI] (CR = 0.938; AVE = 0.835; Cr-α = 0.909)	0.186	−0.093	0.002	0.016	0.131	0.110	0.012	0.053	**0.914**		
Time [TIM] (CR = 0.883; AVE = 0.790; Cr-α = 0.746)	0.152	0.340	0.146	0.192	0.275	0.089	0.508	0.115	−0.059	**0.889**	
Usage [USE] (CR = 1.000; AVE = 1.000; Cr-α = 1.000)	0.107	0.086	0.075	0.242	0.186	0.327	0.239	0.080	0.140	0.136	**1.000**

Considering the convergent validity all constructs exceed the required threshold values—namely: 0.7 or greater for CR, 0.5 or greater for AVE, 0.7 or greater for Cr-α [75, 76]. With regard to the discriminant validity, all square roots of each construct's AVE (bold diagonal values in Table 1) are greater than its highest correlations with any other construct [75, 76]. These results suggest that the measurements of the constructs are reliable and that the constructs are sufficiently distinct to each other. Therefore, the constructs are eligible to be used in the structural model.

4.3 Cluster Variables

In order to investigate the mediating influence of psychometric factors on the behavioral intention, the TRI 2.0 [32] and BFI-10 [33] are used to cluster the subjects into distinctive groups. Following Hair [76] and Anable [25] agglomerative hierarchical clustering is applied to investigate which of the psychometric characteristics provides a significant cluster distinctiveness and cluster size. Applying the explorative driven Silhouette criterion and the Calinski and Harabasz criterion [78, 79] the BFI's attribute of neuroticism is chosen as the eligible criterion for clustering in the sample at hand. With the cluster number of $k = 2$ it satisfies the cluster validity criteria concerning within-cluster homogeneity and between-cluster heterogeneity. It separates the sample into two groups: one with a higher tendency toward neuroticism (HN) with 184 subjects and one with a low tendency toward neuroticism (LN) with 114 subjects. All of the other psychometric dimensions provided a clear distinctiveness as well but failed to have the required size for subsequent model evaluation.

4.4 Structural Model

In order to test the hypothesized model a PLS-SEM approach is applied. This is appropriate since the context adaptation does not allow the assumption of a strong theoretical foundation but has an explorative character [80, 81] and is in line with method used in the baseline model [38]. It also allows an evaluation of the subgroups with a relative small sample size and estimates the appendant parameter. Multicollinearity variance inflation factors (VIFs) are computed and since none of them exceeded the conservative threshold of five, multicollinearity is not seen as a problem in this study [76]. The adapted model investigating the driving forces toward using ITIS is shown in Fig. 1 and illustrates the effect sizes for both subgroups.

With a standardized root mean square residual (SRMR)—explaining the difference between the observed correlation and the predicted correlation—of 0.041 the PLS-model has a good model fit according to Hu and Bentler [82], who see 0.08 as a conservative threshold for acceptance. Further the model accounts for 34%

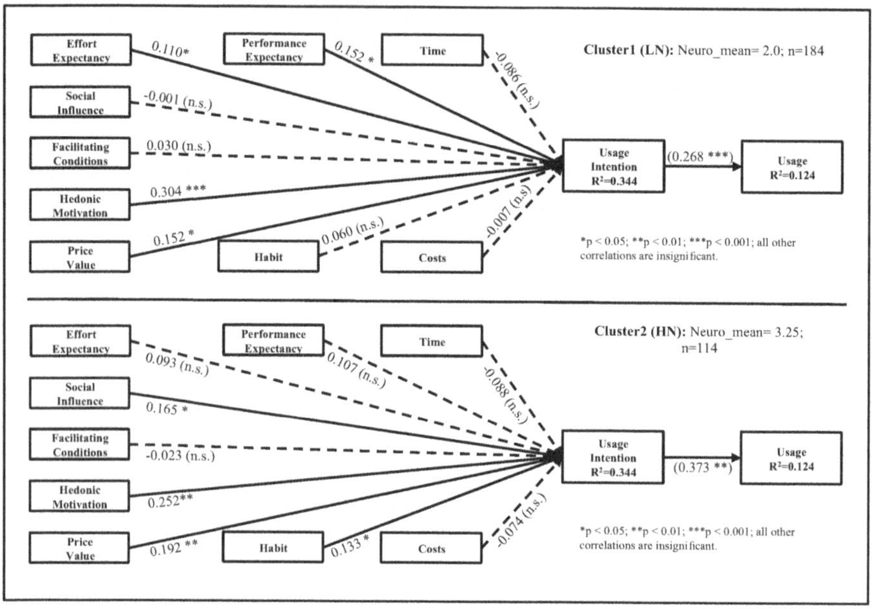

Fig. 1 Structural model of factors influencing the acceptance and usage of ITIS

(R^2 = 0.344) of the observed variance in the behavioral intention toward using an ITIS and 12.5% of the observed variance in the usage behavior (R^2 = 0.124) [44, 83, 84]. Referring to the UTAUT 2 baseline model by Venkatesh et al. [50] this amount of variance explained represents a good result.

The models in Fig. 1 show that neither time nor costs nor facilitating conditions have a significant effect on the behavioral intention in both groups. In contrast the hedonistic motivation has a relatively strong and significant effect in both groups, while the price value has a relative moderate effect in both groups. The models also indicate that the two groups diverge in effects sizes of the factors determining their behavioral intention toward using an ITIS. While effort expectancy (0.110) and performance expectancy (0.152) are strong drivers for LN they are insignificant for HN. On the other hand, are habit (0.133) and social influence (0.165) significant for HNs' usage decisions, but play no significant role in LNs' decisions.

5 Conclusion and Future Research

5.1 Conclusion

This study aims at understanding the driving forces toward the usage of ITIS on an individual's level. Therefore, the UTAUT 2 is transferred to the context of mobility by enriching it with explanatory variables from the area of transportation research.

The results indicate that the driving forces are user-specific and depend on diverse influencing factors that exceed pure economic and socio-demographic dimensions, e.g., habits of past mobility behavior, hedonistic motivations, social influence or psychometric attributes. Thereby it discloses the challenge that a higher sensitivity for individuals' characteristic and complex needs is required to sustainably influence peoples' mobility decisions. Technological innovations can help to exploit this potential. Representing psychological character traits as a source of influence on the decision to use ITIS neuroticism is used in this study to cluster user-types. Diverging effect strengths over the decision triggering factors between these clusters validate the influence of the psychometric measures on the mobility decision. This is important since it legitimates a wider investigation of influencing forces in the mobility context as well as it underpins the necessity for and affectivity of user-adaptation respectively customization.

Neuroticism as a criterion results in two clusters representing the range of high (HN) and lower (LN) tendencies toward neuroticism. Within these groups usage decisions toward using an ITIS are driven by common and group-specific variables. The variables that are active and significant in both groups are the hedonic motivation and price value. While the influence of price value can be explained by the fact that most ITIS are often available for free—seemingly the right pricing strategy —the strong influence of hedonic motivation in both groups is surprising. It shows a stronger effect on the intention to use an ITIS than the 'classic' transport decision factors of time and cost saving, neither of which has an significant effect. This allows two implications: (1) A pleasant design and interaction became crucial service attributes in the ITIS context that can dominate the effects of pure cost and time saving; (2) that the economic benefits of using ITIS are not acknowledged by the users so far. Looking at the potential of sustainable behavioral change the last point reveals a great opportunity and challenge in the development of ITIS. In the end, they have to convince users that using ITIS comes along with economical benefits as well.

Facilitating conditions, costs, and time show no significant impact on the behavioral intention of either group. Considering that the participants in this sample approximate the 'digital native' population it can be interpreted in a way that the handling of ITIS as a mobile app is not perceived as problem at all. As discussed before the investigated population does not perceive these ITIS as being highly economically valuable by now. This is interesting since classical simulation studies on mobility behavior focus on time and costs as main drivers for mobility decisions while neglecting the noneconomic factors. Thus the finding indicates a potential starting-point to improve the quality of simulation studies.

Looking at the group-specific variables in which HN and LN diverge, performance and effort expectancy have a significant effect on behavioral decision in LN while they are insignificant with HN. On the contrary, habit and social influence are significant with LN and do not show any effect with HN. This can be interpreted in a way that LN are more outcome driven when it comes to deciding whether to use an ITIS or not, while an important factor within HN is habit. Together with the strong effect of social influence within HN's decision indicates that they are even more amenable to social pressure than LN. Both insights have consequences on how to approach these customer groups and optimize the ITIS according to their needs. For example, LN seem to be more likely to react to advertisements emphasizing performance and effort reducing benefits, while HN prefer a pronunciation of social and hedonistic aspects.

The insights over the user-specific factor valuations can also be used to adjust and develop groups-specific functionalities and services. For example, an insurance for arriving on time or a group discount can be offered to HN user while an optimization suggestion of the daily trip to work can motivate LN to use ITIS.

5.2 Limitations and Future Research

This study has been conducted with a relatively homogeneous user group. To account for the present heterogeneity in real world consumer behavior the next steps require a cross-sectional representative survey in urban and nonurban areas. This study will gather a wider database about peoples' psychometric characteristics, their typical mobility behavior, and their stated motivations to use an ITIS. This will allow testing robustness of group discriminating features as well as identifying new features since a lager population in the resulting subsamples is required for the application of advanced statistical methods. Further the construct of habit should be extended to more than one item, so that it covers the habitual facets in a more granular way. Since cities diverge in their infrastructures across countries, as does the culture, a multinational survey design is suggested to account for these distinctive features. This is because the available infrastructure and socioeconomic environment of a city have a high impact on the prevailing mobility cultures.

Being able to identify distinct groups of mobility behavior it is important to derive measures and strategies how a change toward sustainable mobility behavior can be initiated. Since pure economic incentives are not sufficient in the long run, the implementation of persuading elements over social influencing systems [85] is a promising approach. Laboratory experiments and deep-interviews with focus groups will support the identification of the decision triggering needs and motivators. Further such settings allow comprising the effect of environmental factors like stress or travel purpose.

An aspect related to service quality that requires further research is the provided functionality. By now many distinct solutions exist to satisfy the divers mobility needs by the customer [13]. Since these solutions are on their own and not connected to a network, where a booking of a service composition is possible in a single app, they do not use their synergetic potential and are not perceived as convenient by the customer. Future research needs to investigate the effects and conditions for service integration, so that the customer recognizes the economic value of ITIS as an overarching network integration platform that serves all needs across the mobility chain.

Altogether this study provides a first step toward a deeper understanding of what drives the usage of ITIS on an individual's level. As a starting point for further investigations of the complex and interdisciplinary topic it contributes to make better mobility decisions and thus increase the quality of life in cities.

References

1. Gehl, J.: Cities for People. Island Press, Washington (2010)
2. Knoflacher, H.: Zurück zur Mobilität!: Anstöße zum Umdenken. Ueberreuter, Wien (2013)
3. Batty, M.: The New Science of Cities. The MIT Press, Cambridge (2013)
4. Larson, K. et al.: Updates from our future city. www.youtube.com/watch?v=pUum3OQuI24c. Accessed 20 June 2016
5. Kenyon, S., Lyons, G.: The value of integrated multimodal traveller information and its potential contribution to modal change. Transp. Res. Part F Traffic Psychol. Behav. **6**(1), 1–21 (2003)
6. Sussman, J.S.: Perspectives on Intelligent Transportation Systems (ITS). Springer, New York (2005)
7. Hilty, L.M., et al.: The relevance of information and communication technologies for environmental sustainability—a prospective simulation study. Environ. Model Softw. **21**(11), 1618–1629 (2006)
8. Eriksson, L., et al.: Interrupting habitual car use: the importance of car habit strength and moral motivation for personal car use reduction. Transp. Res. Part F Traffic Psychol. Behav. **11**(1), 10–23 (2008)
9. Gardner, B.: Modelling motivation and habit in stable travel mode contexts. Transp. Res. Part F Traffic Psychol. Behav **12**(1), 68–76 (2009)
10. Verplanken, B., et al.: Attitude versus general habit: antecedents of travel mode choice. J. Appl. Soc. Psychol. **24**(4), 285–300 (1994)
11. Schneider, R.J.: Theory of routine mode choice decisions: An operational framework to increase sustainable transportation. Transp. Policy **25**, 128–137 (2013)
12. Adler, J.L., Blue, V.J.: Toward the design of intelligent traveler information systems. Transp. Res. Part C Emerg. Technol. **6**(3), 157–172 (1998)
13. Kramers, A.: Designing next generation multimodal traveler information systems to support sustainability-oriented decisions. Environ. Model Softw. **56**, 83–93 (2014)
14. Blau, B. et al.: How to coordinate value generation in service networks, Bus. Inf. Syst. Eng. **5**(1), 343 (2009)
15. Lisson, C. et al.: Decisions in Mobility Service Networks—Coordinating Demand and Supply Using a Mechanism Design Approach. In: Proceedings of the Forty-Ninth Annual Hawaii International Conference on System Sciences (HICSS) (2016)
16. Bamberg, S., et al.: Choice of travel mode in the theory of planned behavior: the roles of past behavior, habit, and reasoned action. Basic Appl. Soc. Psych. **25**(3), 175–187 (2003)
17. Steg, L., et al.: Instrumental-reasoned and symbolic-affective motives for using a motor car. Transp. Res. Part F Traffic Psychol. Behav. **4**(3), 151–169 (2001)
18. Bamberg, S., et al.: Social context, personal norms and the use of public transportation: two field studies. J. Environ. Psychol. **27**(3), 190–203 (2007)
19. Schiefelbusch, M.: Rational planning for emotional mobility? The case of public transport development. Plan. Theory **9**(3), 200–222 (2010)
20. Ben-Akiva, M., Lerman, S.R.: Discrete Choice Analysis: Theory and Application to Travel Demand. The MIT Press, Cambridge (1985)

21. Grotenhuis, J.-W., et al.: The desired quality of integrated multimodal travel information in public transport: customer needs for time and effort savings. Transp. Policy **14**(1), 27–38 (2007)
22. Chorus, C.G., et al.: Use and effects of advanced traveller information services (ATIS): a review of the literature. Transp. Rev. **26**(2), 127–149 (2006)
23. Abou-Zeid, M., Ben-Akiva, M.: Well-being and activity-based models. Transportation (Amst) **39**(6), 1189–1207 (2012)
24. Maslow, A.H.: Motivation and Personality. Harper Row, New York (1970)
25. Anable, J.: "Complacent Car Addicts" or "Aspiring Environmentalists"? Identifying travel behaviour segments using attitude theory. Transp. Policy **12**(1), 65–78 (2005)
26. Vallerand, R. J.: Toward a hierarchical model of intrinsic and extrinsic motivation. Adv. Exp. Soc. Psychol. **29** (1977)
27. Vallerand, R.J., Lalande, D.R.: The MPIC model: the perspective of the hierarchical model of intrinsic and extrinsic motivation. Psychol. Inq. **22**(1), 45–51 (2011)
28. Axhausen, K.W.: Social networks, mobility biographies, and travel: survey challenges. Environ. Plann. B. Plann. Des. **35**(6), 981–997 (2008)
29. Gärling, T., Axhausen, K.: Introduction: habitual travel choice. Transportation (Amst) **30**(1), 1–11 (2003)
30. Götz, K.: Mobilitätsstile. In: Schöller, O. et al. (eds.) Handbuch Verkehrspolitik, VS Verlag für Sozialwissenschaften, 964 (2008)
31. Seebauer, S., et al.: Technophilia as a driver for using advanced traveler information systems. Transp. Res. Part C Emerg. Technol. **60**, 498–510 (2015)
32. Parasuraman, A., Colby, C.L.: An updated and streamlined technology readiness index: TRI 2.0. J. Serv. Res. **18**(1), 59–74 (2014)
33. Rammstedt, B., John, O.P.: Measuring personality in one minute or less: a 10-item short version of the big five inventory in English and German. J. Res. Pers. **41**(1), 203–212 (2007)
34. Chlond, B. et al.: Data quality and completeness issues in multiday or panel surveys. In: Zmud, J. et al. (eds.) Transport Survey Methods: Best Practice for Decision Making, Emerald Group Publishing Limited, pp. 373–392 (2013)
35. Ajzen, I.: The theory of planned behavior. Organ. Behav. Hum. Decis. Process. **50**(2), 179–211 (1991)
36. Taylor, S., Todd, P.A.: Understanding information technology usage: a test of competing models. Inf. Syst. Res. **6**(4), 144–176 (1995)
37. Limayem, M., et al.: How habit limits the predictive power of intention: the case of information systems continuance. MIS Q. **31**(4), 705–737 (2007)
38. Venkatesh, V. et al.: Consumer acceptance and use of information technology: extending the unified theory of acceptance and use of technology. MIS Q. (2012)
39. Grönroos, C.: A service quality model and its marketing implications. Eur. J. Mark. **18**(4), 36–44 (1984)
40. Cronin, J.J., Taylor, S.A.: Measuring quality: a reexamination and extension. J. Mark **56**(3), 55–68 (1992)
41. Diana, M.: Measuring the satisfaction of multimodal travelers for local transit services in different urban contexts. Transp. Res. Part A Policy Pract. **46**(1), 1–11 (2012)
42. Ding, D.X., et al.: e-SELFQUAL: a scale for measuring online self-service quality. J. Bus. Res. **64**(5), 508–515 (2011)
43. MacKenzie, S.B., et al.: Construct measurement and validation procedures in MIS and behavioral research: integrating new and existing techniques. MIS Q. **35**(2), 293–334 (2011)
44. Gefen, D. et al.: An update and extension to SEM guidelines for administrative and social science research. MIS Q. **35**(2), iii–ixiv (2011)
45. Anable, J., Gatersleben, B.: All work and no play? The role of instrumental and affective factors in work and leisure journeys by different travel modes. Transp. Res. Part A Policy Pract. **39**(2–3), 163–181 (2005)
46. Klein, H.K., Myers, M.D.: A set of principles for conducting and evaluating interpretive field studies in information systems. MIS Q. **23**(1), 67–93 (1999)

47. Flick, U.: An introduction to qualitative research. Sage Publications, Thousand Oaks (2014)
48. Glaser, B.G., Strauss, A. L.: The Discovery of Grounded Theory: Strategies for Qualitative Research. Transaction Publishers (2009)
49. Krippendorf, K.: Content Analysis: An Introduction to its Methodology. Sage Publications (2012)
50. Venkatesh, V., et al.: User acceptance of information technology: toward a unified view. MIS Q. **27**(3), 425–478 (2003)
51. Davis, F.D.: Perceived usefulness, perceived ease of use, and user acceptance of information technology. MIS Q **13**(3), 319–339 (1989)
52. Venkatesh, V., Davis, F.D.: A theoretical extension of the technology acceptance model: four field studies. Manage. Sci. **45**(2), 186–204 (2000)
53. Davis, F.D., et al.: Extrinsiacnd intrinsic motivation to use computers in the workplace. J. Appl. Soc. Psychol. **22**(14), 1111–1132 (1992)
54. Rogers, E.: Diffusion of Innovation. Free Press, New York (1995)
55. Bandura, A.: Social Foundations of Thought and Action: A Social Cognitive Theory. Prentice Hall, Englewood Cliffs (1986)
56. Stradling, S.G., et al.: Performance, importance and user disgruntlement: a six-step method for measuring satisfaction with travel modes. Transp. Res. Part A Policy Pract. **41**(1), 98–106 (2007)
57. Davis, F.D.: JSTOR. MIS Q. **13**(3), 319–340 (1989)
58. Thompson, R.L.: Personal computing: toward a conceptual model of utilization. MIS Q. **15**(1), 124–143 (1991)
59. Moore, G.C., Benbasat, I.: Development of an instrument to measure the perceptions of adopting an information technology innovation. Inf. Syst. Res. **2**(3), 192–222 (1991)
60. Xin Ding, D. et al.: The impact of service system design and flow experience on customer satisfaction in online financial services. J. Serv. Res, Vol. **13**(1), 96–110 (2009)
61. Taylor, S., Todd, P.A.: Assessing IT usage: the role of prior experience. MIS Q. **19**(2), 561–570 (1995)
62. Brown, S.A., Venkatesh, V.: Model of adoption and technology in households: a baseline model test and extension incorporating household life cycle. MIS Q. **29**(3), 399–436 (2005)
63. van der Heijden, H.: User acceptance of hedonic information systems. MIS Q. **28**(4), 695–704 (2004)
64. Ben-Akiva, M., Boccara, B.: Discrete choice models with latent choice sets. Int. J. Res. Mark. **12**(1), 9–24 (1995)
65. Dodds, W.B. et al.: Effects of price, brand, and store information on buyers' product evaluations
66. Zeithaml, V.A., et al.: Problems and strategies in service marketing. J. Mark. **49**(2), 33–46 (1985)
67. Kim, S.S. et al.: Research note—two competing perspectives on automatic use: a theoretical and empirical comparison. Inf. Syst. Res (2005)
68. Ajzen, I., Fishbein, M.: The influence of attitudes on behavior. Handb. Attitudes **173**, 173–221 (2005)
69. Groves, R.M., et al.: Survey Methodolog. Wiley, LondonWile (2009)
70. Malhotra, M.: An assessment of survey research in POM: from constructs to theory. J. Oper. Manage. **16**(4), 407–425 (1998)
71. Churchill, G.A.: A paradigm for developing better measures of marketing constructs. J. Mark. Res. **16**(1), 64–73 (1979)
72. Harman, H.: Modern Factor Analysis. University of Chicago Press, Chicago (1976)
73. El Hedhli, K., Chebat, J.-C.: Developing and validating a psychometric shopper-based mall equity measure. J. Bus. Res. **62**(6), 581–587 (2009)
74. Bagozzi, R.P., Baumgartner, H.: The evaluation of structural equation models and hypothesis testing. In: Bagozzi, R. (ed.) Principles of marketing research, pp. 386–422. Blackwell Publishers (1994)

75. Fronell, C., Larcker, D.: Evaluating structural equation models with unobservable variables and measurement error. J. Mark. Res. **18**(1), 39–50 (1981)
76. Hair, J., et al.: Multivariate Data Analysise. Prentice Hall, Englewood Cliffs (1998)
77. Cronbach, L.: Coefficient alpha and the internal structure of tests. Psychometrika **16**(3), 297–334 (1951)
78. Caliński, T., Harabasz, J.: A dendrite method for cluster analysis, Commun. Stat. Theory Methods (2007)
79. Kaufman, L., Rousseeuw, P.J.: Finding Groups in Data: An Introduction to Cluster Analysis. Wiley, New York (2005)
80. Gefen, D. et al.: An update and extension to SEM guidelines for administrative and social science research. Editorial Comment. **35**(2), III–XII (2011)
81. Ringle, C. et al.: Editor's comments: a critical look at the use of PLS-SEM in mis quarterly. http://aisel.aisnet.org/misq/vol36/iss1/2 (2012). Accessed 20 June 2016
82. Hu, L., Bentler, P.M.: Cutoff criteria for fit indexes in covariance structure analysis: conventional criteria versus new alternatives. Struct. Equ. Model. A Multidiscip. J. **6**(1), 1–55 (1999)
83. Chin, W.W., et al.: A partial least squares latent variable modeling approach for measuring interaction effects: results from a Monte Carlo simulation study and an electronic-mail emotion/adoption study. Inf. Syst. Res. **14**(2), 189–217 (2003)
84. Henseler, J., et al.: Common beliefs and reality about PLS: comments on Ronkko and Evermann (2013). Organ. Res. Methods **17**(2), 182–209 (2014)
85. Stibe, A.: Towards a framework for socially influencing systems: meta-analysis of four PLS-SEM Based Studies. In: MacTavish, T., Basapur, S. (eds.) Persuasive Technology, pp. 172–183. Springer International Publishing, New York (2015)

You Are What You Share: Understanding Participation Motives in Peer-to-Peer Carsharing

Mark-Philipp Wilhelms, Sven Henkel and Katrin Merfeld

Abstract Carsharing is changing the automotive industry. Although existing offers fail to fully meet expectations, researchers and mangers agree that these access-based services will strongly affect the industry's existing business model. Employing means-end chain analysis, we provide first insights in car owners' and renters' participation motives to understand the Peer-to-Peer (P2P) carsharing trend. Users are not only concerned about cost savings and altruism but also about expressing their personality through the car they offer or rent. While renters realize particular lifestyle expectations, car owners participate to express their interest in enabling others' mobility. P2P carsharing provides users with sharable life-enriching stories. The 2010s are all about sharing—one's pictures, rooms, and cars.

Keywords Sharing economy · Carsharing · Peer-to-Peer · Consumer behavior · Qualitative method · Means-end chain

1 The Rise of Access-Based Services

We have left the age of ownership and take the first steps in the era of access, where companies are confronted with consumers who no longer want to own products. Those customers prefer to consume at the time, place, and in the quantity and intensity they desire. Hence, numerous so-called access-based services (ABS) are being funded by companies. ABS are defined as a market-mediated subform of the sharing economy, where consumers are willing to pay a fee for access to and

M.-P. Wilhelms (✉) · S. Henkel · K. Merfeld
EBS University for Business and Law, Gustav-Stresemann-Ring 3, 65189 Wiesbaden, Germany
e-mail: mark-philipp.wilhelms@ebs.edu

S. Henkel
e-mail: sven.henkel@ebs.edu

K. Merfeld
e-mail: katrin.merfeld@ebs.edu

G. Meyer and S. Shaheen (eds.), *Disrupting Mobility*,
Lecture Notes in Mobility, DOI 10.1007/978-3-319-51602-8_7

subsequent use of an object [1]. Additionally, currently, consumers become micro-entrepreneurs offering their assets—from a spare room in their apartment to their cars and drills—for rent on one of thousands of online networks. Essentially, these consumers offer ABS, too.

The trend for access will change the automotive industry. Paired with an increasing contempt of the car as a status symbol—especially among younger customers in the triad markets—managers have to act and adapt existing business models. Carsharing has been deemed as a panacea for companies to compete in the age of access. Nevertheless, existing carsharing offers barely meet profit goals and do not yet fully live up to consumers' needs.

Daimler and BMW, on one hand, have founded Business-to-Consumer (B2C) carsharing organizations (e.g., DriveNow, Car2Go) around the globe and flooded cities with thousands of Smarts, Minis, and BMW 1 Series. Ford and Opel, on the other hand, are traveling down a different path: They are experimenting with Peer-to-Peer (P2P) Carsharing. Online communities, such as Turo in the US and GM's CarUnity in Germany, enable car owners to rent their vehicles to other users (renters) who can find the vehicles that they require—it is the AirBnB of cars. P2P carsharing is evolving to fill the gap between classical car rental and carsharing. While the former is typically used for periods of 1 week or more, the latter is used for rental durations of less than a day [2].

By no means are these P2P carsharing experiments the results of marketing frenzies. Top managers, such as Ford's Mark Fields, are well aware of the trend and forecast that in the future, 34% of people will want to rent their vehicle to others. A similar number of people would be waiting to rent a private car to satisfy their mobility needs. Current prognoses expect 9.8 million members of P2P carsharing in 2025 for North America and Europe [2].

Nevertheless, despite the high reported willingness of consumers to partake in P2P carsharing, adoption has been relatively slow and P2P carsharing networks report issues in growing the user base. It is a long way to get a consumer, who intends to participate in such a service, to begin actual usage. This is even more so the case when one expects the consumers to become service providers as is the case with P2P carsharing car owners. Bridging this barrier to provide personal belongings for interpersonal access is a key issue in spreading P2P carsharing. Moreover, it requires a detailed knowledge of the consumer-relevant benefits and a thorough understanding of consumers' motivational drivers to design proper communication campaigns and enhance service acceptance and adaption to attract additional users.

Participation motives have not yet been thoroughly analyzed. Till today, advocates of the sharing economy promote a romantic view and highlight altruistic, prosocial and environmental motives of sharing economy's participants [3, 4]. On the other hand, academia found the domination of utilitarian motives among users [1, 5]. For the specific case of car owners, P2P carsharing networks advertise that they can earn about USD 6000 annually—while car renters can decrease rental-costs by up to 40% [6], further underlining the importance of monetary goals in the participation decision.

To gain a deeper understanding of car owners' and renters' participation motives and motivational patterns in P2P carsharing, we conducted a series of in-depth interviews with German P2P carsharing users. Our findings help managers to grow participation rates through concerted communication campaigns and overcome challenges of differentiation and competition in a growing market. We highlight connections between carsharing-related attributes and identified motives, present prototypes of P2P carsharing users and outline how to address these in communication campaigns.

2 Methodological Background—A Brief Introduction into Means-End Chain Theory

For the current research we employed Means-End Chain (MEC) analysis. MEC is a qualitative research method, suitable to uncover insights on nondirectly observable phenomena (e.g., values, attitudes, and believes), when investigating consumers' motivations for service usage and participation. According to MEC theory, all consumer activities, their directions and strengths, rest on the construct of motivation to explain a consumer's goal-directed behavior [7]. A central assumption of MEC theory is that the hierarchical representation of attributes, consequences, and values represents different goal and motive structures, essentially outlining the consumer decision-making process [8].

The MEC technique allowed us to gain a detailed understanding of a user's P2P carsharing participation motives, as it links participation-relevant attributes (i.e., means), resulting utility components, and underlying individual values (i.e., ends). MEC analysis is widely established in academia and of great value to managers. The results indicate the values that make a service relevant to a specific consumer group and equip managers with an understanding of the functional and psychosocial consequences the customers strive for in service participation. Ultimately, this helps managers to identify market segments and develop concerted positioning and advertising strategies [7].

This can easily be exemplified by the following: in the case of car owners, the attributes *rental income* and *low utilization* describe two participation-relevant characteristics. The consequences can be interpreted as the utility derived from the attributes and distinguished into functional and psychosocial consequences. Functional consequences are direct outcomes regarding the use of a product or service—for car owners these include *reduction of fixed and maintenance costs*, while psychosocial consequences describe the more abstract social and psychological outcomes such as consequences that are outside the core function of a commodity, such as *feels good* [9]. Values represent individual standards or goals, such as *economic interest* [8].

For the data collection process, interview questions were reviewed by independent, experienced qualitative researchers. To uncover the relationships between the MEC elements of the interviewees, we employed the laddering interview technique, as it reflects the assumed hierarchical structure of the MEC elements. In a laddering interview, participants are "forced up" the ladder of abstraction, through continuously asking why questions, uncovering the structural aspects of consumer knowledge as modeled by the MEC. Specifically, we asked participants to provide reasons for their participation motivation, followed by inquiring about each reasons' importance and the underlying rationale until the level of terminal value was achieved or participants were unable to provide further insights.

The data analysis process included a transcription of the interviews' audiotapes, which were reviewed several times by the coders to increase content familiarity and sensitivity. The data was content analyzed in a collaborative effort by a team of coders including some of the authors and independent guest coders. Content analysis followed a three-level process, incorporating the three main phases of description (reliance on verbatim quotes of interviewees), analysis (identification of overarching factors, themes, and relationships), and interpretation (making sense of meanings in the context) [10]. To account for the hierarchical nature of MECs, the elements of the individual laddering interviews were first coded independently and then assigned to the different MEC categories (i.e., attributes, consequences, values). Subsequently, similar codes at the same hierarchical levels were combined to further reduce the number of codes.

For every interviewee, we identified a number of individual MECs. Table 1 provides an explanatory depiction of the uncovered MECs for car owners. The MEC analysis then requires the researcher to aggregate the various individual MECs and depict them in a hierarchical value map (HVM), to derive a set of overarching motivational patterns, relevant in the participation decision [7]. In an HVM, relationship-paths are visualized, starting with attributes at the bottom, which are related to functional and psychosocial consequences, and values at the highest level. Essentially, a HVM graphically indicates the values that make a service relevant to a specific consumer group and the benefits that consumers obtain through participation. In conclusion, a HVM reveals insights into the hierarchical nature of consumers' cognitive structures and derivation of implications regarding the underlying motivational patterns of P2P carsharing participation.

Table 1 Exemplary means-end chains

Respondent	Attribute	Functional consequence	Psychosocial consequence	Value
Achim	Low utilization	Decrease vehicle need	Environmental awareness	Sustainability
Gaby	Rental income	Reduction of fixed costs	Save money	Economic interest
Joern	Interest in sharing	–	Feels good	Help others
Markus	Low utilization	Additional income	Have money for other purposes	Quality of life

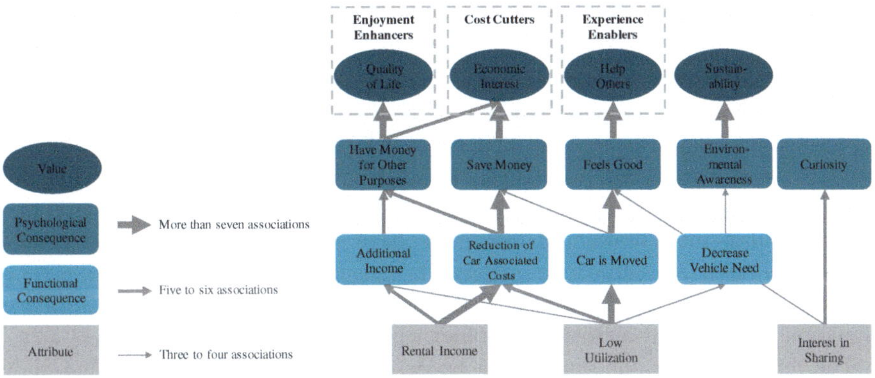

Fig. 1 Car owners' hierarchical value map

Figure 1 depicts the HVM developed for car owners. Looking at the HVM reveals one major theme related to the desire to *save money*, associated with the functional consequence *reduction of car associated costs*. The functional utility *reduction of car associated costs*, is a direct outcome of the associated attributes *rental income* and *low utilization*, as these attributes allow car owners to reduce the costs of car ownership. In conclusion, these associations add to the value *economic interest*.

3 The People Who Share—the Study's Participants

We conducted a total of 31 in-depth interviews with active German P2P carsharers via telephone, covering owners, and renters, to obtain a fundamental understanding of their participation motives. Interviews, each lasting approximately 30 min, were conducted until the point of theoretical saturation had been reached. To determine whether theoretical saturation had been reached, we compared results of new interviews with the already analyzed interviews, until we were unable to identify novel properties or dimensions that would help explain the participation decision [11].

To derive meaningful results and ensure an economic data collection process, participant selection is of paramount importance in qualitative research. Therefore, the interviewee recruitment was organized by a German P2P carsharing network. All interviewees had a proven record of P2P carsharing usage. In detail, the sample covers a broad range of customer types.

The car owners interviewed can be described as follows: On average, they have 2 years of experience in renting their cars on P2P carsharing networks, have an average annual net income of 30,000–40,000 Euros, are 36 years of age and use their car twice a week. In terms of car ownership, seven respondents have more than one car within their household and three have three or more cars. Seventeen respondents own an alternative means of transportation (bike, motorbike, and related forms). Lastly, car owners' P2P carsharing usage behavior is as follows: They rent out their car almost every other week, usually for one to 2 days.

The renters within our sample can be described as follows: On average, they have 2 years of experience in renting cars on P2P carsharing networks, have an average annual net income of 30,000–40,000 Euros and are 39 years of age. Four have at least one car within their household and use it more than twice per week. Nine respondents own an alternative means of transportation (bike, motorbike, and related forms). Finally, renters' P2P carsharing usage behavior is highly similar to that of car owners: They rent cars almost every other week, usually for 1 to 2 days.

4 The Findings—You Share Because You Care?

The goal of this study is to understand overarching participation motives and motive structures in P2P carsharing. In the following sections, we will present our findings—first for car owners, and then for renters. To provide the reader with a deeper understanding of participation motives we developed a set of prototypical P2P carsharing users based on the most dominant motivational patterns that influence P2P carsharing participation. To do so, we considered every MEC displayed in the HVM from bottom (attribute) to top (value) as a potential chain representing a perceptual orientation. To assess the strength of the MECs, intra-chain relations were summarized and evaluated [7]. It is important to note that an individual's P2P carsharing usage will be based on a unique set of different intensities of the motives discussed below. To highlight the unique qualitative insights uncovered through employing MEC analysis, we elaborate on the connections between attributes, consequences as well as values and enrich the discussion with various quotes from the interviews.

4.1 Car Owners

4.1.1 Cost Cutters

For *cost cutters* the attributes *rental income* and *low utilization* are relevant. Those are associated with the consequences *reduction of car associated costs* and *additional income. Rental income* enables cost cutters to reduce their costs of car ownership. For example, Dennis participates to reduce insurance fees and taxes, which are independent of his car's usage: 'I pay insurance, parking, taxes and all that every month—no matter if I use the car or not. When I rent my ride, I can reduce those costs a bit and not pay more.' Additionally, Christian stresses the ability to reduce maintenance costs: 'Renting out my car every now and then does not cover all of the maintenance costs—but that's not the point—all I want is to reduce the costs by some fraction.' Others, such as Michael, aim to reduce costs resulting from low utilization: 'I hardly use my car, and when I do, the battery is often empty. So I thought, I will just rent it. Then, someone else charges the battery for me and I save all of the electricity money and the hassle.' The overarching goal of *saving money* is the psychosocial consequences driving cost cutters. They typically do not plan to save money for a particular objective, but are attracted by the possibility of avoiding costs and increasing savings through participation, such as Jose, who states, 'I just think it [P2P carsharing] is a nice idea to save some money. It allows me to save a bit every time.' Hence, cost cutters differentiate between costs resulting from car ownership, which can be reduced by renting out the car, and costs resulting from low usage of the car.

4.1.2 Enjoyment Enhancers

Through participation in P2P carsharing, *enjoyment enhancers* increase their disposable income and consequently have money for other purposes, which they invest in various activities, enhancing their overall lifestyle. *Low utilization* and *rental income* are attributes related to the consequences *reduction of car associated costs, additional income,* and ultimately having *money for other purposes*. Achim even goes a little bit further: 'When I bought the car I already had in mind to rent it out and use the money for something different. By now, I rent it to pay for these extras.' He proclaims a prior intention to acquire a car for renting purposes and the consequent income, as opposed to the common practice of renting out a car with low utilization. Markus, on the other hand, states, 'I don't use the rental income on my car. I rather go out with friends—it is just extra money.' In conclusion, in our sample, funds are invested in enhancing one's own lifestyle. Car owners want to realize their dreams. Often they have very specific goals in mind including even the enabling of car ownership. Overall, enjoyment enhancement is related to hedonic benefits resulting from the perceived merits of a transaction and the satisfaction and pleasure of obtaining the financial advantages associated with renting out one's car.

4.1.3 Experience Enablers

Experience enablers are mainly driven by the psychosocial consequence that it *feels good* to provide renters with mobility. On an attribute level, *interest in sharing* and *low utilization* are relevant for experience enablers; this is related to the consequences that the *car is moved* and most importantly, that it *feels good* to provide others with mobility. 'I rent out my hardly used white Fiat 500, which comes with red seats. The perfect wedding car. Upon returning the car, the newlyweds always show me the wedding pictures. It's great to be part of these experiences. On top of that, wedding cars are always returned in perfect condition—vacuumed, with the scent of flowers,' states Jan. Joerg mentioned 'You know, it's just nice when you can be part of another person having a great experience. I helped this one guy, who wanted to go to a concert and needed a car—since it was a weekend, I didn't need mine. Afterwards, he told me about the great trip and all of the fun he had. It was nice knowing that I helped create this experience.' These statements underline that experience enablers like to empower others to create lasting memories or to rent cars that they would otherwise not find at a commercial outlet. Lastly, experience enablers do not participate for altruistic motives only, but rather display a stewardship behavior. While they take pride in helping others by granting them the opportunity of mobility, they report to engage in a form of cost-use calculation and say that they would not necessarily participate if no remuneration were paid.

4.2 *Renters*

4.2.1 Savings Seekers

On an attribute level, *distance between renter and car owner* as well as *rental price* matter. These relate primarily to the psychosocial consequence *saving money* in their mobility budget. In this context, participant Anne stated: 'Websites like CarUnity [a P2P network] are simply cheaper than Hertz—especially on the weekend.' Even though this insight is rather evident, it beautifully exposes the following: sharing is not an act of pure altruism. Specifically, users are keen to save money, allowing them to increase their disposable income. One example is Jan A., who states, 'At Hertz, I never know which car I'm going to get. Therefore, I don't know my costs. Doing P2P carsharing I know those exactly. Additionally, the cars are around the corner—hence, I save taxi costs.' Consequently, renters appreciate low and transparent costs of private carsharing, resulting from the transparent presentation of costs and the often short distance between car owner and renter.

4.2.2 Convenience Seekers

The attributes *distance between renter and car owner* and *communication with the car owner* are mainly associated with the value convenience. Jan B. highlights: 'Ideally, I rent a car from my neighbor; often it is even in front of my door. That's easy', referring to the consequence *easy accessibility*. Furthermore, renters such as Andreas are stressing the ability to *plan precisely*, and state, 'I communicate with the car owner to get to know him. Is he easy-going or a nerd about cars? I want to know this to have an easy renting process and plan accordingly.' Another consequence, frequently associated with convenience, is *saving money*, as the short distance between *place of residence and car* reduces commuting costs. In conclusion, the psychosocial consequence *saving stress, effort and time* drives renters' desire for convenience. Participation enables them to save money, plan ahead, and easily reach the cars' location.

4.2.3 Expression Seekers

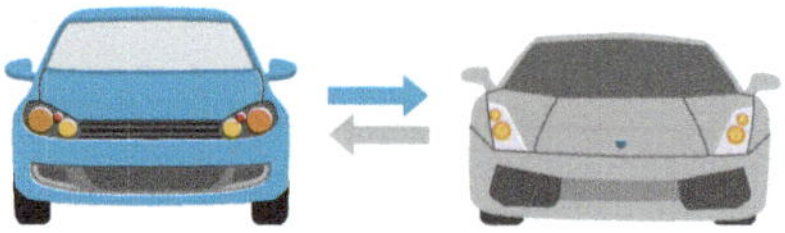

Expression seekers appreciate the lower *rental prices* compared to commercial outlets and the *variety of vehicles*. These are associated with the consequence of being well equipped for a symbolic (e.g., status) purpose and investing the *saved funds* into specific activities, enhancing overall well-being. Merlin appreciates that he can rent average cars, with local license plates, so he does not get unwanted attention from his friends. On the contrary, Herbert rents "nice looking salesman cars when I need to drive to clients instead of using my old Honda." Hence, renters like Merlin and Herbert use P2P carsharing to signal status. Due to their desire to signal status, they value knowing beforehand which car they will get. Others use the saved funds for activities enhancing their overall well-being such as Valentin, who invests these in his hobby as means for self-expression. Conclusively, expression seekers value P2P carsharing for the ability to express themselves by signaling status with the rented car and investing the saved money in alternative, life-improving measures.

4.2.4 Certitude Seekers

On the attribute level, elements such as the *car owner's profile*, *communication with car owner*, *distance between renter and car owner*, and *car type and features* are relevant for certitude seekers. The resulting consequences are *planning certainty*, being *well equipped* but also *getting to know the car owner*. Renter Sarah elaborated, 'I wanted to surprise my boyfriend with a special convertible for the weekend. At Hertz we got a VW Tiguan. On CarUnity, I found a great convertible with a children's seat on the second try. It was a great weekend!' She wonderfully shows how P2P carsharing allows for meeting specific functional needs, combining certain knowledge to obtain a specific car with distinct features and enabling exactly the experience the participant desires. Martin, another renter, has a more altruistic need: 'Shelby, my Labrador, loves to take walks in the woods. Therefore, I need a car with a dog-box. Only on CarUnity you can be sure to get that.' Those renters seek certitude because they want the specific mobility solution that they require. In addition, renters like Lukas value that they get to know the car owner, his preferences and attitude toward the car and can thus handle the car as well as the transaction better. Concluding, certitude seekers seek to be aware of all relevant product and service features prior to the first and for all subsequent usages which allows a repeated utilization under known conditions.

5 Management Takeaways

The analysis of the MECs' strengths revealed some indications about the relative importance of the derived prototypes. A sense of this importance enables managers to organize communication activities strategically. In the case of car owners, the motives of *cost cutters* and, to a slightly lesser extent, *experience enablers* were dominant. For car renters, *savings seekers,* and *certitude seekers* were the most prominent outcomes in our case. Opposing the idealistic view of the sharing economy, we find that environmental benefits are rather perceived as a side effect of P2P carsharing participation. Users share a form of environmental consciousness, as do users of B2C carsharing, and perceive carsharing as a sustainable mobility option. But, they do not consciously decide to participate in P2P carsharing to fulfill environmental needs or meet correlated concerns. They rather appreciably accept the beneficial connotation. In this context, Martin states: 'If P2P carsharing would adversely affect the environment I would reconsider participation.' He thereby shows that sustainability is a factor, but it does not primarily trigger his usage. This stands in opposition with some P2P carsharing network's current communication messages and implies a rather limited applicability of this trigger to the otherwise widely used communications of sustainability benefits.

Throughout the various interviews, it became evident that users are highly satisfied with their P2P carsharing experiences. As consumers are generally interested in the service, increasing user numbers is predominantly a marketing communication issue. This study shows that a variety of basic customer needs are fulfilled through P2P carsharing. These beneficial aspects need to be highlighted in communication campaigns to grow participation rates.

As social media is one of the core drivers and enablers of the sharing economy and mobility activities are the most frequently shared content, it is the most adequate communication channel. In the P2P carsharing context, campaign messages need to include themes, which are framed in a way that can be easily shared in social media. In conclusion, marketing professionals are asked to embed carsharing-related benefits into stories to be liked, commented, and shared on the mediums of Facebook and Instagram.

5.1 How to Better Attract Car Owners

Car owners are motivated by a set of values, each of which can be used to identify a specific customer segment and develop targeted marketing campaigns. Connecting the main participation motives is a 'me-first' logic: Car owners put their personal benefits first and postpone the benefits for renters. Consequently, advantages for others are perceived as a side product of P2P carsharing, even for experience enablers, as they aim to balance their cost-benefit calculations.

Cost cutters want to achieve one main goal through participation: lowering the direct and indirect costs of car ownership. To attract them, P2P networks need to clearly outline the ability to save money by reducing fixed costs, and, at the same time, highlight the minimal effort required to participate in P2P carsharing. Therefore, it is insufficient to highlight unspecified savings potential. Instead, communication campaigns should adapt the indirect costs resulting from car ownership and highlight the possibility to reduce these costs. This problem-solving approach is common in advertising communication and is likely to increase user numbers. Moreover, the desire for minimal effort is essential to attracting car owners. In today′s world, consumers are used to managing their lives with the touch of a button on their smartphone screens while simultaneously perceiving disposable time as rather limited. Therefore, apps facilitating the entire rental process, including rental contract management, communication hubs, immediate access to insurance services in case of an accident, authentic reviews, and an easy means to contact the P2P network are of paramount importance to lower especially mental barriers to access. A further digitalization of the rental process, for instance, by substituting the personal key hand over process through technology mechanisms, would foster a seamless rental experience. For example, in an effort to optimize the user experience, French P2P carsharing network Koolicar, introduced KoolBox, which enables keyless transactions between car owners and renters. If implemented correctly, P2P networks could then follow Airbnb's footsteps and serve the demands of business travelers as well as move closer to the benefits of B2C solutions.

Enjoyment enhancers participate to generate additional income through P2P carsharing. These extra funds are spent on very special things in life such as premium opera tickets instead of the average seats or a night out with friends instead of watching Netflix at home. Therefore, the motivation for enjoyment enhancers is not to save money but to rather spend it in fields they could otherwise not afford. The existence of this car owner group offers potential for automotive manufacturers, such as Ford and Opel, operating P2P networks, too. Communication campaigns addressing this consumer group can raise upselling and post-purchase sales potential. Providing consumers with a means to privately generate additional funds can be an effective tool to lower their perceived costs at the time of car purchase, while simultaneously increasing sales. This can be particularly effective in leasing scenarios. In a sales pitch, dealers could show consumers how a single one-day rental transaction can create the necessary funds needed to pay for months of a sunroof, entertainment or driver assistance system.

Experience enablers take pride in empowering renters with a unique experience and in being part of this experience. They value both: meeting the renters and hearing stories of the experience they helped to create by renting out their cars. At the same time, experience enablers are rational; they engage in cost-benefit calculations, too, just like the cost cutters do. Communication campaigns should emphasize the good feeling that car owners experience when their car is generating a purpose by offering mobility to others. Doing so will result in a number of benefits for P2P carsharing organizations: First, when a car becomes an essential part in an experience a renter wishes to have, it will be treated as such. Car owners value knowing beforehand how their car will be used and for what purpose, which creates a certain level of perceived security. Second, experience enablers will help create stories. Stories are a cornerstone in today's social media driven world. If exciting, unexpected, and with a twinkle in the eye, they provide great potential to be shared by others, thus increasing the awareness of P2P carsharing. Lastly, it has to be mentioned that P2P organizations, wanting to attract experience enablers, should by no means exclusively address customers owning iconic cars. Average cars can create experiences too. Whether it is a car with a tow-bar, a ski-box, or a convertible for a drive along the ocean coast, virtually any car, if marketed correctly, can have a feature that renters desire.

5.2 *How to Grow the Renter Base*

First, renters do not follow the romantic mantra 'sharing is caring.' They participate for a number of personal benefits and are not predominantly motivated by altruistic attitudes. As is the case for car owners, environmental concerns are rather perceived as a side effect of P2P carsharing by renters too. The participation motives identified in our study highlight a number of P2P carsharing features that should be capitalized on in communication campaigns. Doing so will better promote the consumer-relevant benefits P2P carsharing offers, thereby increasing renter participation in the service.

Current campaigns stress the benefits for *savings seekers* and the environment. While the former is an important participation motive and should be adapted in future communication campaigns, the latter is only a side effect of participation, hence, should be promoted less prominently. To attract *convenience seekers*, P2P networks should capitalize on the ability to save stress, effort, and time because the physical distance between car owner and renter is shorter and the pick-up as well as drop-off times can be freely negotiated. Furthermore, the technology related aspects identified as growth drivers for cost cutters (i.e., car owners) will also enhance service convenience for renters, thus enabling a more intuitive and swift P2P experience.

Expression seekers participate in P2P carsharing to express themselves whether by investing saved funds in hobbies or to signal status with the rented car. Hence, communication campaigns should further emphasize these aspects. P2P carsharing, as a way of increasing disposable income, can be easily incentivized via the pure monetary aspect but also can be communicated through the manifestations of the more tangible outcomes, such as attainment of desired activities or accessible products. Moreover, the connotation of lifestyle, the key visuals, and media neutral messages that can be created around this concept are more appealing to today's consumers than a sole focus on costs. In particular, Generation Y is more attracted to lifestyle motives; they crave content that can be shared on social media. In this context, elements of self-actualization such as having the disposable income to engage in one's hobby and renting the car one desires to express status are easily communicated through advertising, social media, or word-of-mouth. Keeping in mind that cost savings are not sharable but lifestyle, enabled by service participation is, allows for a focus on wholly new aspects in communicating P2P carsharing.

The greatest opportunity for communication campaigns is offered by capitalizing on the benefits *certitude seekers* value in P2P carsharing: getting exactly the mobility solution renters require and book. This level of certainty, combined with the variety of options, can neither be provided via classical car renting nor through B2C carsharing. Getting what they booked does not only refer to the rented car itself—it extends to service features, too. Certitude seekers have a deeply rooted desire to gain trust and reduce risks correlated with wrong decisions. They want to know, prior to a transaction, how the car owner will behave, how the transaction process will be facilitated, and what they can accomplish with the car. These are features that are rather unique in the P2P context and are unlikely to be satisfied in a rental or B2C sharing context, due to the impersonal interaction and exchangeability of service providers. The importance of the motive also has significant implications for P2P carsharing networks. To grow the adoption of service usage and follow Airbnb's strategy of providing solutions for corporate travel arrangements, P2P networks have to implement measures that help renters easily book the desired mobility option. At the same time, car owners need to be conditioned to provide the necessary information and the service level that experience seekers desire. This transformation will not be accomplished overnight, but requires significant insights into a network's users, development of incentive schemes, educational material, and monitoring mechanisms.

References

1. Bardhi, F., Eckhardt, G.: Access-based consumption: the case of car sharing. J. Consum. Res. **39**(4), 881–898 (2012)
2. Frost & Sullivan.: Strategic Analysis of the European and North American Peer-to-Peer Carsharing Market (2015)
3. Belk, R.: Sharing. J. Consum. Res. **36**(5), 715–734 (2010)
4. Botsman, R., Rogers, R.: What's Mine is Yours: The Rise of Collaborative Consumption. Harper Collins, New York (2010)
5. Lamberton, C., Rose, R.: When is ours better than mine? A framework for understanding and altering participation in commercial sharing systems. J. Mark. **76**(4), 109–125 (2012)
6. Frost & Sullivan.: European P2P Carsharing Market, Retrieved 18 Feb, 12 Oct 2015. www.prnewswire.com/news-releases/european-p2p-carsharing-market-300157657.html. Accessed 18 June 2016
7. Reynolds, T., Gutman, J.: Laddering theory, method, analysis, and interpretation. J. Advertising Res. **28**(1), 11–31 (1988)
8. Pieters, R., Baumgartner, H., Allen, D.: A means-end chain approach to consumer goal structures. Int. J. Res. Mark. **12**(3), 227–244 (1995)
9. Herrmann, A., Huber, F.: Value-oriented brand positioning. Int. Rev. Retail Distrib. Consum. Res. **10**(1), 95–112 (2000)

10. Wolcott, H.E.: Transforming Qualitative Data: Description. Analysis and Interpretation, Sage, Thousand Oaks (1994)
11. Holton, J.: The coding process and its challenges. Grounded Theory Rev. **9**(1), 21–40 (2010)

Multimodal Transportation Payments Convergence—Key to Mobility

Michael Dinning and Timothy Weisenberger

Abstract Travelers have more choices than ever before and information about these travel choices is being integrated and delivered to individuals, making trip planning easy. But each of these transportation options may require its own method of payment, requiring travelers to use different payment devices and establish multiple payment accounts. Multimodal payments convergence promises to make the travel experience truly seamless. With payments convergence, travelers will be able to plan, book, access, and pay for their trips with minimal effort. Payments convergence will enable transportation service providers to better market their services, incentivize mode choice, manage demand, and may reduce costs. Multimodal payment will be part of bundled mobility services, which will provide travelers access to many mobility options. Both public and private sector organizations are developing, demonstrating, and implementing various forms of multimodal payments. This paper will outline three approaches to payments convergence, and will discuss the challenges to successful implementation of each approach.

Keywords Payment · Multimodal payments · Payment media · Mobile apps · Mobile payments · Accounts · Co-marketing · Incentives · Gamification · Mobility-on-demand · Mobility-as-a-service

1 Introduction to Transportation Payment Convergence

There are more mobility options for consumers than ever before. These options include traditional modes of public transportation (rail, bus, paratransit, ferry), as well as private and nonprofit-delivered mobility services. The transportation

M. Dinning (✉) · T. Weisenberger
U.S. Department of Transportation, Volpe National Transportation Systems Center, 55 Broadway, Cambridge, MA 02142, USA
e-mail: michael.dinning@dot.gov

T. Weisenberger
e-mail: timothy.weisenberger@dot.gov

G. Meyer and S. Shaheen (eds.), *Disrupting Mobility*,
Lecture Notes in Mobility, DOI 10.1007/978-3-319-51602-8_8

industry is exploring partnerships among these providers, called "Transportation Service Providers" (TSPs) in this paper, and a new mobility ecosystem is emerging.

This mobility ecosystem has been fueled by advancements in mobile communications, networking, and computing, and the rise of the shared use economy in transportation.

This emerging mobility ecosystem can deliver many benefits for TSPs, including:

- Improving customer service
- Developing better first-mile/last-mile and on-demand services
- Improving transportation demand management
- Establishing cross-promotions and incentives
- Fine tuning operations
- Enhancing the richness of transportation data.

Consumers have been quick to adopt these new mobility services. Consumers now expect exceptionally convenient ways to engage mobility services and are being rewarded with compelling benefits, including:

- Increasing the ease of end-to-end trip planning
- Overcoming barriers to using multiple transportation modes
- Enhancing mode choice
- Receiving incentives, discounts, and loyalty program promotions.

To date, partnerships among TSPs have not often included convergence of payment methods, focusing more on common mobile portals for trip planning, cross-promotional programs, or agreements to co-locate services (e.g., providing spaces in train station parking garages for car share vehicles).

Many public sector leaders and private sector developers of mobility services have stated that transportation payment must not be a barrier to achieving a multimodal mobility ecosystem. In fact, many mobility innovators have called the integration of trip planning with payments the "holy grail" of seamless mobility apps. The consumer must be able to plan, travel, access, and pay for mobility in much the same way consumers can access goods and services through an online shopping mall. With payments convergence, there will no need to pay in multiple places for those services. Consumers simply add mobility services to the online shopping cart, check out, and pay for them all.

Raymond Telles, the Executive Director of the Camino Real Regional Mobility Authority, described his vision for multimodal payments: "This is where the idea of having a unified system really starts to make sense," Telles said. "Take the idea of the card with the bike share program the [mobility authority] is developing and you could use the card at a parking meter and also at a bike share station. So with it, you can get a bike, pay for parking, take toll roads and cross the bridges. We are now talking about a unified system where you could pay into one account to use for all these mobility uses" [1].

2 Approaches to Payment Convergence

Past transportation payment systems have been modally stove piped, designed for use on one provider's services. But these payment systems have been increasingly modernized over the last several decades, making them more ready to accept multimodal options. Inefficiencies in older methods, such as cash payments and paper tickets, have given rise to a variety of new payments methods: radio-frequency identification (RFID) devices, contactless smart cards, transponders and license plate reading technologies, acceptance of financial payment cards, account-based systems, and mobile payments and mobile ticketing applications (apps). Modern transportation payment technologies create the opportunity to converge payments across the individual systems, modes, and providers.

Three types of multimodal payments convergence are emerging:

- Using the same payment media or technology to pay for or access services on multiple modes of transportation, such as using a transit contactless smart card to pay for many types of transit and to unlock a shared bike or car
- Linking payment accounts for different modes or services, or creating bundled multimodal mobility accounts
- Creating linkages between trip planning, booking and payment mobile portals and apps.

Figure 1 shows that most journeys involve four activities: (1) the available mobility choices are "offered" to the traveler through a mobile trip planning app, kiosk or website (2) the traveler selects or "books" their choice of mobility services (3) the traveler pays for the service using a payment media and (sometimes) account (4) the traveler accesses the service (i.e., goes through a transit turnstile or removes a bicycle from a docking station). The integration of payment for any service is an emerging trend in this new mobility ecosystem.

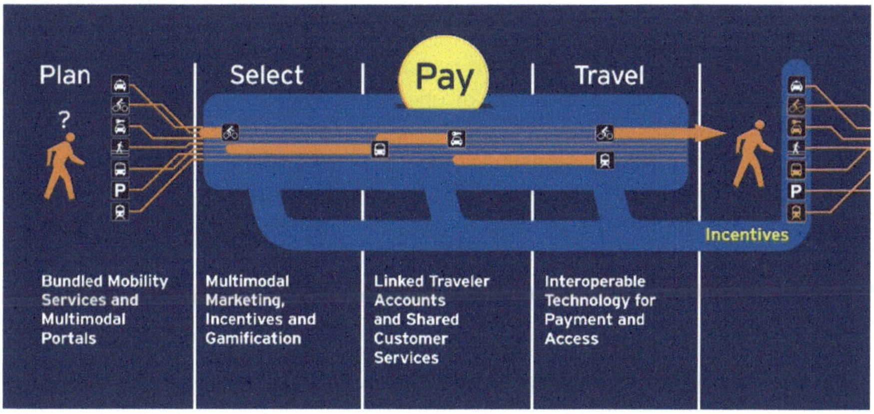

Fig. 1 Payments as part of the multimodal travel experience [2]

As the examples in this paper illustrate, many public and private sector organizations are developing, demonstrating and implementing various forms of multimodal payments. The various approaches to convergence are examined below.

2.1 Common Payment Media

TSPs sometimes agree to use common or compatible payment media so customers can travel on different types of services using a single payment instrument. For example, many transit agencies in a region may agree to use the same contactless smart card for fare payment. In the Washington, D.C. area the SmarTrip card issued by the Washington Metropolitan Area Transit Authority (WMATA) can be used to travel on several regional rail and bus systems. The card can also be used to pay for parking at transit park-and-ride garages. In the San Francisco Bay Area, the Clipper card is accepted for transit payment on over 20 transit agencies for bus, rail, ferry, and cable car travel, and in some parking facilities.

This type of approach is also being used in some examples of multimodal convergence, where the payment media of the larger operator (usually the transit provider) is used as the payment media for the smaller providers in other modes. As an example, the Los Angeles Metro Bike Share program may use the fare card or identification number from the transit system payment media, the TAP card, to access shared bikes.

Similarly, 37 toll road operators in 16 states are members of the E-ZPass Group, and have agreed to use the same electronic toll transponder. The tag can be used on any member toll road, and at selected airport and intercity rail parking garages [3].

When common payment media are accepted by multiple transportation agencies, the traveler purchases the media and may have an account with one agency, and a settlement process is used to reimburse other agencies for services provided.

Differences in the types of payment technology used on different types of mobility services can make it challenging to use the same payment media. Solutions that bridge transit and non-transit mobility often take the form of hybrid devices (e.g., stickers on fare cards, sticker chips on mobile phone cases, attaching a fare card to an RFID device) or insertion of one device into another (e.g., transit card inserted into a toll transponder).

In Chicago, the Chicago Transit Authority demonstrated using their contactless transit fare card to access carsharing vehicles. Because different technologies were used for transit and carsharing, two different radio-frequency antennae and processing chips had to be married to one plastic card.

Many transit authorities are considering partnering with bike share programs to provide first and last mile options, but the technologies used to pay for transit and access the bike share bikes are generally not compatible.

In many European cities "city cards" are issued that allow access to several forms of public and private transportation, as well as tourist attractions. Participating organizations have agreed to use the same technology, and often do co-branding in promotions and discounting. The Bremen "All-in-one-card" is an example of a city card which is accepted for payment on trams, buses, shared bikes and cars, and retail purchases. The program was branded with a distinctive image of a mythical creature called the "egg-laying-wool-milk-sow," representing one product that could be used for many uses [4]. To date, this approach has not been implemented in the United States, and requires challenging coordination among participants' programs and technology investments.

Transit systems in some regions have adopted open payment system architectures which enable a wide range of media to be accepted for fare payment, including: contactless smart cards issued by the transit authority or a third-party, contactless payment cards issued by a financial institution, mobile payment apps linked to a credit card account, mobile ticketing apps, or contactless identification cards issued by an employer or university. These open payment architectures are usually account-based, so the value available to the traveler is held in an account similar to the way that electronic toll payments are managed.

A recent development in some cities is to use the power of mobile apps and communications to work around the front-end technology differences. As an example, users can unlock bicycles in the Bixi bike share in Montreal by using an access code that is sent to their smart phone. This initial use in Montreal is expanding to other cities, such as Aspen, Chicago, Chattanooga, and Chicago [5].

2.2 Common or Linked Accounts and Bundled Mobility Services

Another form of convergence is to create a linkage between the payment system accounts in different modes and services. In account-based systems the customer's value is not stored on the device on a "purse." Rather, the front-end device acts as a credential or token to identify the user to their payment account, as in. roadway toll electronic fee collection (EFC). Account-based transit payment systems have been implemented at Utah Transit Authority, Chicago Transit Authority, and Southeast Pennsylvania Transportation Authority. The Miami-Dade County Department of Transportation and Public Works and other agencies are also planning account-based payment systems.

The convergence across systems that use an account-based approach may be more technically straightforward than hardware convergence. Operators can use different payment devices and reader hardware, and link the accounts in their central system "back-end." Integration is more software-intensive, and while still complex, is potentially less cumbersome and more easily customized than hardware. Linked accounts also facilitate giving multimodal incentives to consumers, which help TSPs manage travel demand.

A potential advantage of this approach is that, since the integration is more in software than in hardware, it is more likely to be open architecture, unlike proprietary front-end devices. This approach can entail the development of an application programming interface (API) and/or data exchange protocols that can be open and flexible enough to allow future partners to join. Further, by remaining independent of front end devices, the system is open to technological advances in the payment space. Unlike hardware integration, account-based approaches may be a flexible solution that can be used in other regions and by other partners.

Several examples of this approach of linking multimodal accounts exist or are planned. In 2013, the Los Angeles ExpressLanes/Express Bus program was created which linked data from consumers' Fast Trak toll accounts with data from their LA Metro TAP Card account. The linked account data tracks the number of trips taken on each type of service, and incentives are given to consumers based on use of the Express Bus. The one-year pilot program was so successful it was made permanent, and is called the Metro ExpressLanes Transit Rewards Program, with over 5000 active users [6]. LA Metro is using customer relationship management software to manage transit customer data, and these customer accounts could be extended to manage payments and use in bike share and other modes [7].

Mobility options may also be bundled together, allowing customers to pay for and receive a variety of mobility services from a common mobility payment account. This tighter linkage gives the consumer an experience very similar to an online shopping site, where the goods purchased are provided by a variety of companies, but the consumer never leaves the shopping site and pays for all goods they have added to their online shopping cart.

Some public agencies and transportation demand management (TDM) organizations are considering offering bundled mobility options. A demonstration of a multimodal transport broker subscription service was conducted in Gothenburg, Sweden under the UbiGo demonstration. Participants were able to subscribe to receive a range of transportation services, which they ordered through a mobile app [9]. A similar initiative, Mobility-as-a-Service, has been proposed in Helsinki, Finland [10]. A key challenge of implementing the bundled mobility model is reaching agreement on how the services will be coordinated and administered, and what organization would play the role of mobility manager.

In many regions, urban developers are working with transportation agencies to integrate mobility options into land use designs. The TDM organization may partner with developers to make bundled mobility packages available to residents, and use discounts and incentives as part of strategies to reduce automobile use. Integrated, multimodal mobility accounts can facilitate these bundled mobility offerings [11].

Travel and payment data from mobile devices, payment systems, and other sources can help planners understand travel demand, and optimize mobility services. Travel data may also be valuable to retailers and event planners. Data integrators are developing APIs and analytical tools to access and interpret mobility information.

The International Organization for Standardization (ISO) is currently developing a "Common Transport Service Account" technical report to show in detail how account linkage can occur in linked and common account approaches. The technical report, Intelligent Transport Systems—Common Transport Service Account Systems - Part 1: Framework and Use Cases, is being jointly developed by the ISO Public Transit and ISO Electronic Fee Collection working groups, and is anticipated to be published in 2016.

2.3 Multimodal Portals and Apps

With many travelers now using mobile devices, public transportation agencies and commercial data integrators are developing mobile apps and web portals which present information on a wide range of mobility options. Travelers can use these multimodal apps to plan their trips, to book services, and to pay for these services. Where mobile ticketing is used, the mobile device may be accepted as proof-of-payment for access to the service (Fig. 2).

Multimodal portals depend on transit authorities, shared use mobility, and other TSPs making their data available to integrators. The integrators must have data sharing agreements with each service provider, and APIs can facilitate data exchange.

Several models of linking to the payment process are emerging. In one approach the trip planning portal is linked to the TSP's app, which is used to book and pay for the trip. In another approach, payment is made to the planning portal app, which passes information on the transaction to the TSP for processing. In a third approach, the traveler has an account with the data integrator or mobility manager, who collects payments and reimburses the TSPs as part of a payment settlement process.

Pilot implementations of multimodal trip planners that are linked to payment are being tested in Austin, Texas, and Portland, Oregon, and an app linked to payment

Fig. 2 Multimodal transportation apps are becoming commonplace [8]

has been implemented in Stuttgart, Germany. Similar deployments are expected in several other regions. API's and software development kits have been developed to facilitate the interfaces among mobile apps and data.

2.4 Co-marketing, Incentives, and Gamification

Public and private TSPs and non-transportation organizations are engaging in cross promotions, co-branding, and other joint marketing programs to make customers aware of complementary services. These programs may include incentives to influence travel behavior, such as encouraging use of public transit to get to major events. Multimodal payment systems can facilitate these initiatives.

Transit providers have implemented successful co-branding programs with e-hail services, special events, retailers and car sharing services. For example, the Dallas Area Rapid Transit (DART) and the Metropolitan Atlanta Rapid Transit Authority (MARTA) have partnered with an e-hail provider to give customers information on services that provide first and last mile connections to transit routes. Both DART [12] and MARTA [13] allow customers to link from the transit agency apps to the e-hail app to arrange for a ride to or from transit service.

DART has used their GoPass^{SM} mobile ticketing app to host promotions for Dallas-area special events to encourage transit usage and ease traffic congestion at the events [14]. For example, DART partnered with the Texas State Fair to give a ten per cent discount when consumers bought fair tickets on the DART trip planning app [15].

In Montreal, the Societe de Transport de Montreal (STM) has created the Merci program to encourage transit use [16]. STM customers can show their Opus transit fare card and receive discounts and other special offers at over 1300 retailers, theaters, arenas and Montreal's Bixi bike share system.

In Portland Oregon, a promotion called "Orange Marks the Spot" used gamification to stimulate interest in a new Tri-County Metropolitan Transit Authority (TriMet) transit line. Bluetooth Low Energy beacon technology was placed along the new line, and travelers could earn discounts at local merchants by bringing their mobile ticketing app near the beacons [17].

There are many examples of partnerships between public transit providers and car share companies. Many transit authorities give preferred parking spaces to car chare vehicles at park-and-ride garages and lots, and include marketing for car sharing on the transit trip planning apps. One example of such a partnership was established in January 2014 at TriMet in Portland, OR [18].

Other types of shared-use mobility providers are also co-marketing and giving incentives. In a campaign to get lower-income patients to try the bike sharing system, the Boston Medical Center has arranged discounts for bike share memberships under the Prescribe-a-Bike program [19].

Transit and toll road authorities are beginning to work together to offer multimodal incentives. As part of the Metro ExpressLanes initiative to reduce freeway congestion in Los Angeles, the Los Angeles County Metropolitan Transportation Authority has used gamification and incentives, combined with improved transit service, dynamic toll pricing, and real-time information on congestion, to encourage travelers to switch from single-occupancy vehicles at the most congested times. For example, frequent riders of the express bus rapid transit system are given credits to use on toll roads. In addition, members of car-pools are given chances to win gift cards, depending on how often they car-pool during the month [20].

In some regions, transportation demand management organizations are hosting competitions to encourage travel behavior which reduces congestion and emissions. The Clear the Air Challenge in Salt Lake City gave travelers chances to win prizes when they recorded trips using alternatives to the automobile [21]. If financial incentives are provided, these initiatives could be linked to travelers' payment accounts.

The UbiGo demonstration in Sweden offered rewards to participants who used more sustainable means of travel, but the rewards appeared to play only minor roles in changing behavior compared to convenience and economy [22].

3 Benefits of Payments Convergence

This paper has provided many examples of how public and private organizations are using seamless payment for travel services as a key link to unlocking a true multimodal, public-private mobility ecosystem in regions. Mobility on demand services will allow consumers to create personalized journeys and make changes based on day of week, time of day, service outages, weather issues, or for sporting and other special events. Payments convergence is a key element that can make travel more convenient for customers by providing easy ways to pay for any type of transportation. Consumers will have new abilities to track their usage and budget their travel funds as they see fit.

Multimodal payments enable TSPs and regional planning organizations to manage travel demand by giving consumers pricing cues and incentives to modify their travel habits. By combining services in seamless and logical ways, TSPs can create more efficient services tuned to consumer.

Public transportation providers can potentially reduce costs and enhance revenues. Private and nonprofit providers can grow their businesses and enhance revenues by linking to the much larger public providers in regions. Combining payment-related services for multiple modes, like account management or customer service centers, may enable transportation agencies to reduce costs and streamline procurements and contracts by pooling their requirements. This integration would require collaboration among agencies, however, and presents challenges in coordinating policy, governance, financial management, and acquisition strategies.

Payment data is rich data. By linking payment accounts, payment data from various TSPs can be analyzed together. This data can be combined with other data from Intelligent Transportation Systems, such as automated vehicle location data, to give TSPs very granular information to fine tune operations and thus provide better customer service. Transportation data, combined with data from other sensors in "smart cities," can provide valuable information to city planners, utilities providers, and real-estate developers to enhance mobility and optimize services.

While many public agencies are interested in multimodal payments, their primary goal is to improve mobility options and services. It is important that payments convergence and incentive programs complement multimodal service integration. To provide first mile and last mile connections, public transit agencies are developing agreements to align car and bike sharing and e-hail services with transit services. In Boston, San Francisco, Kansas City, and other cities, transit agencies are evaluating new mobility services, like dynamically routed vehicles, to provide better transit coverage. A priority for public transit agencies is to provide service to everyone who needs it, particularly transit-dependent populations. For example, the Metro ExpressLanes project in Los Angeles and the Prescribe-a-Bike project in Boston gave discounts to lower-income travelers. The challenges to fare integration programs will be to identify which services are eligible for transit benefits, and to comply with legal requirements to maintain separation of the funds allocated to each type of service [23].

4 Opportunities to Develop and Implement Multimodal Payments

The interest in multimodal payments continues to grow, and both the public and private sectors have initiatives to develop, demonstrate, and implement different types of multimodal payments. Public agencies and data integrators are continuing to develop multimodal trip planning apps that are linked to payment. In some regions, agencies are exploring the concept of a universal transportation or mobility account for travelers. Public agencies are also evaluating the feasibility and potential benefits of integrating processes and services for different modes, such as having an integrated customer service center [24].

The U.S. DOT's ITS Joint Program Office and the FTA are planning to sponsor demonstrations in a Mobility on Demand (MOD) Sandbox. The goal of the MOD Sandbox is to leverage innovative technology to improve mobility using a traveler-centric approach. The MOD Sandbox aims to provide opportunities where integrated MOD concepts and solutions, supported through key local partnerships, are demonstrated in real-world settings. The intent of the program is to demonstrate both technical and institutional innovations that improve mobility [25]. The U.S. DOT is also sponsoring a demonstration of innovative technologies as part of the DOT Smart City Challenge. Seventy-eight cities submitted proposals to the challenge, and Columbus, Ohio was selected for the demonstration [26]. The types of payments convergence described in this paper could be included as part of future smart city demonstrations and initiatives.

To help develop an understanding and awareness of the potential opportunities for multimodal payments, the Transportation Council of the Smart Card Alliance and the Association for Commuter Transportation are developing a Multimodal Payments Convergence white paper, which is expected to be completed in 2017. The paper describes multimodal payment concepts, provides use cases of innovative deployments, and outlines challenges and opportunities for future development.

The U.S. DOT Volpe Center and many transportation industry associations are organizing discussions of multimodal payments at key events to keep public and private stakeholders aware of the latest developments, and to stimulate additional interest and innovation. Multimodal payment is a featured topic at the Payments Summit of the Smart Card Alliance, the Fare Collection and Revenue Management Summit of the American Public Transportation Association, and national fora sponsored by ITS America, the Association for Commuter Transportation, the Transportation Research Board and the Shared Use Mobility Center.

5 Conclusion

Mobility options are converging into an ecosystem that includes public sector-delivered transit, toll, and parking services, as well as private and nonprofit sector-delivered services, such as e-hail, shared use, micro-transit, and on-demand services. People have come to expect exceptional ease of use through the explosion of transportation apps for trip planning and booking.

Payments must not be a barrier to the continued development of the mobility ecosystem. Initiatives in private and public sector transportation industries, as well as in academia and government, are contributing to research, development, and implementation of convergent multimodal payments. In the past this convergence has taken the form of hardware integration of the payment devices used in different transportation services.

More recently, software integration of mobile apps and payment accounts have been fielded, and are poised to fuel a true mobility ecosystem that allows consumers to plan and select their preferred travel options, and pay for the services in a more streamlined and convenient way. This account-based approach eases the integration burden, while creating more open and replicable platforms that allow new partners to join, and keeps the payment system flexible to accept emerging payment technologies.

Without achieving true multimodal payments convergence, the mobility ecosystem would be lacking the ability to maximize customer convenience and mode choice, enhance transportation planning, manage travel demand, reduce costs, and increase revenues.

Multimodal payments convergence is being explored and advanced through outreach and thought leadership activities in industry associations, government programs funding field demonstrations and deployments, and industry pilots of innovative payments convergence capabilities.

The full achievement of integrated mobility ecosystems and their many benefits will hinge on not only technology innovation, but also development of collaborative relationships among government, the private sector, and academia to enable them to work together to address institutional, business case, and policy issues.

Authors' note: this material is disseminated under the sponsorship of the Department of Transportation in the interest of information exchange. The United States Government assumes no liability for its contents or use thereof. The U.S. Government is not endorsing any manufacturers, products, or services citied herein and any trade name that may appear in the work has been included only because it is essential to the contents of the work.

References

1. Las Cruces Sun-News, May 24, 2014
2. Weisenberger, T., Dinning, M.: Multimodal payments convergence—key to mobility. Poster presented at the MIT Disrupting Mobility Summit, 12 Nov 2015
3. E-ZPass Group Website.: http://e-zpassiag.com. Accessed 23 Feb 2016
4. Rhys, T.: The amazing egg laying wool milk sow. July 23, 2008, http://thecityfix.com/blog/the-amazing-egg-laying-wool-milk-sow. Accessed 28 June 2016
5. Sion, J.: Cities beyond driving, Presentation at Transportation Research Board Webinar, 7 Mar 2016
6. Metro Expresslanes Transit Rewards Program Website.: https://www.metroexpresslanes.net/en/about/transit.shtml
7. O'Hara, R.: Collaboration + cooperation = connectivity: equation for integration of 26 fare systems in LA County. In: APTA 2016 Fare Collection Summit and TransiTech Conference, April 12, 2016
8. Weisenberger, T., Dinning, M.: Multimodal payments convergence—key to mobility. Poster Presented at the MIT Disrupting Mobility Summit, 12 Nov 2015
9. Sochor, J., Strömberg, H.: Travelers' motives for adopting a new, innovative travel service: insights from the Ubigo Field operational test in Gothenburg, I.C. Mari Anne Karlsson Chalmers University of Technology. Presented to 21st World Congress on Intelligent Transportation Systems, Detroit, 7–11 Sept 2014
10. Heikkila, S.: Mobility as a service—a proposal for action for the public administration, Case Helsinki, 28 April 2014
11. Hiatt, R.: San Francisco County transportation authority, treasure island mobility management: congestion toll policies to support new, comprehensive transportation system for a redevelopment area, 2016. TRB Annual Meeting, Jan 2016
12. DART News Release: DART, Uber stepping up "complete trip" efforts, 14 April 2015. http://www.dart.org/news/news.asp?ID=1179. Accessed 28 June 28 2016
13. MARTA Press Release: MARTA partners with Uber, adds Wi-Fi to buses, 23 July 2015. www.wsbtv.com/news/news/local/marta-partners-uber-adds-wi-fi-buses/nm5p3. Accessed 28 June 2016
14. DART Website: DART to special events. www.dart.org/about/riderinsider/specialevents.asp. Accessed 28 June 2016
15. DART Website: Ride DART to the 2015 State Fair of Texas® 25 Sept–18 Oct 2015. https://dart.org/statefair/statefair.asp. Accessed 28 June 2016
16. STM Website: Merci Program, Exclusive offers and benefits. www.stm.info/en/offers-and-outings/Merci-program/discounts-and-offers-merci/discounts-and-offers-merci/discounts-and. Accessed 28 June 2016

17. Murphy, A.: TriMet News Release, TriMet and Zipcar join together to increase transportation options, 22 Jan 2014. http://news.trimet.org/2014/01/trimet-and-zipcar-join-together-to-increase-transportation-options. Accessed 28 June 2016
18. Mass Transit Magazine: TriMet launches game to explore the new max orange line, 11 Sept 2015.
19. Gaitan, C.: New city program lets doctors 'prescribe' bike-sharing memberships. Boston Globe, 27 Mar 2014
20. Los Angeles Congestion Reduction Demonstration ExpressLanes Program: National Evaluation Report, FHWA-JPO-14–126
21. Clear the Air Challenge fact sheet. http://cleartheairchallenge.org/wp-content/uploads/Factsheet-ClearTheAirChallenge.pdf. Accessed June 28, 2016
22. Sochor, J., Stromberg, H., Karlsson, I.C.M.: The added value of a new, innovative travel service: Insights from the UbiGo field operational test in Gothenburg, Sweden, Chalmers University, International Conference on Mobility and Smart Cities, Rome, 27–28 Oct 2014
23. Feigon, S., Murphy, C.: Shared mobility and the transformation of public transit. TCRP Report 188, pre-publication draft, June 2016
24. Kuester, C.: Presentation at ITS America 2016, Payment as an Organizing Principle in Urban Transportation panel, 15 June 2016
25. ITS/FTA Mobility on Demand Fact Sheet. http://www.its.dot.gov/factsheets/pdf/MobilityonDemand.pdf. Accessed 28 June 2016
26. U.S. DOT Smart City Challenge, www.transportation.gov/smartcity. Accessed 31 Aug 2016

System Effects of Widespread Use of Fully Automated Vehicles—Three Scenarios

Wolfgang Gruel and Joseph M. Stanford

Abstract As a promising technology close to entering the market, fully automated vehicles (AVs) have been widely analyzed from many angles. In spite of many discussions and articles published on the subject, there has been little structured analysis of the broader system-level effects of full adoption of AVs, and therefore a lack of consideration of the potential effects of multiple simultaneous causal relationships and feedback loops. This paper utilizes an existing system dynamics model that conceptually illustrates many important dynamics in surface transportation systems. It builds on this model, incorporating changes that are likely to occur under three different scenarios of AV adoption.

Keywords Autonomous driving · Self-driving vehicles · Automation · Automated driving · Driverless cars · Long-Term effects · System dynamics

1 Introduction

Fully automated vehicles (i.e., self-driving, driverless, or robotic vehicles) are well on their way to becoming a reality. It is expected that they will transform existing mobility systems as we know them [1]. They hold great promise for people who are currently unable to drive, e.g., due to infirmity or age [2], and they are expected to increase the efficiency of traffic flow and dramatically improve safety [1, 3, 4]. With fully automated operation, travelers will be able to use the time spent in a car in ways that are potentially far more productive or entertaining. Significant effort has

W. Gruel (✉)
Massachusetts Institute of Technology| Car2go Group GmbH,
Fasanenweg, 13-15, 70771 Leinfelden-Echterdingen, Germany
e-mail: wgruel@mit.edu

J.M. Stanford
Massachusetts Institute of Technology, 77 Massachusetts Avenue,
Cambridge, MA, USA
e-mail: joseph.stanford@sdm.mit.edu

G. Meyer and S. Shaheen (eds.), *Disrupting Mobility*,
Lecture Notes in Mobility, DOI 10.1007/978-3-319-51602-8_9

been invested in examining the technological and legal requirements related to widespread adoption of fully automated vehicles (AVs). There have also been plenty of informal discussions about the consequences of this emerging technology, which usually consider only one or two effects of the use of AVs (e.g., workforce changes, traffic flow improvements, etc.). However, there remains a lack of research into the longer term outcomes on a system level.

The goal of our article is to address that gap and to systematically investigate the long-term effects of AVs. Therefore, our work focuses on the interdependence and interaction of different direct and indirect effects of automated driving. We consider potential benefits made possible by AVs, while also investigating how they could change mobility behavior and thus traffic volume, congestion, land use, and mode choice. For our analysis, we use a structured qualitative approach to identify and better understand the relations between different factors. As there are considerable uncertainties related to this topic, we apply a scenario-based approach. To capture the interlinked effects, we develop conceptual system dynamics models for the different scenarios. Based on these holistic models, we discuss ideas on how to influence the system in order to create desired outcomes.

2 Literature Review

Many publications about AVs address very specific aspects of vehicle automation. They usually focus on only a few factors. As long as AVs are not widely in use, research on the effects of AVs is limited to proxies like simulations. Several studies have investigated the effects of AVs on traffic flow and human behavior (e.g., [5, 6]). Others have examined the optimal management of robotic vehicles (e.g., [7–10]). Some research has also been performed on attitudes toward AVs, behavior, and impacts on land use and urban planning (see [11–14]). In [1] a thorough discussion of the expected benefits of AVs is provided—e.g., improvements in safety (up to 99% reduction in fatalities) and congestion (coordinated platoons, better route choices, shorter headways).

Generally, however, most research to date has focused on only few variables and only very few studies draw a more holistic picture. In one of these [3], the authors examine potential opportunities and risks of AVs, considering not only direct consequences like impacts on safety, congestion, energy use, pollution, and mobility for underserved populations. They also consider longer term effects on land use, such as decreased urban density that is caused by an increased dispersion of destinations (due to improved highway speeds and an increased willingness to travel). They also refer to a reduced need for parking and new opportunities, this opens up for changing urban form.

The authors of [15] investigate the effects of a mobility system that builds on shared AVs using analytical and simulation approaches. They conclude that the use of shared AVs potentially provides "better mobility experiences at radically lower cost" and could be beneficial for customers, cities, and the environment. In [9, 10, 16] it is argued that shared vehicle systems can satisfy mobility demands with significantly fewer vehicles, but with more trips in order to rebalance the fleet.

The implications of AVs are also investigated with the help of scenario approaches [17, 18]. The scenarios in [17] focus on the implications on transportation planning. They take a closer look on policies and technological development. The scenarios in [18] consider different factors that could influence the adoption of AVs, as well as financing schemes, the role of planning, and the effects of automated driving on land use and other transportation systems. A wide range of expected outcomes from these scenarios are a reflection of how much uncertainty remains about how AVs will be used, and what their impacts on our towns and cities will be.

Most work on the longer term effects of AVs falls into two categories: The first category consists of rigorous studies considering a small number of variables and narrow sets of data. While their results are well supported, they are limited to a small set of implications. The second category consists of broader speculations. These rarely build on existing literature, models, and frameworks and are thus difficult to connect to well-known mechanisms and established phenomena. Furthermore, the complex system interactions—e.g., multiple variables interacting simultaneously, feedback effects, etc.—have not been considered. However, an examination of transportation dynamics suggests that these system effects could be of great importance. For example, many studies acknowledge that AVs could increase vehicle kilometers traveled (VKT) and automobile dependency and that this is likely to further encourage urban sprawl [1]. However, they neglect to consider that an increase in sprawl is likely to further increase VKT and automobile dependency, creating a feedback effect. The effects of such "vicious cycles" have not yet garnered significant attention.

The goal of this paper is to identify direct and indirect effects relating to AVs, the existing feedback mechanisms they may influence, and potential new feedback effects that may arise. It aims to illustrate these relationships in a compelling conceptual model. Such a model should help to better understand the potential effects of AVs as they relate to user behavior, traffic volumes, public transit ridership, and land use. Furthermore, such a model will be a useful tool for the discussion about desired outputs of AVs and potential levers to achieve those results.

3 Approach

The primary goal of our work is to conduct a structured examination of the basic effects that come with vehicle automation at a system level. Since there are a number of well-known models of transportation systems, we have chosen to build on a simple existing base model that describes known relationships among several of the important variables involved in transportation systems. We conducted qualitative interviews to investigate how AVs might change behavior. This section describes the modeling technique and data collection, the development of scenarios, and the specific base model chosen.

3.1 Modeling Technique and Interviews

To investigate the effects of vehicle automation on a system level, we have chosen to use system dynamics—an effective tool for analyzing complex, dynamic systems [19]. At its core is a mapping process, which identifies key variables and the causal relationships among them. By considering many variables in a system at once, instead of focusing on relations between just a few, system dynamics can help to build "a better understanding of the complex transportation system problems" [20] (for further discussion of the use of system dynamics to model problems in transportation see [21]).

Because we focus on the relationships among variables and the overall structure of the system, our work is limited to the use of causal loop diagrams (CLDs). CLDs can help to identify key variables and causal relationships. They make perceived system structures visible—and thus open them for analysis and discussion [21]. One of the benefits of CLDs is that they can generate valuable insights before the relevant system behaviors can be observed or measured. They shed light on the systemic implications of known or hypothetical relationships by revealing the structure and potential behaviors that they entail [22].

CLDs consist of only two main elements: variables, which are indicated by their names, and causal links, which are indicated by arrows pointing from the independent variable to the dependent variable. Causal links have positive or negative polarity, indicating the nature of the relationship—e.g., a link from variable A to B with positive polarity means that an increase in A will cause B to be larger than it would be without A's impact, and vice versa (a decrease in A will cause B to be smaller than it would otherwise be). A negative polarity indicates opposing changes. Marks can also be added to causal links to indicate delays. Once CLDs are assembled, "loops" will arise when the causal links from one variable connect back to itself, after connecting to other variables [23]. These loops play an important role in system dynamics modeling, as they drive many system behaviors.

The structure of system dynamics models is largely based on qualitative data [24, 25]. In order to add formality to our data collection, we used very common qualitative techniques: interviews and group workshops [26, 27]. We conducted semi-structured interviews and workshops with 30 individuals, with expertise in a range of fields, including transportation planning, the automotive industry, the sharing economy, urban planning, and government policy.

In these sessions, we presented a base model (Fig. 1) that describes key aspects of current mobility dynamics and we asked the participants to suggest changes that might be caused by the introduction of AVs, and to identify how they might affect the structure of the model. With each session, we analyzed the collected information and used the results as input for further sessions [25]. The summarized results are reflected in the models shown in Sect. 4.

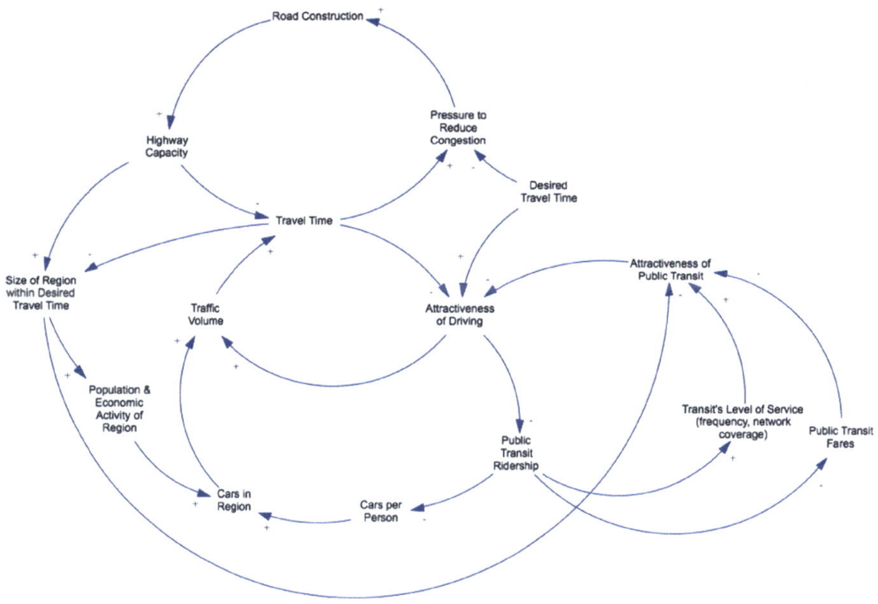

Fig. 1 The baseline model (adapted from [23])

3.2 Scenario Development

Given the amount of uncertainty in the potential effects of AVs, we chose to apply scenario analysis methods [28, 29]. Rather than trying to determine the most likely outcomes, this approach aims to develop useful insights by exploring a range of different futures. It tries to identify the main forces that will affect outcomes in question, followed by speculation about the development of these forces. To counter the inherent subjectivity of this process and steer our analysis toward more objective outcomes, we used interviews and group discussions with a diverse group of participants with widely varying perspectives [26].

The public discussion at the moment rarely questions *if* AVs will become a reality—but focuses more on *when* they will be available. That is why we assumed in our scenarios that AVs are successfully developed and adopted. Technological challenges, legal obstacles, factors that may spur or hinder adoption, and the adoption process itself were not considered.

Based on existing literature [18, 30] and our interviews and workshop discussions, two major dimensions of uncertainty emerged that seem to drive the main outcomes of vehicle automation: How will mode choice change due to this technology, and how much new demand for travel will be caused by the changes in total costs of driving (including costs of time spent driving and other disincentives)? Based on these sources of uncertainty, we developed three scenarios that are examined in Sect. 4.

3.3 Baseline Model

As a base model for our research, we used an established CLD of traffic and congestion developed by Sterman [23] and slightly simplified it in order to adapt it to our research question. The starting point of the model is the question of how to reduce traffic congestion, and it considers the system-wide effects of road building, a common intervention that aims to reduce congestion. The model (Fig. 1) is adequate for our purposes because we can treat the use of AVs as somewhat similar to road building, as they are widely seen as a way to increase road capacity without expanding infrastructure. This increase would be achieved by better coordination, platooning, and shorter headways between vehicles [1]. Similar to other transportation models, Sterman's model considers variables relating to road capacity, travel demand, land use, and public transit.

The core element of the model is the variable *Attractiveness of Driving*. It is influenced by *Desired Travel Time* and *Actual Travel Time*, and the *Attractiveness of Public Transit*. *Attractiveness of Driving* is a crucial variable, because as it increases, it tends to counteract and balance out the initial benefits of new roads: as *Attractiveness of Driving* increases, more and longer trips are taken, due to both the latent demand for travel and the mode shift from transit, resulting in higher *Traffic Volumes*. These effects tend to lead to a new equilibrium state with similar levels of congestion and higher levels of VKT. Another effect of new roads is that people can travel farther in less time. That means that the size of the region that can be reached within a desired travel time increases. As a consequence, people are able to move farther away from their work and other destinations, and settlement density decreases. There are also potentially powerful feedback effects on public transit (referred to as the "mass transit death spiral"): if public transit ridership falls, revenues of transit operators fall, which leads to higher fares or lower service quality (or both), which in turn lead to a reduced *Attractiveness of Public Transit*, which further reduces ridership, and so on.

While our model is simplified from Sterman's model, it indicates the same fundamental outcomes from increasing capacity: (1) additional trips will be induced as travel times go down (which will drive traffic volumes to a higher equilibrium level); (2) some riders will be drawn away from transit (which could trigger the "death spiral of transit," described above); and (3) longer trips will occur due to higher speeds and thus shorter travel times, which will increase sprawl (reduce density of land use), which makes public transit service more difficult to provide, ultimately reducing the attractiveness of transit and potentially triggering the "death spiral".

4 Introducing Automated Driving: Three Scenarios

In the following sections, we discuss three scenarios for a future mobility system with full adoption of AVs (see Sect. 3.2). In each section, we first describe the scenario, then explain the major changes made to our base model to reflect the assumptions that define each scenario. All scenarios assume complete adoption of AVs, and assume AVs also have the benefits of connected-vehicle technologies.

4.1 Scenario 1: Technology Changes, but We Do Not

Scenario description: This scenario assumes that AVs replace our current vehicles, but the adoption of AVs does not result in significant behavioral changes. In this scenario, the adoption of AVs comes with improvements in safety, efficiency, and utility of travel time. These are repeatedly mentioned in the literature [1, 3] and were also mirrored in the comments of our interview subjects (e.g., "driving will be so much safer," "my commute will be so much more pleasant—and productive"). The assumption that the improvements brought by AVs will make driving more pleasant and less costly, but will not cause additional travel, was often expressed in our interviews. However, as indicated in the baseline model, the initial reduction in congestion due to AVs is expected to generate some additional trips, cause some people to switch from public transit to automobile trips, and encourage some longer trips due to higher speeds (and shorter travel time). Furthermore, there is assumed to be significant latent demand for travel pent-up in elderly and disabled people who currently cannot drive, so the introduction of AVs will increase overall demand for travel.

Modifications to base model: To consider the effects of AVs, we replace *Road Construction* in our base model with *Use of AVs*. This is reasonable, as AVs are assumed to increase road capacity (by improving efficiency of driving and traffic flow, both through coordinated driving using connected-vehicle technologies and crash reduction through automation). We renamed the variable *Effective Highway Capacity* to make it clear that no actual road expansion takes place.

The basic assumption of the scenario is that the benefits provided by AVs do not change people's behavior—i.e., length of trips, their number, and mode choice stay the same. To reflect this assumption and the benefits of automated driving at the same time, we introduce the variable *Benefits of Automated Driving* into the model but do not show any causal links emerging from it. Another effect of AVs is the *Ability to Use Automobiles by People Unable to Drive,* which has the effect of increasing the number of *Cars* and *Traffic Volume* (Fig. 2).

Expected system-level outcomes: In scenario 1, without major behavioral changes, a number of substantial benefits can be expected: traveling by car will be safer, less expensive, less energy consuming, more environmentally friendly, and time spent in the car will be used in better ways. Furthermore, the situation of millions of people who currently have limited access to mobility (e.g., elderly and disabled) will be able to use automobiles. Normally, the increase in capacity would cause higher average speeds by reducing congestion, which would trigger the three main effects in our baseline model. However since our Scenario adds a lot more automobile users (elderly, disabled), they are likely to consume most or all of the additional effective highway capacity resulting from AVs. Therefore, the usual

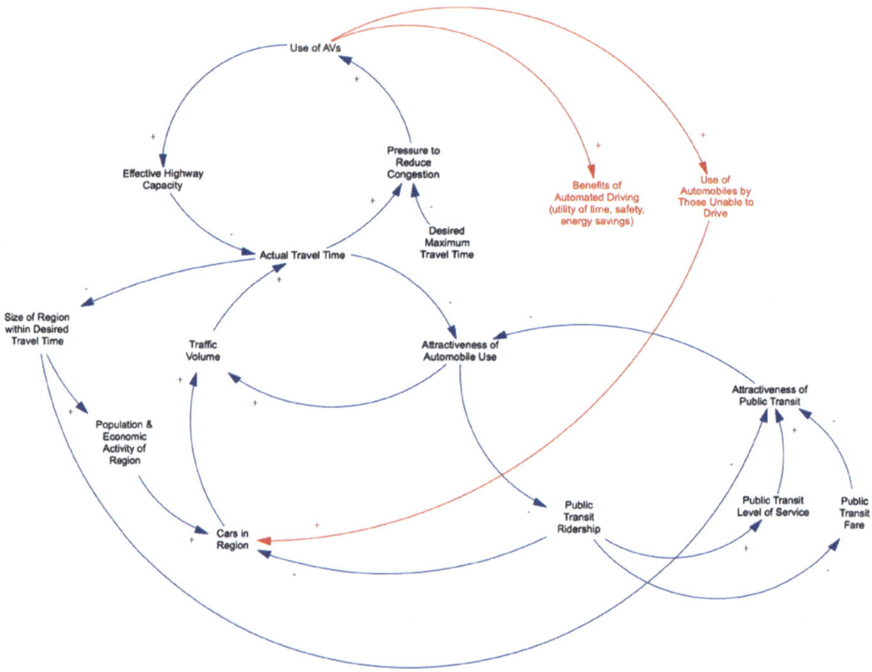

Fig. 2 Changes to the baseline model resulting from assumptions of Scenario 1 are shown in *red*

effects of the baseline model are likely to be weak or nonexistent. There will be more drivers, more VKT, but probably equal levels of congestion, no major increases in sprawl, and relatively unchanged levels of transit use.

4.2 Scenario 2: Technology Changes Our Mode Choice

Scenario description: In contrast to Scenario 1, this Scenario assumes that the benefits of automated driving will increase the attractiveness of automobile use and this will influence mode choice behavior: as automobile use becomes more attractive, public transit becomes less attractive in comparison. This scenario reflects a widely held belief that improvements in the experience of automobile use may cause people to change modes. However, the scenario maintains the assumption that these improvements will not increase the overall demand for travel. This reflects statements from some interviews, where subjects were confident that they already make all the trips they desire to and that—regardless of the mode—the number and the length of the trips they take would not change.

Modifications to existing model: In this scenario, we allow the *Benefits of Automated Driving* to increase *Attractiveness of Automobile Use*. This increase

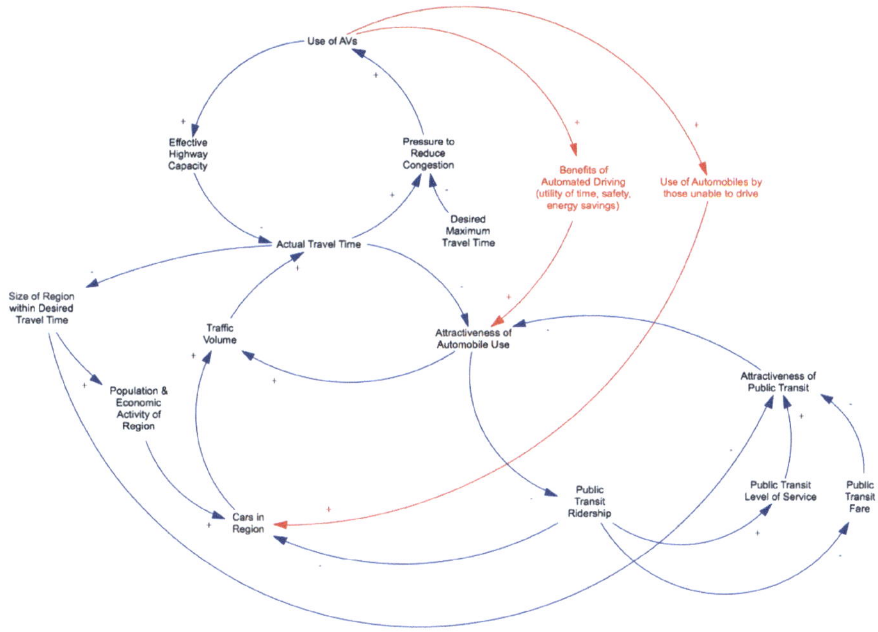

Fig. 3 Changes to the baseline model for Scenario 2 are shown in *red*

often makes the automobile a better choice compared to public transit. This reduces *Public Transit Ridership*, and increases *Traffic Volume*. In keeping with the fact that overall travel demand by current drivers is unchanged, for every new trip by car resulting from a change in mode choice, there is one less trip by public transit (Fig. 3).

Expected system-level outcomes: Similar to Scenario 1, there will be substantial benefits on an individual basis from the adoption of AVs. We take into account that increased *Attractiveness of Automobile Use* has further consequences: increasing this variable will reduce *Public Transit Ridership* and drive the two challenging feedback loops described in the baseline model that result in potentially much lower levels of *Public Transit Ridership*. It is uncertain how far the mode shift caused by AVs will alter the equilibrium state. However, given the frailty of many existing public transit systems, it may not take much to drive the reinforcing feedback loops described above beyond a point where the systems will not be able to recover, or will have to shrink dramatically to stay in operation. The fundamental change here is that travel by automobile will become even more attractive relative to transit (people will tolerate higher levels of congestion to meet their current demands, because traveling by car will have become so much more attractive). Ultimately, in this scenario, we can expect more VKT, more congestion, and less transit use. Sprawl is not expected to increase much in this scenario, because road speeds are not likely to go up, as congestion will probably settle out at similar levels to the original conditions before the dynamics driving sprawl can take effect.

4.3 Scenario 3: New Technologies Offer New Opportunities

Scenario description: This scenario builds on scenario 2, and adds the assumption that people will exploit new opportunities for using AVs. This reflects the notion that there is significant latent demand for travel, currently held back by the disutility of time spent driving, and this is supported by our interviews and also by the literature. For example, some interview subjects observed: "I could work in my car, so a longer commute wouldn't be that bad, and I could live farther away," or "I could travel from Boston to New York every week to see my family—if I could sleep or work in the car, I wouldn't care how much time it takes". These statements indicate that people would be willing to travel more and to make longer trips than they did in conventional cars. In addition, this scenario also allows for entirely new behaviors and business models, made possible by the ability to operate unmanned automobiles. These could include individuals sending the car home for parking, unmanned delivery services, mobile meeting spaces, low-cost billboard trucks, errand-running services, etc.

Modifications to existing model: To reflect these potential new behaviors, we introduce *New Uses for AVs*. This new variable has a causal link leading directly to *Traffic Volume,* as the new uses will directly increase the number of trips taken. Furthermore, although it is not directly shown in the model, it is assumed that the increased *Attractiveness of Automobile Use* will lead to higher *Traffic Volume* not only due to mode shift away from transit, but also through entirely new trip generation, resulting from the *Benefits of Automated Driving*, particularly the increased utility of time spent in an automobile (Fig. 4).

Expected system-level outcomes: As observed above and in Scenario 2, increasing *Attractiveness of Automobile Use* will increase *Traffic Volume*, both through a mode shift away from transit and through generation of entirely new trips. Furthermore, the new uses in Scenario 3 enabled by unmanned vehicle operation will further increase the traffic volume. As in Scenario 2, increasing the *Attractiveness of Automobile Use* will trigger dynamics within the system that lead to a new equilibrium with higher traffic volumes at higher levels of congestion and with decreased *Public Transit Ridership.*

Increased *Effective Highway Capacity* (and a more pleasant driving experience—not explicitly shown in this model, but reflected in *Benefits of Automated Driving*)—will increase the maximum desired travel time and thereby will increase the *Size of Region within Desired Travel Time.* This will also drive the dynamics that increase sprawl, making it more difficult to provide adequate public transit and thus reducing the *Attractiveness of Public Transit* even further. This scenario appears to further amplify the effects seen in Scenario 2, with higher *Traffic Volumes*, increased sprawl, and the triggering of vicious cycles that could undermine public transit systems.

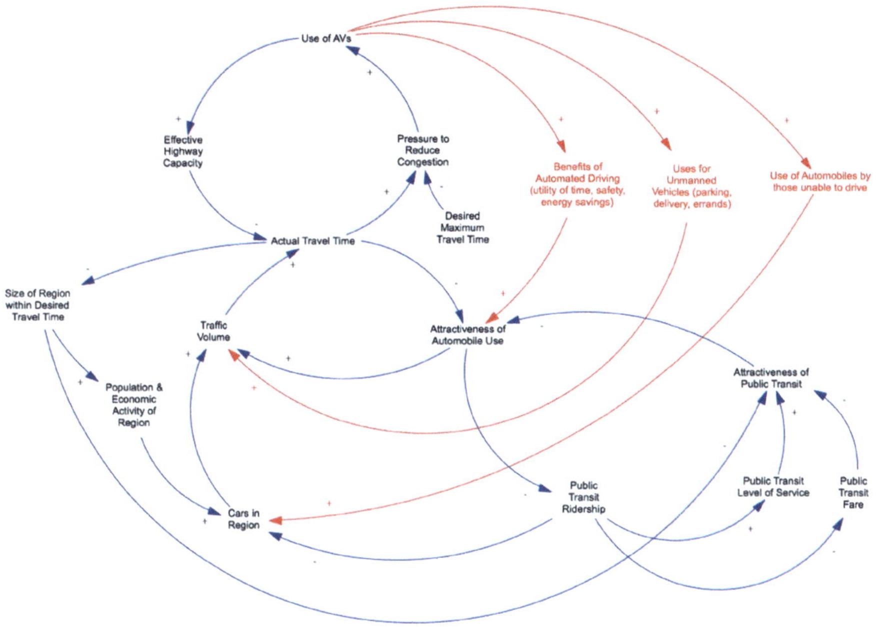

Fig. 4 Changes to the baseline model for Scenario 3 are shown in *red*

5 Discussion

The scenarios examined here illustrate a variety of potential outcomes from the adoption of AVs. They are not meant to predict future conditions but rather show a range of possibilities. In all three scenarios, driving becomes safer, time spent in the car is used more productively or enjoyably, and mobility for millions of elderly and infirm people improves. Furthermore, the cost and energy consumption associated with VKT would decrease. However, our models suggest that many effects could be undesirable: in all three scenarios, VKT is likely to increase, leading to increases in energy consumption and emissions. The structure of the models suggests that these changes could vary significantly across the scenarios.

Scenario 1 shows that if people's behavior does not change, AVs will most likely greatly improve our transportation systems. If, however, behavioral changes occur as depicted in Scenario 2, AVs would increase the attractiveness of automobile use and mode choices would change to favor automobile use. These effects would increase traffic volumes and potentially trigger vicious cycles that could undermine public transit, thereby further increasing VKT. In Scenario 3, when we allow both mode choice and trip generation behavior to be affected by increased *Attractiveness of Automobile Use,* and this is combined with additional trips generated by *New Uses for AVs*, the model suggests substantially higher levels of VKT, which may outweigh any traffic flow efficiency benefits, resulting in more congestion. Also, sprawl would be further encouraged, and public transit would become less and less adequate.

One important factor that we did not consider in the scenarios is the effect of AV sharing. A cursory analysis suggests that widespread vehicle sharing could play a role in mitigating the potentially harmful impacts of AV use. Multiple factors might cause a shift to increased vehicle sharing—e.g., increased appeal of AV-sharing services when vehicles can provide door-to-door service, lower prices due to higher vehicle utilization, or enforcement by cities that ban private cars from certain areas.

The use of shared AVs could reduce *Traffic Volume* by improving price transparency for automobile trips, which might discourage some trips. On the other hand, we can expect some additional VKT due to the zero occupancy trips needed to rebalance the supply of vehicles. Additional benefits could arise if shared AVs were used to improve first-/last-mile connectivity to transit, supplement existing services, or even replace high-cost underutilized transit routes [31]. And even more substantial benefits may be possible in dense urban areas, where ridesharing could play a large role in reducing the total number of vehicle trips [32, 33]. Furthermore, vehicle turnover will be faster (due to higher utilization), so vehicles on the road will tend to be newer and more efficient overall; less parking will be required, which will enable denser land use in some places; and large-scale vehicle sharing systems could provide more appropriate vehicle choices for different uses, saving energy in the process.

A number of interesting effects could emerge from the different cost structure involved with vehicle sharing. The greater the imbalance between demand and supply in an area, the more empty rebalancing trips are needed, and the more vehicles are required in a fleet to fulfill all travel requests. The number of empty trips and idle vehicles will increase the per-mile cost for all users. This means travel costs in sprawled-out areas will be higher than in small, dense areas. For people using shared AVs, this would fundamentally change the cost structure of automobile use, as the current per-mile cost advantage of long car trips would be eliminated. This could put downward pressure on sprawl, which could drive some of the positive reinforcing effects of public transit: if vehicle sharing encourages denser land use and enables first-/last-mile connections, then public transit will gain attractiveness, resulting in higher ridership, which over time translates to better transit service, which improves the attractiveness of transit, and so on.

In addition to vehicle sharing, policies and other interventions are likely to be needed to steer the system toward more desirable outcomes. To reach those outcomes, it may be essential to develop ways to reduce the *Attractiveness of Automobile Use,* increase the *Attractiveness of Public Transit,* discourage urban sprawl, limit the amount of driving that people can do, or some combination of these. Measures could include increasing financial costs (e.g., road pricing), limiting driving (e.g., through some form of rationing), or making other modes substantially more attractive.

Considering potential interventions, we are of the opinion that the longer term, systemic effects of AVs need to be discussed and analyzed more on a broader basis, in order to enable proper preparations for the changes to come. Taking a reactive stance and merely waiting to see what happens when AVs are in use might come with serious consequences. Once widespread adoption occurs, it might be too late to exert sufficient influence on the new mobility system and push the adoption and use of AVs in a desired direction. A more proactive approach could help to shape a framework for AVs to become an integral part of a more sustainable transportation system. For example, it may be possible to get AVs widely accepted as shared-use vehicles and public transit tools, before they become entrenched as a new form of private personal mobility. They might also be used in new ways that connect different modes or replace inefficient trips [31].

Our work has identified some powerful forces that may come into play on the dynamics of our transportation system–and thus provides a valuable starting point for future policy discussions. AVs are a promising technology, but to achieve their desired benefits, the public discourse needs to be focused more on the desired—and undesired—long-term outcomes of their use.

References

1. Fagnant, D., Kockelman, K.M.: Preparing a Nation for Autonomous Vehicles. Eno Center for Transportation, Washington, DC (2013)
2. Google Inc.: Just press go: designing a self-driving vehicle, 2014. https://googleblog.blogspot.de/2014/05/just-press-go-designing-self-driving.html. Accessed 30 June 2016
3. Anderson, J.M., Kalra, N., Stanley, K.D., Sorensen, P., Samaras, C., Oluwatola, O.A.: Autonomous Vehicle Technology: A Guide for Policymakers. RAND Coorperation, Santa Monica (2014)
4. Hayes, B.: Leave the driving to it. Am. Sci. **99**(5), 362 (2011)
5. Hoogendoorn, R., van Arem, B., Hoogendoorn, S.: Automated driving, traffic flow efficiency, and human factors, transportation research record. J. Transp. Res. Board **2422**, 113–120 (2014)
6. Ioannou, P.A., Chien, C.C.: Autonomous intelligent cruise control. IEEE Trans. Veh. Technol. **42**(4), 657–672 (1993)
7. Bösch P.M., Ciari F., Axhausen K.: Required autonomous vehicle fleet sizes to serve different levels of demand. In: Transportation Research Board 96th Annual Meeting (2016)
8. Pavone, M., Smith, S., Frazzoli, E., Rus, D.: Robotic load balancing for mobility-on-demand systems. Int. J. Robot. Res. **31**(7), 839–854 (2012)
9. Spieser K., Samaranayake S., Gruel W., Frazolli E.: Shared-vehicle mobility-on-demand systems: a fleet operator's guide to rebalancing empty vehicles. Transportation Research Board 96th Annual Meeting (2016)
10. Spieser K., Treleaven K., Zhang R., Frazzoli E., Morton D., Pavone M.: Toward a systematic approach to the design and evaluation of automated mobility-on-demand systems: a case study in Singapore. In: Meyer, G., Beiker, S. (eds.) Road Vehicle Automation. Springer (2014)
11. Coughlin J., Yoquinto L.: The long road home, autonomous cars could lead people to live almost 200 miles from their workplaces, 2015. www.slate.com/articles/technology/future_tense/2015/05/autonomous_cars_and_the_future_of_the_commute.html. Accessed 30 June 2016
12. Brustein J.: Self-driving cars will mean more traffic, 2014. www.bloomberg.com/bw/articles/2014-07-21/self-driving-cars-will-mean-more-traffic. Accessed 30 June 2016
13. Chin R.: Driverless cars—the future of transport in cities?, 2014. www.theguardian.com/sustainable-business/driverless-vehicles-future-car-sharing. Accessed 30 June 2016

14. Madrigal A.C.: Driverless cars would reshape automobiles *and* the transit system, 2012. www.theatlantic.com/technology/archive/2012/09/driverless-cars-would-reshape-automobiles-and-the-transit-system/262953. Accessed 30 June 2016
15. Burns L., Jordan W., Scarborough B.: Transforming personal mobility, 2013. http://sustainablemobility.ei.columbia.edu/files/2012/12/Transforming-Personal-Mobility-Jan-27-20132.pdf. Accessed 30 June 2016
16. Fagnant, D., Kockelman, K.M.: The travel and environmental implications of shared autonomous vehicles using an agent-based model. Transp. Res. Part C **40**, 1–13 (2014)
17. Milakis D., Snelder M., van Arem B., et al.: Development of automated vehicles in the Netherlands: scenarios for 2030 and 2050 (2015)
18. Townsend A.: Re-programming mobility: the digital transformation of transportation in the United States (2014)
19. Forrester, J.: Industrial dynamics: a major breakthrough for decision makers. Harv. Bus. Rev. **36**(4), 37–66 (1958)
20. Abbas K.A.: The use of system dynamics in modelling transportation systems with respect to new cities in Egypt. In: System Dynamics Society Conference Proceedings, 1990, www.systemdynamics.org/conferences/1990/proceed/pdfs/abbas016.pdf. Accessed 30 June 2016
21. Shepherd S.P.: A review of system dynamics models applied in transportation, Transportmetrica B: Transp. Dyn. (2014)
22. Forrester, J.W.: System dynamics—the next fifty years. Syst. Dyn. Rev. (Wiley) **23**(2/3), 359–370 (2007)
23. Sterman, J.: Business Dynamics. McGraw-Hill, Boston (2000)
24. Forrester, J.W.: Industrial dynamics—a response to Ansoff and Slevin. Manage. Sci. **14**(9), 601–618 (1968)
25. Luna-Reyes, L.F., Andersen, D.L.: Collecting and analyzing qualitative data for system dynamics. Methods Models Syst. Dyn. Rev. **19**(4), 271–296 (2003)
26. Babbie, E.R.: The Basics of Social Research. Wadsworth Publishing, Belmont (2014)
27. Vennix, J.: Group Model Building: Facilitating Team Learning Using System Dynamics. Wiley, Chichester (1996)
28. Schwartz, P.: The Art of the Long View. Doubleday/Currency, New York (1991)
29. Wilkinson A., Kupers R.: Living in the futures. Harv. Bus. Rev. (2013)
30. Jonas A., Shanker R., Liu J. et al.: Shared autonomy: put this chart on your wall, it's my sad life, 2015. http://orfe.princeton.edu/~alaink/SmartDrivingCars/PDFs/MorganStanley%20040715ReportJonas.pdf. Accessed 30 June 2016
31. Gruel, W., Piller, F.T.: A new vision for personal transportation. MIT Sloan Manage. Rev. **57** (2), 20–24 (2016)
32. Fagnant, D.J., Kockelman, K.M., Bansal, P.: Operations of shared autonomous vehicle fleet for Austin, Texas, market, transportation research record. J. Transp. Res. Board **2536**, 98–106 (2015)
33. Herbawi W., Weber M.: Modeling the multihop ridematching problem with time windows and solving it using genetic algorithms. In: IEEE 24th International Conference on Tools with Artificial Intelligence (ICTAI) 1, 89–96 (2012)

Smartphone App Evolution and Early Understanding from a Multimodal App User Survey

Susan Shaheen, Adam Cohen and Elliot Martin

Abstract Travelers are increasingly turning to smartphone applications for an array of transportation functions. Four types of transportation apps have emerged: (1) mobility apps; (2) connected vehicle apps; (3) smart-parking apps; and (4) courier network service (CNS) apps. This chapter discusses the history and trends leading to the growth and development of transportation apps and summarizes key characteristics of 83 transportation apps identified through an Internet search cataloging transportation apps with more than 10,000 downloads each. Seventy-one percent of the 83 apps identified incorporated a real-time data function (e.g., traffic conditions, roadway incidents, parking availability, and public transit wait times). Additionally, the chapter reports on findings from a survey, conducted in spring 2016, of 130 app users who downloaded the RideScout mobility aggregator app (which ceased operations in August 2016). The survey, which asked respondents questions about their use of mobility aggregators more generally, sought to understand how multi-modal information apps shift travel behavior. The findings showed that most users of such apps would walk, drive alone, and carpool during a typical month. Fifty percent of respondents drove alone once or more per

S. Shaheen (✉)
Transportation Sustainability Research Center (TSRC), University of California, 408 McLaughlin Hall, Berkeley, CA 94720, USA
e-mail: sshaheen@berkeley.edu

A. Cohen · E. Martin
Transportation Sustainability Research Center (TSRC), University of California, 1301 S. 46th StreetBldg. 190, Richmond, CA 94804, USA
e-mail: apcohen@berkeley.edu

E. Martin
e-mail: elliot@berkeley.edu

G. Meyer and S. Shaheen (eds.), *Disrupting Mobility*,
Lecture Notes in Mobility, DOI 10.1007/978-3-319-51602-8_10

day. Twenty-five percent owned one vehicle, and 75% owned two or more vehicles. Thirty-nine percent of respondents reported that they drove less or much less due to the apps. Findings from the survey suggest that multi-modal app users do change their travel behavior in response to information provided, and they may contribute to a reduction in vehicle use.

Keywords Smartphone · Applications · Multimodal apps · Survey · Transportation apps · Travel behavior

1 Introduction

Smartphones represent one of the most important transportation innovations of the twenty-first century. A variety of factors are changing the way people think about mobility including demographic shifts, advancements in geospatial routing and computing power, the use of cloud technologies, faster wireless networks capable of carrying greater bandwidth, congestion, and heightened awareness about the environment and climate change. Mobility consumers are increasingly using smartphone applications, dubbed "apps" for an array of transportation use cases. More people are starting their trips with smartphones to plan routes, seek departure information for the next bus or railcar, find a taxi via an eHail app, or source a private driver through services, such as Lyft or uberX. Factors driving transportation app growth include: time savings (e.g., high occupancy vehicle lanes available to users of dynamic ridesharing); financial savings (e.g., dynamic pricing providing discounts for peak and off-peak travel and for choosing low-volume routes); incentives (e.g., offering points, discounts, or lotteries); and gamification (e.g., use of game design elements in a nongame context) [1].

For mobility consumers, transportation apps offer instant access to real-time information previously unavailable (such as estimated departure and arrival times). This makes trip planning more convenient by aggregating modal and information feeds to provide users with a comparison of routes, departure times, and modal options. For public agencies, transportation apps can aid network management functions, such as disseminating roadway and public transportation information on incidents, delays, congestion, and service disruptions. Transportation apps enable mobility consumers to make more informed transportation decisions, which can aid public agencies in network management. This chapter includes five key sections. First, we provide an overview of the history and evolution of smartphone applications. Second, we present a synopsis of smartphone apps impacting transportation and results of our 2015 North American transportation app benchmarking analysis. Next, we present some behavioral understanding from a 2016 survey of 130 multi-modal app users. Fourth, challenges to the adoption and mainstreaming of transportation apps are discussed. Finally, we conclude with a summary of key findings.

2 History and Evolution of Smartphone Applications

To understand how mobile apps have evolved and are impacting travel behavior, we discuss the history and trends leading to their growth and development. Smartphone apps have progressed through five key phases:

1. Basic Applications
2. Wireless Application Protocol (WAP)
3. The Rise of Proprietary Platforms
4. Platform Wars
5. The Rise of Multi-Platform Advanced Features.

These phases are summarized in Fig. 1.

Phase 1: Basic Hardware and Applications: Early-1980s to Late-1990s
Mobile applications trace their origins to basic devices of the mid-1990s. These applications were extremely limited by rudimentary processors, simple user interfaces, and few features, almost entirely due to limited hardware capability. The Motorola DynaTac 8000X was the first commercially available cellular phone. First marketed in 1983, it had a talk time of about 30 min and retailed for approximately US$4000, slightly less than a new car. The Motorola DynaTac placed calls and included a basic app to manage contacts [2]. Early apps emphasized basic functions, such as arcade games, ring tone editors, calculators, and calendars. During Phase 1, software, application features, and design were facilitated by the original equipment manufacturers (OEMs). As cellular hardware began to advance, new multi-functional applications started to emerge. These developments began to change the way users viewed their phones, transforming them from a single-purpose calling device to a multi-purpose business tool and personal assistant as consumers increasingly requested more features [2].

Phase 2: Emergence of Mobile Data: Mid-1990s to the Mid-2000s
Beginning in the mid-1990s, equipment manufacturers started turning to the Internet to deliver mobile content while limiting third-party access to proprietary software and hardware developed by OEMs. Because early hardware was not directly compatible with the Internet due to limitations in screen size, bandwidth, and processing power, manufacturers developed the Wireless Application Protocol, known as WAP. WAP was a technical standard for accessing information over a cellular network and represented a lower bandwidth, which is a more simplified form of the hypertext transfer protocol (HTTP)—the foundation for the World Wide Web [2]. WAP was designed to operate within the hardware and bandwidth limitations of cellular networks. WAP offered equipment manufacturers the ability to develop a single mobile browser and enabled developers to create third-party content. This set the stage for the development of third-party content, including future app marketplaces. However, the lack of a direct interface with HTTP, limited user interfaces, and technological limitations (screen size, bandwidth, etc.) were common criticisms of WAP [2]. WAP browsers were notoriously known to be

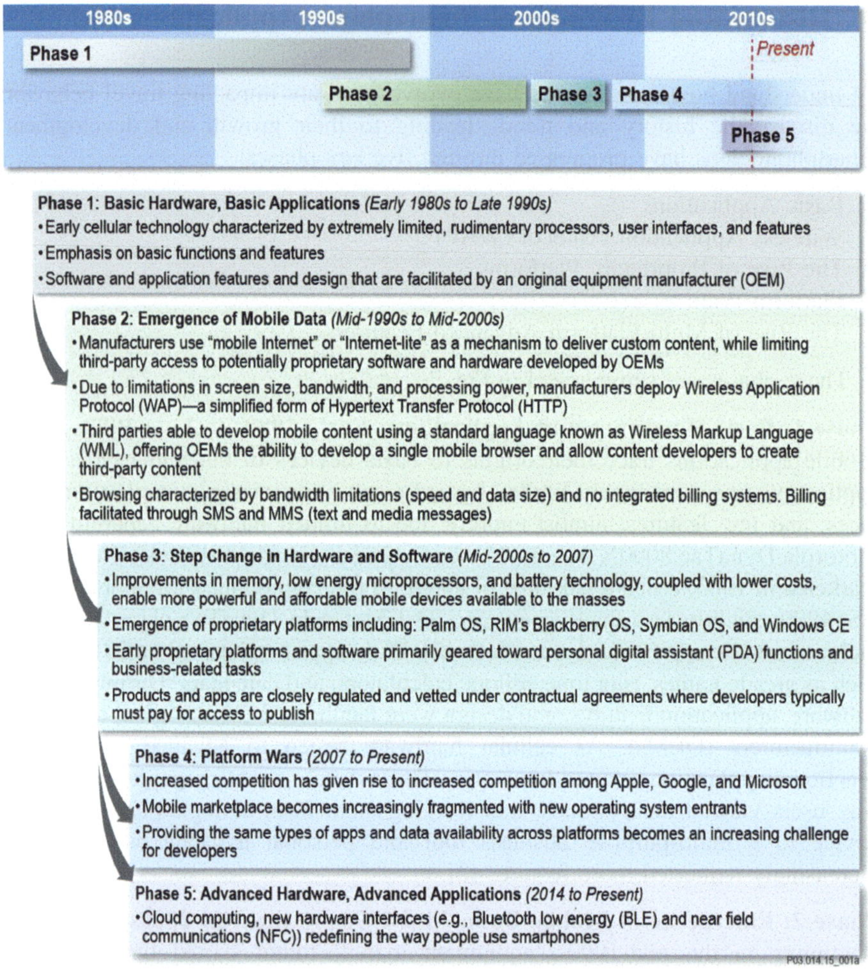

Fig. 1 Five key phases in the evolution of smartphone apps [6]

slow, tedious, and lacked an integrated billing system. Thus, early mobile payments had to be awkwardly facilitated through either Short Message Services (SMS—text messages) or Multimedia Messaging Services (MMS—a picture or multimedia messages). Additionally, users found it tedious to type on numeric keypads and small screens resulted in content that was hard to read. Moreover, many users found it frustrating to load fragmented sentences and then wait for the next sentence fragment to download. Broadly, the poor user experience, due largely to early technological (hardware and bandwidth) limitations, curbed commercial viability.

Phase 3: Step Change in Hardware and Software: Mid-2000s to 2007
Improvements in memory, microprocessors, and batteries coupled with lower hardware costs led to the development of more powerful mobile devices that could run more sophisticated operating systems, such as Windows and Linux. Most of these proprietary, "closed ecosystems" were regulated by the handset maker and/or operating system developer. Desktop computer developers, previously nonparticipants in mobile app development, could now create content for new devices [2]. During this phase, a variety of proprietary platforms emerged including the following:

- *Palm Operating System (OS)*—A mobile operating system developed by Palm Inc. for personal digital assistants (PDAs) in 1996. Later versions were extended to support smartphones. As of 2008, there were 50,000 third-party applications for Palm OS [3].
- *Blackberry OS*—A mobile operating system developed by Research In Motion (RIM), also known as BlackBerry Ltd. for its line of BlackBerry smartphones. BlackBerry OS was discontinued in January 2013.
- *Symbian OS*—A mobile operating system developed by Symbian Ltd. used on numerous basic mobile phones and smartphones produced by Motorola, Nokia, Samsung, and Sony Ericsson in the mid-2000s. In 2006, Symbian was estimated to have 73% of the market share of all smartphone operating systems [4].
- *Windows CE*—A modular mobile operating system developed by Microsoft that was designed to support several types of devices, including smartphones. Originally released in 1996 (v. 1.0), the most recent version was issued in 2013 and will be supported through 2023.

These early proprietary platforms were primarily geared toward personal digital assistant (PDA) functions and business-related tasks.

In the late-1990s and early-2000s, mobile manufacturers began to bridge the gap between business-use mobile computing and cellular phones. In 1996, Nokia launched its communicator series meant to serve as a mobile phone and computer with facsimile (fax), email, text messaging, and web browsing functionality. Similarly, Microsoft's Pocket PC (later renamed Windows Mobile) was designed to mirror the experience of Windows XP, offering users a mobile start button. Pocket PC/Windows Mobile was intended to bridge hardware gaps by operating on smartphones with touchscreens, mobile phones without touchscreens, and on PDAs with stylus functionality. These early hardware and software platforms laid the foundation for contemporary smartphone technologies but were often restricted by hardware, storage, and data bandwidth limitations.

Apple's launch of its iPhone in 2007, also represented a significant advancement in the user experience, as well as hardware and software. The iPhone quickly became the first mass-marketed smartphone device supporting third-party applications and cloud computing using a mobile broadband connection. Incorporating global navigation satellite systems (later coupled with assistance using cellular triangulation) allowed iPhone devices to quickly locate and lock onto satellite signals setting the stage for a variety of mobility functions. This changed not only how smartphones were used but also how users traveled. Another key development was full Web site compatibility on the iPhone. Web sites no longer needed to be concerned with bandwidth limitations, special mobile sites, and protocols. Rather, full webpages could be readily displayed on a smartphone screen. This was critical in bridging the hardware and software digital divide that had previously limited the delivery of products and services to mobile users. iPhone's success was quickly replicated by Google's Android and an updated version of Windows Mobile, known as "Windows Phone."

With the advent of proprietary mobile platforms, developers and their apps became closely regulated and vetted under contractual agreements. Under this framework, developers began to pay to publish their apps to marketplaces. The marketplace model has often been criticized as limiting innovation, app availability, and compatibility across platforms, largely due to the lengthy app screening process for posting and updating apps.

Phase 4: Platform Wars: 2007–Present
Increased competition has resulted in Phase 4 or "platform wars," evidenced by increased competition among Apple, Blackberry, Google, and Microsoft [2]. As new OS entrants launch, the marketplace becomes increasingly fragmented. Developing and maintaining apps across multiple platforms becomes an increasing challenge for developers, particularly for individual developers and less resourced companies that cannot afford to develop application versions for multiple platforms. This lack of open-source standardization has created a complex and challenging marketplace for new entrants (entrepreneurs and developers) with limited resources to make their content available for all mobile users across a growing array of operating systems.

Phase 5: Advanced Hardware, Advanced Applications: 2014–Present
New advanced hardware interfaces, including cloud computing, Bluetooth Low Energy (BLE),[1] and near field communications (NFC),[2] are changing the way people use mobile devices [6]. These innovative technologies offer a number of practical uses for mobility functions (e.g., mobile fare payment and integration) and

---

[1]*Bluetooth Low Energy (BLE)*: With BLE, wireless transmitters, known as BLE beacons (approximately the size of a matchbox with a coverage radius measured in feet) send Bluetooth signals to smartphones and other Bluetooth-enabled mobile devices. BLE communicates with many users, allowing notifications for coupons, offers, and promotional information when entering the Bluetooth range. For example, a user walking past a bikesharing kiosk, public transit station, or a bus stop could be notified of bicycle availability, special rates, or the departure time of the next public transit vehicle. BLE also supports beacon-based navigation, which can assist in guiding users to destinations. San Francisco International Airport is using BLE-beacon technology to assist the visually impaired in navigating its terminals. BLE also supports mobile payment [5].

[2]*Near Field Communications (NFC)*: With NFC, smartphones communicate with postage-stamp sized NFC tags. NFC has a range of about seven inches and communicates with a single user. NFC is best suited for settings requiring one-on-one secure data delivery. NFC can be used for mobile payment, transportation passes, and access cards (e.g., entering a carsharing vehicle).

are changing how users interact with transportation apps. These trends include a variety of data sharing, aggregation and disaggregation, such as

1. **Wider, more integral use of data**: Apps, such as Google Maps, aggregate disparate data feeds (traffic sensors, device satellite tracks, and self-reported roadway incidents) to provide more integrated and accurate predictions of user travel time [6].
2. **Increased data sharing among services**: Apps, such as Google Now, pull data from multiple sources. Third-party apps can provide summaries of important information from multiple apps and data sources. For example, a calendar app may integrate with a map app to display optimal trip routing [6].
3. **Functional disaggregation**: Apps are becoming less multi-functional and are instead focusing more on one or two key functions [6].
4. **Bundled apps as services**: As data become more open and functions become more dispersed, new aggregator services—either new apps or native functions of operating systems—are creating innovative services, or "cards," from a grouping of apps. For example, a card notification on a smartphone informs the user of a new email, allowing the user to quickly respond via email or a texting app and to add an event to their calendar through the calendar app—all without opening a single dedicated app [6].

These trends are leading to a seamless, integrated, and narrowly tailored user experience. Many transportation apps are responding to these trends. For example, Lyft and uberX ridesourcing vehicles can be hailed from inside Google Maps, and delivery services are embedded into restaurant apps. In the future, app users can expect that the basic function of mode (or multimodal) selection to a destination involves a single app that is integrated with multiple, separate apps (routing, booking, payment, social media, and more) to deliver a personalized route recommendation, so the user is not burdened by referencing multiple apps. Together, this technological evolution is driving the development of new app-based services that will continue to impact the transportation sector.

3 Transportation Smartphone Apps

In this section, we define four key areas of transportation apps. We also present our United States (U.S.) smartphone transportation apps benchmarking analysis. To provide some context for this discussion, we present analysis from a 2015 comScore study of the U.S. smartphone market. This research estimates that Android has the largest mobile operating system market share, accounting for 53.2%, followed by Apple iOS at 41.3%. In contrast, Microsoft and BlackBerry each had 3.6 and 1.8% of the market share, respectively [7]. A limitation of the comScore study; nevertheless, is that it only reflects smartphone users. It excludes other mobile devices users (e.g., tablets, 2 in 1 notebooks, and wearable devices). While Microsoft Windows is widely recognized to have a small percentage of the smartphone marketplace, they have a relatively large mobile PC and tablet presence in the app marketplace. Further, it is important to note that in 2015, Windows 8 was phased out in anticipation of the Windows 10 Mobile release in the third quarter. Other mobile devices, such as wearable technology, tablets, and notebooks, can serve a transportation function but may have more limited app availability as many transportation apps are designed for smartphones. The increasing use of universal apps (single apps that can run on different size devices), such as Windows 10 universal apps, may expand the availability of transportation apps on a wider array of devices.

The main categories of transportation apps include: (1) mobility apps, (2) connected vehicle apps, (3) smart parking apps, and (4) courier network services (CNS) apps. There are also nontransportation apps that may impact the transportation network. Broadly, these apps are changing how people travel, interact with privately owned automobiles, and ship merchandise.

Mobility apps assist users in planning or understanding their transportation choices and may enhance access to alternative modes. Mobility apps can include a variety of apps including the following: (1) business-to-consumer sharing (e.g., carsharing, bikesharing); (2) peer-to-peer sharing (e.g., peer-to-peer carsharing, bikesharing); (3) mobility trackers (e.g., Moves); (4) apps for real-time public transportation information; (5) ridesourcing (e.g., uberX and Lyft); (6) eHail taxi apps (e.g., Flywheel); and (7) multi-modal trip aggregator apps (e.g., Swiftly, Moovit).

Connected vehicle apps allow remote access to a vehicle through an integrated electronic system. Generally, connected vehicle apps are designed for emergency situations (e.g., vehicle lockouts, dispatching assistance during an accident, etc.) and may also provide other vehicle services (e.g., diagnostic information, geolocating a vehicle, etc.). Many connected vehicle apps are developed by vehicle OEMs (e.g., General Motor's OnStar).

Smart parking apps provide information on the cost and availability of parking. Some smart parking apps may facilitate electronic payment. Generally, smart parking apps are paired with public or private parking systems or both (e.g., SFpark). Broadly, smart parking apps include *eParking* (apps that streamline the parking process and *eValet*, such as Luxe (for-hire parking services used to dispatch valet drivers to pick-up, park, and return vehicles).

Courier network services (CNSs) provide for-hire delivery services for monetary compensation using an online application or platform (such as a web site or smartphone app) to connect couriers using their personal vehicles, bicycles, or scooters with freight (e.g., packages, food).

Additionally, three categories of nontransportation apps that may impact the transportation network include as follows: (1) Health Apps; (2) Environmental/Energy Consumption Apps; and (3) Insurance Apps. Health apps can assist users with monitoring their health (e.g., calories burned, heart rate, etc.) and changing their behavior (e.g., exercising more and eating less). Health apps can also be employed to help users understand the health impacts of their transportation choices (e.g., Map My Walk). Environment/Energy Consumption Apps track environmental impacts and the energy consumption of user behavior. These apps may predict a user's greenhouse gas (GHG) emission consumption and may also include apps that encourage environmentally conscious behavior, such as eco-driving and eco-routing apps (e.g., Refill and greenMeter). Finally, insurance apps provide a variety of coverage and claims functions for users. These apps can also contain transportation functions, such as pay-per-mile automobile insurance (e.g., Metromile) and other usage-based pricing and incentives related to distance, travel time, and safe driving (e.g., Allstate's usage-based insurance app).

While transportation apps are readily available on app marketplaces, basic data benchmarks, such as the number of downloads, and usage characteristics, are often difficult to identify and catalog. For example, a walking app may be listed in the "health and fitness" category yet serve an important transportation function. Thus, one could miss a transportation-related app in such an analysis due to how apps are cataloged in a marketplace. Further, it is not possible to uniformly compare the number of downloads and user ratings across major app marketplaces. For instance, only the Google Play store publicly provides an approximate number of downloads, while the Apple's iTunes store alone distinguishes customer ratings among the current and earlier versions of an app. Finally, no marketplace has developed a metric to determine the frequency of app use (e.g., whether an app is downloaded a million times and used once or twice before it is uninstalled versus an app downloaded half as frequently but used daily on average). Consequently, it can be challenging to assess which apps have the greatest impacts on the transportation ecosystem.

North American Transportation App Review

Between January and February 2015, we conducted a review of smartphone applications on four major North American app marketplaces (Apple, Blackberry, Google, and Microsoft). We excluded public transit agency apps from the review because many of these apps are available for direct download from public transit agency web sites, which makes such apps more challenging to catalog due to the vast number of public transit agencies across North America. As part of this review, we benchmarked key qualitative functions among transportation apps with more than 10,000 total downloads. Key qualitative characteristics identified include: (1) operating system (OS), (2) real-time information availability, and (3) use of gamification and incentives.

Operating System

We identified 83 transportation apps across all four marketplaces that had 10,000 total downloads or more. We found that the majority of transportation apps were only available on Android and iOS and frequently they were unavailable on Windows and Blackberry. As noted earlier, Windows 8 was being phased out at this time in anticipation of the Windows 10 Mobile release in Q3 of 2015.

Real-Time Data

We calculated that 86 and 80% of transportation apps were available on Android and iOS, compared to just 36 and 23% on Windows and Blackberry, respectively. Seventy-one percent of the 83 apps identified incorporated a real-time data function (e.g., traffic conditions, roadway incidents, parking availability, and public transit wait times).

Gamification

Gamification is the use of game theory and game mechanics in a mobile app to engage users. Apps that employ gamification configure the user as a "player" within a gamified app design. For example, the use of leaderboards, badges, levels, progress bars, and points are intended to encourage and/or discourage particular user behaviors [1, 8]. In a gamified context, app users may receive points, increased rankings, or other rewards for environmentally conscious behaviors, such as carpooling or riding public transportation instead of driving alone. Particularly "bad" behaviors may be penalized by the loss of points or rankings, including driving alone on a "spare-the-air" day.

Gamification tends to leverage social aspects of competition to encourage socially and environmentally preferable outcomes. Some of the most successful behavior change mechanisms pair gamification with social pressure. For example, the Waze and GasBuddy apps use competition and status seeking behaviors to encourage desired behavioral change. In this vein, gamification is often paired with incentives. Reporting roadway incidents and gas prices, in the case of each of these apps (respectively), can lead to the accumulation of points or statuses that can be redeemed for lottery entries, prizes, and leaderboard rankings to further increase gamified elements, competition, and social pressure. The urge to compete, rank highly, and conform to community norms can be a powerful motivational tool.

We found that 23% of the apps incorporate a gamified incentive, such as raffles or special badges (also known as favicons), symbolizing an achievement level. Some apps employ loyalty points that can be redeemed for rewards (e.g., discounts, gift cards, etc.). Loyalty points were the most common mechanism employed, accounting for approximately 21% of all incentives.

Gamification, social pressure, and incentives can be an effective way for apps to promote use and adoption, encourage certain types of transportation behaviors (e.g., ridesharing, cycling, etc.), and provide a mechanism for disbursing a variety of transportation demand management incentives. The user impacts of gamification and incentives in transportation apps have not been extensively studied and are not well understood.

4 Impacts of Multi-modal Apps on Travel Behavior: 2016 Exploratory Survey of Multi-modal Transportation Information App Users

Multi-modal transportation information apps are smartphone apps that provide users with trip planning information on surrounding mobility options. RideScout was an example of an app that integrated an array of public and private transportation services. On August 3, 2016, RideScout ceased operations and the company transitioned to a different business model oriented toward mobile payments for travel. As an app, RideScout offered a single interface for users to compare transportation options (e.g., cost, mode, departure and journey time, etc.). Some of the services formerly integrated into RideScout's interface included public transportation, ridesourcing, taxis, carpooling, ridematching, carsharing, bikesharing, scooter sharing, and eParking. In the summer of 2015, RideScout merged with GlobeSherpa and rebranded itself as Moovel, which is a public transit platform facilitating mobile ticketing and back-end transit agency operations. Moovel also offers an application programming interface (API), known as RideTap that allows third-party developers to integrate ridesharing, carsharing, bikesharing, and other transportation modes into a single interface.

In March 2016, the authors surveyed 130 people that had downloaded the RideScout app. We designed the survey to evaluate the attitudes and perceptions of respondents to apps like the RideScout mobile app, as well as evaluate the travel behavior and modal shift that might result from using multi-modal information apps. The population sampled included people who downloaded the RideScout app across the United States. It is important to note that we asked about the impacts of multi-modal apps more broadly and not specifically about RideScout. Hence, the impacts reported are not necessarily due to RideScout specifically but rather to multi-modal apps more generally. That is, those that downloaded RideScout were also users of other multi-modal apps, and their responses were given in the context of all such apps that they used. To recruit participants, RideScout sent an email on behalf of our study team with a survey link to an estimated 3000 randomly selected users. This population consisted of those who had agreed in the summer of 2015 to be contacted by RideScout for future surveys. The sample contained 56.2% (73) male respondents, 42.3% (55) female respondents, with 1.5% (2) respondents declining to state their gender. Respondents were balanced toward younger ages, as 42% were between 24 and 36 years of age. Forty-six percent of respondents had completed a post-graduate degree program. Eighty-six of the respondents were Caucasian/White, compared to 7% Asian, 5% Hispanic/Latino, and 4% African American. The remaining sample declined to provide their race/ethnicity (percentages do not add up to 100 because respondents could select more than one race/ethnicity). Finally, 72% of respondents earned more than $50,000 annually.

Nearly all respondents reported that they use transportation apps for multiple functions. The most common application was the use of mapping apps for driving, as 85% of respondents reported using transportation apps for this purpose. Other common applications included obtaining access to on-demand ride services (84%), obtaining information about public transportation services (77%), and receiving information about multi-modal transportation options (60%). Among those that used apps with multi-modal information, 22% used them every day, and 66% reported using them at least once a week.

The survey contained questions that evaluated how use of multi-modal apps changed respondent travel behavior. Respondents were asked "How has your use of multi-modal transportation app(s) influenced your ⟨use of mode X⟩…" where a number of modes were described in this context within separate questions. Fifty-eight percent reported that they did not change their driving behavior due to the availability of a multi-modal app. Of the respondent pool that did change their driving behavior, 38% stated they decreased their driving because of the multi-modal app in contrast to 4% that reported increasing it. Respondents generally reported shifts in travel behavior that led to less-energy intensive mobility. For example, 36% of respondents reported walking more versus 8% walking less. At the same time, 22% reported bicycling more, while only 5% reported bicycling less. For public transit modes, multi-modal apps seem to be even more beneficial. For urban rail, 43% reported that multi-modal apps increased their use of this mode, while only 8% said that they decreased their rail use as a result (the rest indicating no change). In the case of bus, 56% reported increasing their use, while 5% reported decreasing their bus use due to multi-modal apps. Respondents reported that emerging shared mobility options, such as Uber and Lyft, as well as one-way carsharing were positively influenced by the use of multi-modal apps. Forty percent increased their use of ridesourcing (e.g., Lyft, Uber) due to these apps versus only 8% of respondents reporting a decline. For one-way carsharing, 22% reported an increase, while only 4% reported using the mode less. Other modes, like commuter rail, also experienced a reported increase in use as aided by the multi-modal app. Roundtrip carsharing experienced little change, positive or negative, due to multi-modal apps, while modes such as taxis, driving alone, and riding as a passenger in a car all had more respondents reporting a decrease versus an increase in use. The impact on taxi use was modestly negative, as 17% of respondents reported a decline in their taxi use, while 9% reported that the apps had increased their taxi use. Hence, the results suggest that multi-modal information apps more broadly are enabling people to use public transit and nonmotorized modes more. Overall, the effect on shared mobility modes is more mixed, with ridesourcing and one-way carsharing benefiting the most from such apps.

Survey respondents also reported that obtaining real-time information about public transit arrivals, trip planning, and the convenience of having multiple modes in a single interface were the most common reasons for using multi-modal apps. Cost savings was rated as the least popular reason for using multi-modal apps. The top-rated reasons for not using a multi-modal app include as follows: (1) users know in advance which mode they want to use, and (2) users do not like the uncertainty associated with real-time planning and mobility. Respondents were also asked about the impact of multi-modal apps on wait times, with 41% reporting a decrease in wait times, 6% reporting an increase, and the remaining sample reporting no change.

In summary, this early user survey suggests that mobility aggregators could have the potential to improve real-time information and convenience for app users, while reducing the frequency of vehicle use. Nevertheless, more research is needed with a larger sample and additional mobility aggregators to determine if these results are applicable to a wider user population.

5 Challenges and Opportunities for Adoption and Mainstreaming of App-Based Services

Despite the growing prevalence of transportation apps, some challenges impact the adoption and effectiveness of app-based transportation services. These challenges and opportunities include: (1) privacy challenges, (2) accessibility considerations, and (3) open data standards and data sharing.

Privacy policies for most apps, app marketplaces, software, and operating systems (e.g., Apple, Google) are often written in legalese making user agreements opaque, long, confusing, and difficult to understand for the vast majority of users. Apps, operating systems, and app marketplaces typically have multi-page user agreements with fine print that software companies "expect" users to read and consent. For the vast majority of users, this text is challenging to read on mobile devices. Many users may not understand what information they are consenting to share or are unaware of what private information they are exposing to third parties through app use. Smartphone apps may intentionally or unintentionally collect a wide array of sensitive information, such as email addresses, phone numbers, financial and location information, and usage history of the apps installed on their phone and mobile browsing history. Location history may represent some of the most sensitive data collected and stored by transportation apps and shared with third parties to offer users additional products and services. Privacy and security concerns are complicated by this type of data sharing because this is often facilitated through third-party APIs,[3] which may contain security vulnerabilities in addition to the cloud, software, and hardware security protocols. App developers, marketplaces, and OEMs have a continuing obligation to enhance security features and monitor their apps for potential security vulnerabilities. App marketplaces, in particular, play a critical role in ensuring that apps distributed on their sites are secure and free of malware.

In addition to privacy challenges, it is important that public agencies and app developers ensure accessibility for all users. Smartphones and data packages are often expensive (if not out of reach) of low-income individuals. Additionally, data availability and bandwidth speeds can limit smartphone app use in less urbanized and rural locations, which can also limit access.

Developers and public agencies interested in launching smartphone apps can address service quality and bandwidth limitations by allowing the caching of data when larger bandwidth is available and by designing "lite" versions of smartphone apps. Lite app versions and functionalities can provide users with a more functional and enjoyable user experience in times of lower bandwidth and poor data coverage. Additionally, app-based services that facilitate electronic payment may not be

[3]An API, short for "Application Programming Interface" is a set of routines, protocols, and tools for building software and applications. APIs can help developers and smartphone apps share data and information between apps and make it easier for third parties to develop apps and incorporate features from existing apps.

usable by unbanked users (e.g., users without a bank account or credit/debit card). Some of these apps may require fare payment via credit/debit cards or mobile/Internet banking. Unbanked users may find it challenging to use mobile apps requiring electronic fare payment. Public entities and app developers can address this challenge by allowing alternative payment methods in conjunction with paperless transactions or establishing programs that offer banking products and services for these users. Capital Bikeshare, for example, has established the "Bank On DC" program to assist prospective unbanked users open an account at local financial institutions. Finally, public agencies and app developers should give special consideration to users with special needs and ensure that disabled users have the ability to use all of the app features.

Public–private partnerships represent one of the greatest opportunities to enhance transportation access for all travelers. Fundamentally, smartphone apps can help to bridge an information divide and make multi-modal transportation more convenient, cost effective, and desirable by aggregating information and simplifying user choices. Offering open data allows public agencies and local governments to disseminate real-time transportation information to their communities, without the cost or responsibility of developing or maintaining their own smartphone apps. Establishing policies that facilitate real-time and static data sharing for APIs and other data is critical. Local governments can support data sharing by adopting acceptable use policies and developing terms and conditions for their data use. Efforts aimed at opening data and developing sharing standards will improve transparency and accessibility, while simultaneously encouraging the private sector to develop new features and apps that take advantage of these data feeds. Local governments and public agencies can meet future data needs by establishing a technology or data officer to manage the collection, sharing, and dissemination of transportation data, as well as the creation of a data dashboard to process and track travel behavior data.

6 Conclusion

Increasingly, mobility consumers are using smartphone applications (apps) for an array of transportation functions, such as vehicle routing, real-time data on congestion, information regarding roadway incidents and construction, parking availability, and real-time transit arrival predictions. Over the years, smartphone apps have evolved from early basic applications to apps with advanced features and functionality that is common today.

Four types of transportation apps have emerged: (1) mobility apps, (2) connected vehicle apps, (3) smart parking apps, and (4) courier network service (CNS) apps. In addition to these core transportation apps, a number of other apps can assist with transportation functions (e.g., health apps, environment and energy consumption apps, and insurance apps).

Findings from a user survey of 130 multi-modal app users showed that respondents are generally using public transit and nonmotorized modes more in response to the information provided by the apps. They are also driving less, while the impact on shared mobility modes is mixed depending on the service. Thirty-eight percent of respondents indicated driving less or much less due to the apps. In addition, respondents reported that the apps facilitated reduced wait times. In the future, more research is needed with a larger sample and across a larger number of mobility aggregators to determine if these results are applicable to a wider user population. Additionally, more research is needed to understand user behavior in response to transportation apps and to fully understand their impacts on travel behavior choices, modal split, and other factors impacting the transportation network.

Furthermore, public–private partnerships can help users overcome the information divide and make multi-modal transportation more convenient, cost effective, and desirable by aggregating information and simplifying user choices. Establishing policies that facilitate data sharing, adopting acceptable use policies, and developing terms and conditions for data use represent key opportunities for public–private collaboration.

Acknowledgments The authors would like to thank Apaar Bansal, Nelson Chan, Corwin Bell, Apoorva Musunuri, and Teddy Forscher of the UC Berkeley Transportation Sustainability Research Center; Ismail Zohdy of Booz Allen Hamilton; Beaudry Kock and Regina Clewlow of RideScout (now Moovel); Wayne Bermand and Allen Greenberg of the Federal Highway Administration; and UCCONNECT for their generous support of this research. The authors would also like to thank the many specialists and practitioners that provided invaluable expertise on smartphone apps at a 1-day workshop in July 2015. Finally, we thank the RideScout survey respondents for their time in responding to our questionnaire. The contents of this chapter reflect the views of the authors and do not necessarily indicate sponsor acceptance.

References

1. Marczewski, A.: Gamification: A Simple Introduction & A Bit More. Amazon Digital Services, Seattle (2012)
2. Clark, J.: History of Mobile Applications, MAS 490: Theory Practice of Mobile Applications. University of Kentucky, Lexington (2012)
3. Treo and Centro: Treo and Centro Software titles. http://blog.treonauts.com/2008/01/treo-and-cent-1.html (2006). Accessed 15 Sept 2016
4. Litchfield, S.: Symbian Market share analysis from ABI. All About Symbian. www.allaboutsymbian.com/news/item/5059_Symbian_market_share_analysis_.php (2007). Accessed 15 Sept 2016
5. Mogg, T.: SFO trials beacon-based navigation system to help blind visitors explore terminal. Digital Trends, 4 Aug 2014. www.digitaltrends.com/mobile/sfo-trials-beacon-based-navigation-system. Accessed 15 Sept 2016
6. Shaheen, S., Cohen, A., Zohdy, I., Kock, B.: Smartphone applications to influence travel choices: practices and policies. FHWA-HOP-16-023, U.S. Department of Transportation, Washington, DC, www.ops.fhwa.dot.gov/publications/fhwahop16023/index.htm (2016). Accessed 15 Sept 2016

7. Soomro, D.: Android OS still the top mobile OS in U.S. Retrieved from Android Central: http://androidcentral.us/2015/04/android-os-still-the-top-mobile-os-in-u-s. Accessed Apr 2015
8. Herger, M.: Resources. The Enterprise Gamification Consultany. www.enterprise-gamification.com (2012)

Getting Around with Maps and Apps: How ICT Sways Mode Choice

Adam L. Davidson

Abstract As computers have become cheap and mobile—most notably in the form of the smartphone—the data and information that they convey has become increasingly practical and spatial for a critical mass of consumers. In short, through mobile devices and GPS functionality, a substantial connected class can now find information that is relevant to them based on where they are, and when they are there. One of the most practical and common behaviors exhibited is the ability to check digital maps and transport information while mobile, often with real-time data. What influence might this have on travel behavior? This study uses focus groups to gain an exploratory understanding of the qualitative influence that information communication technology (ICT) exerts on travel mode choice.

Keywords Interactive communication technology · Travel behavior · Mode choice · New york · Transportation · Public transit · GPS · Apps · Smartphones · Focus group · Urban planning

1 Introduction

One of the most practical and common behaviors to stem from smartphones is the ability to check digital maps and transport information while mobile, often with real-time data. As information and communication technology (ICT) infrastructure and real-time communication expands, so to do the opportunities for communication with users about schedules, routing, delays, congestion, cost, mode options, and other transport metrics. Incumbent transportation agencies are responding to this new technology by gradually adopting schedule and real-time information feeds through standards, such as the general transit feed specification (GTFS), or construction, incident, and congestion information through RSS or Twitter.

A.L. Davidson (✉)
City University of New York, Graduate Center, 365 Fifth Avenue,
Room 4306, New York, NY 10016, USA
e-mail: ADavidson@gradcenter.cuny.edu

G. Meyer and S. Shaheen (eds.), *Disrupting Mobility*,
Lecture Notes in Mobility, DOI 10.1007/978-3-319-51602-8_11

Agencies frequently cite high marks of approval from the public and their relatively low cost as reason to invest in information technology. However, a less explored topic is that with these information improvements in place could they actually influence key travel decisions?

The purpose of this exploratory study is to find out, through series of qualitative focus groups, how ICT delivered mobility information is influencing the transportation decisions of users who face mode and/or routing decisions in their typical week. It is hypothesized that the personalized, dynamic, consumer-oriented information that is delivered through smartphones, GPS devices, and Internet portals could influence the travel decision by mode choice, routing, timing, or even whether to attempt the trip at all by changing the confidence, understanding, or perception of the traveler toward their travel options.

This chapter seeks to summarize some of the main literature suggesting that personalized information could have an influence on changing travel behavior. It then presents the results of a series of focus groups that gave people from three neighborhoods in the New York City region an opportunity to discuss how their devices influence their use of transport.

2 Literature

How one chooses to travel is made up of a complex matrix of utility, purpose, constraint, safety, convenience, infrastructure, and perception—yet the decision is made quite routinely and quickly. Travel decisions become travel habits where common behaviors, patterns, and cues are utilized and embedded into everyday behavior—complex information is not sought and analyzed for each trip [1]. A growing body of the literature suggests that higher quality information tailored to individual circumstances and delivered via ICTs could alter this behavior.

One of the strongest studies to suggest that personalized information leads to changes in transport behavior comes from a pilot program tested repeatedly and with laudable success in Australia. Starting in 1997 various municipalities and a firm named *SocialData* began piloting a program called *TravelSmart*. The big idea of the program was to encourage people to walk, bike, carpool, and take public transit more. Society would benefit from a reduction in carbon emissions and a reduced strain on infrastructure, while individuals would achieve a more active and flexible lifestyle. What was unique about the program was that it was going to first find individuals who were willing to consider alternative transportation and then provide them customized information and support, and finally measure the effects of their intervention. It did this by selecting a neighborhood with multimodal transport options, canvasing the neighborhood with a general transport survey, and then followed up one-on-one with respondents who indicated that they were willing to consider different transport options. At the time, the project was considered a great success. For example, in 2000 the program for South Perth reported a 14% drop in driving and 25% increase in walking amongst participants. Amazingly, participants bicycled 61% more. Similar results were seen in other Australian cities throughout the late 1990s and early 2000s [2].

While those results may seem dramatic, they actually had several caveats. The most important one being that these projects were small in the number of households they reached and were resource intensive. The South Perth example was one of the larger samples and it included only 1434 participating households in the final report. Each community required a series of interventions, including several surveys and an ability to discuss transport needs one-on-one with interested households. This intervention was also scheduled in advance and must be recalled by the participant when seeking mobility rather than providing contextual information when the participant was dealing with the actual task of transport. The programs' estimated costs were approximately $2 million to conduct in an area with 10,000 households. Though *TravelSmart* comes with heavy resource costs it demonstrates that in many cases the problem is not that the provision of alternative transportation is not sufficient—it is that it is not understood.

TravelSmart was also designed at a time when travel information was still relatively inflexible. Historically transportation information had been static, with updates happening typically at the scale of months and years. Information is presented via fabricated signs, printed maps, and time-tables. These mediums are also very general—needing to reach a broad audience. Travel times might only be posted from specific places assuming "normal" conditions, signs might focus more heavily on tourist or mass-market destinations. Ultimately, the narrative power of the information is necessarily limited and determined by a central authority [3]. Only a few years after *TravelSmart* was conducted ICTs evolved enough to give a fundamentally new perspective to the presentation of travel information: yours. Jain (2006) explores how ICTs in the form of mobile trip planners can shift the relationship from mass market to individual by targeting information to specific circumstances—such as traffic conditions right now on your route home, or directions from where you are to your friends' favorite restaurant. Mobile trip planners, she argues, "exploit the need to personalize the timetable and reduce its complexity by reconfiguring its presentation for individual needs," [4].

Less than a decade after *TravelSmart* the smartphone market was growing immensely, mobile trip planners had seen several generations of development, and a number of transport agencies had advanced Intelligent Transportation System (ITS) projects. In short, travel information could easily be accessed and personalized at a large scale. In line with the *TravelSmart* experience, a series of 2012 Chicago-based studies found that real-time arrival information made it psychologically easier for people to consider public transport by reducing anxiety and increasing the sense of control [5], which lead to an actual increase in ridership [6]. Brakewood found a similar increase in New York after the BusTime project went live in 2014 [7] (this occurred after data collection for the study conducted in this chapter). Other studies found an increase in customer satisfaction [8], and a decrease in wait times [9]. These studies suggest that the existence of real-time information present a gradual easing of attitudes that allow habits to be reformed.

A 2011 qualitative travel study [10] would agree with the concept of gradual behavior change due to ICT usage. The authors conducted a travel survey and interviews of purposive sample consisting of part-time working mothers and university students, due to their varied schedules and constraints on time. Both groups of participants identified as tech savvy, but were not professional technologists, which oriented the study towards future potential habits of the broader population. The study concluded that ICT use was strongly embedded in the lives of the participants, often encouraging a blurring of the boundaries between work, home, and school. ICT devices were used both to augment official travel information and to allow on-the-fly reconfiguring of scheduled activities. Whether, it was the ability to use a mobile app or make a mobile phone call or SMS, ICTs were "seen to compensate for the unreliability or unpredictability in both the transport system and people's schedules of activities," [10]. Since the role of the ICT device seemed to be to compensate for behavior rather than outright change it, researchers were led to the conclusion that the incorporation of ICTs into travel behavior was a gradual process, but one in which people developed increasing reliance. In short, "ICT behaviors are adopted not because they are revolutionary, but because they merely accentuate practices that already exist" [11].

In summary, though reliance on ICTs is growing, its impact is still in its infancy. New networks and patterns are developing over time where "the devices will permit new practices and innovation in our relationship with space and travel," [11]. As the ability of ICTs to present information dynamically and individually centered increases, opportunities are created to alter travel decisions by reconfiguring habits [4]. This happens as new ICT regulated systems become available thus adding to, refining, or clarifying pertinent spatial, economic, personal, or temporal aspects of travel. This information allows insight into existing systems, like real-time transit routing [12] or congestion avoidance systems, or creates new systems that utilize excess capacity like car-sharing or ride-sourcing [13, 14]. Presenting people with improved information on their travel options has been shown to improve use of those options [2], thereby indicating investments in information can improve return-on-investment in infrastructure. Thus, ICT systems seem to offer opportunities to gradually reconfigure transportation habits by changing transport practices.

3 Methodology

To explore these emerging habits three purposive focus groups were commissioned in late November and early December of 2013 to discuss their habits of using travel-related ICTs in their everyday lives. The focus group methodology was chosen for the depth of opinion and detail that it can offer for subject exploration. The main advantage of a focus group as an exploratory tool is to give both the respondents greater ability to direct discussion and give the moderator greater ability to probe reactions. This allows for greater discovery of perceptions, emotions, and behaviors that can easily miss articulation in delineated surveys. A strong disadvantage of the focus group option is the small sample size which limits the ability to confidently extrapolate towards broad populations.

3.1 Respondent Requirements and Data Collection

The requirements for the respondents in each group were that:

- They had to be from the same or neighboring zip codes
- Taken at least one round trip on public transit in the previous week
- Must have an active mobile phone line and access to the Internet at home
- A majority needed to have a smartphone with an active data plan
- Reasonably reflect the demographics of their neighborhood, thus there needed to be a gender and minority balance
- Could not work for a transportation provider
- Between the ages of 18 and 85

In addition to the above requirements, other demographic and behavioral questions were asked during recruitment to provide some initial data about the composition of the groups.

The groups only materially differed by the neighborhood from which they came from. Each group had to be respondents from the same neighborhood so that there was a baseline of similar transportation options, modes, and routes to support the discussion. Since neighborhoods had to be targeted, they were chosen by a zip code level analysis of census journey-to-work information that sought modal split between driving and public transportation. This information worked as a proxy to suggest that the neighborhood population in general was regularly engaged in decisions about mode choice. The three neighborhoods that were chosen from this analysis were Yonkers (Westchester County), Bay Ridge (South Brooklyn), and Greenpoint (North Brooklyn). All three neighborhoods featured a mix of bus, train, and highway access in peripheral locations. The sessions were videotaped and transcribed for analysis.

4 Pre-survey Results

During recruitment respondents were asked demographic data such as age, gender, household income, travel modes used in the past week, the kinds of travel information that they had looked up using a device, and their preferences for choosing a travel mode and route. Table 1 is a summary of their responses. Given the small sample sizes this data was not tested for significance, but rather stands as a descriptive measure of the participants.

Table 1 Demographic, transport, search, and preference characteristics of the focus group sample

Categories	*n*	Transit rides prior week	Number of transport modes used[a]	Different transport query types[b]	HHI ($1 k)	Age	Directness	Cost	Speed	Comfort	Reliability
Total sample	30	2.9	4.1	6.2	$102.7	40.8	2.5	3.1	3.2	3.9	2.4
Rank							2	3	4	5	1
Female	15	3.0	4.1	6.0	$98.5	40.4	2.2	2.7	3.5	4.3	2.4
Male	15	2.8	4.1	*6.3*	$106.8	41.3	2.8	3.5	2.8	3.5	2.3
Non-drivers	15	2.6	4.4	6.4	$92.5	45.3	2.6	3.4	3.0	3.5	2.4
Drivers	15	3.2	3.8	5.9	$112.8	36.3	2.4	2.8	3.3	4.4	2.3
Non-white	7	3.3	4.8	6.0	$109.6	45.7	1.7	2.9	3.4	4.6	2.4
White	23	2.8	3.9	*6.2*	$100.5	*39.3*	2.7	3.2	3.1	3.7	2.3
College	22	2.8	4.3	6.0	$105.3	43.2	2.0	2.8	3.5	4.4	2.5
Some college	8	3.1	*3.5*	*6.5*	$95.3	34.4	3.9	3.9	2.3	2.8	2.1
Bay ridge	10	2.4	4.1	5.8	$93.8	44.8	2.8	3.7	2.2	3.7	2.6
Greenpoint	10	3.2	4.4	6.9	$103.0	38.4	2.6	3.0	3.8	3.8	2.2
Yonkers	10	3.1	3.8	5.8	$111.3	39.3	2.1	2.6	3.5	4.3	2.3

[a]Score out of 6. Modes were: private auto, taxi, bike, bus, subway, commuter rail

[b]Score out of 9. Query types were: map views, addresses, transit routing, transit schedules, driving directions, walking directions, biking directions, taxi e-hail, other

Half the participants were male and half were female ($n = 30$). Half of the participants also indicated that they had driven a car in the past week ($n = 15$). However, 67% of those drivers were female ($n = 10$). As a group participants indicated that they used a median of 3 (out of 6 listed) travel modes in the prior week. The ages ranged from 21 to 70 with a mean of 40.8 years. The sample was generally middle and upper middle-class with a mean household income estimated at \$102 k per year with a range from \$27 k to a max reportable value of \$150 k per year. The average participant took about 4 one-way public transit trips in the prior week and looked up 6 out of 9 different kinds of transport information. Our non-white sample (23%, $n = 7$) earned \$10 k more per year than our white sample but took one more transit trip in the prior week than the white population. All participants had at least some college education with only 26% ($n = 8$) not having graduated. Despite having equal access to cars, the college graduate group took nearly one more transit trip in the week prior to the survey than the some college group.

Transport preferences included asking respondents to rank on a Lichert scale (1 = most important, 5 = least important) five factors used when making a transport decision: directness, cost, speed, comfort, and reliability. Rank numbers were then averaged across groups in the summary table. Reliability or directness often came as the first or second most important, while comfort or speed was indicated as least important. This leaves cost as the middle concern.

5 Discussion Summaries

The requirements of the 90 min discussions had two distinct parts covering three broad topics. One part asked respondents about their prior behaviors and perceptions, the other part had respondents attempt travel decision tasks with varying degrees of constraint and information. The discussion topics were centered around their expectations, the influence of data, and their behavior.

Expectations included discussion about planned outcomes based on factors such as speed, cost, ease, health, and others that were important in making personal travel decisions. An important part of this area was explaining the information needed to make those determinations.

The *Influence of Data* of discussion asked respondents about the reliability of the information that they received and their confidence in following its resulting advice, especially if it may lead to a different outcome than predicted. They also were tasked with planning trips that had particular constraints or information sources.

Behavior questions dealt with methodology and frequency of use. The moderator encouraged respondents to discuss their actions in detail both in the general discussion and when undergoing tasks.

The aforementioned tasks also provided a comparison point against the broad discussion. It not only asked them to demonstrate their behaviors, but it also sought

95th Street @ 5th Ave to
Spumoni Gardens, Bensonhurst

	Dollars	Minutes	Calories	Lbs/CO2
	$0.00	65	443	0.0
	$0.13	23	130	0.0
	$2.38	36	105	0.9
	$15.00	13	14	4.0
	$4.20	13	15	4.0

Fig. 1 A sample of the ‘Decision Grid’ which informs the traveler about cost, time, calories, and carbon dioxide emissions by mode for a particular trip

to witness decision modifications based on constrained options or improved information.

One task attempted by all three groups presented in an easy-to-read matrix that was called a Decision Grid in the discussions (Fig. 1). It was a comparison of possible personal impacts by travel mode. The rows were modes of transportation (walking, bicycle, transit, taxi, and driving), while the columns were impacts (cost, time, health/calories, and environment/CO_2). This task was unique because it differed from most travel planning tools which require the person to determine their mode before seeking information.

5.1 Bay Ridge—A Case of Too Few Options

Bay Ridge is located at the end of the R train in southern Brooklyn and with no other trains serving the neighborhood a significant amount of the discussion was directed by participants toward their trials and tribulations with that line. Though express and local buses do provide alternatives, many did not see them as consistent substitutes. Cars and bikes were discussed in context to the R—often as a tool to avoid that train.

Discussing their transportation perspectives revealed that they had many issues—the long journey from the end of the line, the high frequency of stops, off-peak infrequency and unreliability, and the inability of the bus to act as a speedy substitute.

The perceived lack of reliability on the R train led respondents to seek different solutions. One solution was avoid the transit system whenever possible, substituting cars or bikes where they could. Another solution was to just cope by padding their trips with extra time or resigning themselves to the whims of the system. The solution focused on in this study—utilizing information technology to make wiser use of the system—was an emerging solution. Given the limited substitutes, few people were regularly proactive in seeking information for alternatives or updates unless their destination and routing would have been uncommon for them. However, they felt that information could help them and should exist, whether or not they knew if it did.

The information that the respondents most wanted was real-time arrival information. The R train does not currently feature it, though it is found on some lines in the transit system. Due to technical limitations plans to install it on that line in the future are murky, but this situation did not appear to be communicated to these riders by the MTA. One respondent was very enthusiastic about an app called "Roadify," which combines posted schedules and user updates to 'hack' real-time information. However most respondents did not proactively check applications unless there was a problem. A few people were even unaware that schedule or construction information was available online.

One of the more animated discussions involved just getting next train information on the platform. An older gentleman who felt that the R train was reasonably good and a younger gentleman who felt just the opposite had differing views on the necessity.

> Older Male: "So if you're on the platform whether the next train is coming in five minutes or fifteen minutes it doesn't matter. You still have to stand there and wait for it."

> Younger Male: "At least you know though… You're correct, 100% correct but I'd rather be on the platform and see like, half an hour for the train and sit there than just be like, what am I doing? And yeah, if I have no idea if it's coming, I'm just standing there. I could get out and take a cab that would be so much quicker. Like if I saw that and it said a long time until the train I would just get out and take a cab because I don't want to wait… If I don't know it is 30 min and I've been waiting 20 I'll be going crazy."

Noticeably absent from the discussion was any mention of the B63 bus by the participants. This was the MTA's first bus route with real-time information and it serves Bay Ridge on a route parallel to the R train to downtown Brooklyn. However, this is consistent with the view that the bus system did not provide a reasonable alternative to the R train, despite their issues with it. In addition, the real-time info was a pilot project on that line only at the time of the interview. It has since expanded to the entire city-wide bus network and thus currently more visible.

The Bay Ridge group was tasked with finding the best way from their homes to a local restaurant in an adjacent neighborhood on Saturday night. While most respondents turned to their phone for guidance, an older respondent requested a laptop. Some people combined sources of information and local knowledge to make a travel decision.

> "I opened up my Google Maps app. I typed in Spumoni Gardens. I'm very familiar with the place. Then on my Google Maps app I have it set up so that it shows the colors of the train lines and where they go through the streets and stuff. I just zoomed out a little bit, saw where the closest train stations were, and then I looked on HopStop to see what the delays were and then in my head I would have taken the D train because it's faster."

Other people preferred to get information directly from the establishment. One respondent found the restaurant website and assumed he could find directions on there, however they were not available and he did not complete the task. He expected them to have a clear section with written transit and driving directions.

The respondent who used Roadify had a two-step process of finding the address using Google, and then entering the address into the Roadify app. While she found that to be the best method for her, she expressed a desire to be able to search for places by name within the Roadify app. She identified this two-step process as a barrier.

After the group made their initial choices the Decision Grid was handed out and the respondents were asked if this information would have affected their choices. Four of the nine respondents said that it would. An older gentleman said that it would encourage him to take a cab, since it revealed that he could easily afford the fare. A middle-aged woman appreciated the encouragement it offered her to walk by showing the calorie count. A male who had been a bike messenger said that it showed a competitive time for biking and he would probably beat that, so it would have reminded him to bike. Another female revealed in the course of discussion that it got her to reconsider biking as transportation:

> Female: "I don't have other choices then the train. Maybe bike… So then I would just need bike and train [information], but nothing else.
>
> Moderator: Why would you say maybe bike?"
>
> Female: "Because I have a bike but I don't really use it as transportation, more like exercise. But now that I know this thing and the calories, I would consider this as well."

5.2 Greenpoint—A Case of Too Many (Mediocre) Options

Greenpoint is located in Northern Brooklyn with the popular neighborhood of Williamsburg to the south, Manhattan to the west across the East River and Queens to the north and east across Newtown Creek. Its only subway line, the Crosstown G, connects Brooklyn and Queens but does not enter Manhattan without a transfer. However, two popular subway lines (the L and 7) in adjacent neighborhoods are a 10–20 min walk from most parts of the neighborhood.

This group appeared to be very accustomed to Internet-based travel planning, particularly for public transit—though all modes were represented. The pre-survey revealed that they made the most trips by public transit, used the most variety of modes, and used the most online travel tools. Given that they generally had a positive and experienced viewpoint on their use of smartphones and Internet sites to coordinate travel. In fact, most respondents said that they often sought local travel information before making a trip.

Reliability was also cited as an important factor in making travel decisions and one which they readily acknowledged was influenced by the accessibility of information, but only to the extent that the service matched the broadcasted schedule. In one case a young male expressed confidence and relief at having easy access to the schedule[1] of the G train for late night commutes home:

> Male: "There is a lot of hate for the G—but I gotta say. Google Maps—there's been times where I work late nights [and] I have to take a dreaded cross town and take the G at four in the morning but my phone in many cases has been pretty accurate, where within two or three minutes. It says 4:33 in the morning and the train shows up at 4:34. There I am—I'm heading home."

Later another respondent agreed but qualified the extent of the information's influence only to the trains:

> Moderator: "Do you think [the information in these apps is] reliable and trustworthy?"
>
> Female: "I think the trains are much more reliable, for the bus I absolutely wouldn't depend on it to save my life."

This translated into a desire for real-time bus information among the group as an improvement to the existing information options.[2] Thus indicating that good information communicated to them through their personal devices was a positive and perhaps expected service. One respondent surmised that these information systems gave her more control in public transportation:

> Moderator: "But have [travel apps] changed your behavior, what value do they really serve?"
>
> Female: "Theoretically they would make it more likely to rely upon a public method like subway or bus versus thinking I have no control over it [which is when] I'll go take an individualized method like a taxi, your car or bike or something."
>
> Moderator: "So there is a possibility that they are giving you more control?
>
> Female: Yes."

Another theme was the predictability that accessible information technology gave the respondents. One male explained it as "*you can see into the future,*" while

---

[1]Only static planned schedule information is available. The line does not feature real-time reporting.

[2]Since the study took place real-time bus arrival information has become available for all bus lines in the city.

a female talked about how it allowed her to organize her time better, while yet another male said that it altered his life because "*it means that I am generally never late.*" However, as a dissent the guy who noted that it was like seeing into the future also noted that "*we are not as free roaming as we might be... we're kind of going from point A to point B rather than exploring.*" Despite agreement about that another respondent said that if they are going somewhere that they "*don't go to that often there is not a way I'm going to leave without checking it because it's a lot easier.*"

Participants also frequently turned to Google Maps for trips of all modes, but would occasionally use HopStop (which has since been integrated into Apple Maps) if they knew they wanted public transit or Waze if they would drive as these apps give cultivated and targeted information towards these modes.

When this group was tasked with planning a journey to a restaurant in Bushwick, a nearby neighborhood, they were presented with Decision Grid right away. While they chose a variety of modes based on their preferences and their precise routing from their homes, they found the comparisons in the grid helpful. Two people cited being presented with CO_2 emissions as having an effect on their decision making. One respondent said that the time and cost comparison between bike and subway helped to pick the slightly faster bike, especially since it would give her more control, while another respondent choosing between cab and subway went with the subway due to a larger than expected cost savings. A suggested improvement to the Decision Grid information would have been weather conditions.

5.3 Yonkers—When Driving Is the Default Option

The Yonkers participants seemed to provide a middle ground between the engagement of Greenpoint and the apathy of Bay Ridge when it came to ICT engagement. Yonkers was the only neighborhood in this study not inside New York City. Though the small city has reasonable access to transit services, the population makes more trips via automobile than the city residents. As such, most respondents used public transit only for certain trips to avoid using their car rather than the other way around. Rather than use technology to decide between modes, they most often used it to achieve confidence and reliability in their driving experience. Because of this orientation they spent a significant amount of the discussion focused on vehicular GPS.

A consensus quickly emerged that the GPS made their lives easier, while also gave them confidence to go places that they might not have otherwise. One respondent, a freelancer, claimed that it made him much more capable of taking jobs that are located in different areas. Another respondent liked how GPS turned wrong turns into minor delays instead of big problems due to their ability to recalculate routes—it gave her a sense of security. However, even in this group vehicular GPS was not the sole device used to gather information. The smartphone, tablet, and computer all had roles in helping people get around. As an example, the moderator asked each person about how they got to the focus group site. Notably, some combined personal knowledge with the electronic information that they sought, either by double checking familiar routes for delays, timing, or specific maneuvers before beginning the journey, or altering their journey mid-route when presented with unexpected conditions.

By and large the group expected the information they received to be accurate, and were not forgiving when it was not. One respondent claimed that her GPS unit tried to direct her into a brick wall—a potential artifact of bad data—and it tempered her confidence in her use of the device. Another gentleman bemoaned what he saw as a confusing online trip planner provided by the Westchester County bus system, coupled with a call center staff that closed at 4 pm, leaving no acceptable information alternative.[3] They also had growing expectations about the availability of personalized transport information. One respondent put it succinctly—when leaving the house she wants to know "*what you are going to encounter.*" This included tolls, fares, delays, detours, congestion, journey time, and if there would be any material difference on the return journey. A respondent summed it up as anything unusual that could "*get in your way or hinder you getting to your destination on time.*"

When tasked with finding a restaurant using the Decision Grid many expressed a modal preference based on their potential decision to drink, to travel with friends, or avoid headaches such as parking or delayed trains. However that does not discount that the Decision Grid may have weakly influenced the choice of mode for some respondents. One respondent said "*in all honesty I usually drive, but being that it's so close, and it seems like it's short distance, I can just take the train, so it's not a big deal.*"[4] This group revealed that there are many considerations in choosing a mode for a social journey.

In tasked with describing the somewhat more stressful journey of heading to JFK Airport, one of the more interesting insights came from a woman who chose to drive herself. If faced with traffic her first instinct would be to turn to "her friend" the traffic reporter on the radio. Despite, referring to this radio personality warmly, when asked if she felt that method was reliable she responded "*No, not at all… By the time they get that information the traffic's either over or there's a new traffic situation.*" While she felt the radio fostered a better human connection, the information was not recent or tailored to her situation.

[3]The BeeLine system did not provide GTFS data at this time, which as a standardized data feed allows third party applications to present schedule information. This limits the applications and information sources that the rider can utilize.

[4]The train journey was recorded to be 4 min longer than the car journey.

6 Key Themes and Conclusions

Travel decisions are an expression of a group of preferences meant to balance factors like time, reliability, cost, and comfort. For many people, those factors are not considered every time a trip is taken—having been set in the past familiarity is often the dominant determinant on how a trip is made. However, this does not mean that opportunities to change habits are rare. As evidenced by the members of these panels, common challenges like traveling to new parts of the region, avoiding congestion, or dealing with delayed or infrequent services are all reasons to seek a better trip. Electronic devices—in-car GPS, smartphones, and computers—were utilized as tools to help people on our panels match their transportation options with their transportation preferences throughout their week.

Based on the above discussions several themes emerged about how ICT was being integrated into the transport decisions of the participants:

Usage: Respondents found accessing travel information via interactive communication technologies most valuable for travel involving off-peak, unfamiliar routes, multiple options, or where delays are encountered or expected.

Options matter: Having the perception of choice or constraint may be a condition for more active use of travel-related information technologies.

Reliance: Most of our respondents have integrated requesting and using this information into their life.

Control: Many respondents cited increased confidence and control in their travel plans as result of accessing dynamic travel information.

Reliability of information: Many respondents combined multiple information sources with personal knowledge to make a travel decision.

Current/base expectations: Viable app-based schedule information for transit is an expected service; as is accurate and safe driving directions.

Future expectations: Real-time information of all kinds is an expected service for the future. There is also a desire for more seamless application systems that involve fewer steps to retrieve relevant information and predictive abilities.

Application diversity: Though Google dominates, customers use a variety of applications to access this information.

Modal bias: The applications are mode based, and even Google requires you to pick a mode first.

The Decision Grid system, which presents an agnostic view towards mode, may encourage mode switch.

These themes suggest that ICT technology is influencing behavior when it comes to transportation choices. The behaviors exhibited propose that ICT usage is becoming an integral part of the transportation experience by helping our participants to increase perceived reliability. Reliability was cited as one of their most important transport needs. Reliable and tailored information allows users to make active decisions, thereby exercising a form of control on their personal outcomes. As these tools expand in coverage and capability, so too will reliance on this information.

References

1. Iftekhar, S., Tapsuwan, S.: Review of transportation choice research in Australia: Implications for sustainable urban transport design. Nat Resour Forum **34**(4), 255–265 (2010)
2. SocialData, Evaluation of Australian TravelSmart Projects in the ACT, South Australia, Queensland,Victoria and Western Australia: 2001–2005, Government Report, Australian Greenhouse Office in the Department of the Environment and Heritage, Canberra, 2005
3. Harley, J. B.: Deconstructing the map. Cartographica: Int. J. Geogr. Inf. Geovisualization. **26**(2), 1–20 (1989)
4. Jain, J.C.: Bypassing and WAPing: reconfiguring timetables for real time mobility. In: Sheller, M., UrryMobile, J. (eds.) Technologies of the city, pp. 79–101. Routledge, London (2006)
5. Tang, L., Thakuriah, P.: Ridership effects of real-time bus information system: a case study in the City of Chicago. Transp. Res. Part C Emerg. Technol. **22**, 146–161 (2012)
6. Tang, L., Thakuriah, P.: Will the psychological effects of real-time transit information systems lead to ridership gain? Transp. Res. Rec. **2216**, 67–74 (2012)
7. Brakewood, C., Macfarlane, G.S., Watkins, K.E.: The impact of real-time information on bus ridership in New York City. Transp. Res. Part C Emerg. Technol. **53**, 59–75 (2015)
8. Zhang, F., Shen, Q., Clifton, K.: Examination of traveler responses to real-time information about bus arrivals using panel data. Transp. Res. Rec. J. Transp. Res. Board **2082**, 107–115 (2008)
9. Watkins, K.E., Ferris, B., Borning, A., Rutherford, G.S., Layton, D.: Where is my bus? Impact of mobile real-time information on the perceived and actual wait time of transit riders. Transp. Res. Part A Policy Pract. **45**(8), 839–848 (2011)
10. Line, T., Jain, J., Lyons, G.: The role of ICTs in everyday mobile lives. J. Transp. Geogr. **19**(6), 1490–1499 (2011)
11. Aguilera, A., Guillot, C., Rallet, A.: Mobile ICTs and physical mobility: review and research agenda. Transp. Res. Part A **46**(4), 664–672 (2012)
12. Tang, L., Thakuriah, P.: Ridership effects of real-time bus information system: a case study in the City of Chicago. Transp. Res. Part C Emerg. Technol. **22**, 146–161 (2012)
13. Bottsman, R., Woo, R.: What's mine is yours: the rise of collaboartive consumption. Harper Collins Publishers, New York (2010)
14. Benkler, Y.: Sharing nicely: on shareable goods and the emergence of sharing as a modality of economic production. Yale Law J. **114**(2), 273–358 (2004)

Online and App-Based Carpooling in France: Analyzing Users and Practices—A Study of BlaBlaCar

Susan Shaheen, Adam Stocker and Marie Mundler

Abstract This paper examines the characteristics and practices of ridesharing users in France. In May 2013, the authors surveyed members of BlaBlaCar, the largest online and app-based carpooling service in France, to analyze the sociodemographic characteristics and usage patterns of the respondents. The survey results identify correlations between socio-demographic characteristics and usage elements. Notably, users with a lower-income level are more inclined to be passengers, while higher-income users employ carpooling mainly as drivers. Students are shown to be more frequent users as well. These findings indicate some equity balancing effects, which may be unique to this shared mobility mode.

Keywords Ridesharing · Carpooling · Travel behavior · Socio-demographics · Ordinal regression

1 Introduction

Carpooling first appeared in France in 1958 under the name of "organized hitch-hiking" [1]. The original objective of carpooling was to provide mobility to young people with little money who could not afford to drive a personal vehicle. Despite

S. Shaheen (✉)
Transportation Sustainability Research Center, University of California, Berkeley, 408 McLaughlin Hall, Berkeley, CA 94720, USA
e-mail: sshaheen@berkeley.edu

A. Stocker
Transportation Sustainability Research Center, University of California, Berkeley, 2150 Allston Way, Suite 280, Berkeley, CA 94704, USA
e-mail: adstocker@berkeley.edu

M. Mundler
Institut de Géographie et Durabilité, University of Lausanne, NIL Mouline—Géopolis, 1015 Lausanne, Switzerland
e-mail: marie.mundler@unil.ch

G. Meyer and S. Shaheen (eds.), *Disrupting Mobility*,
Lecture Notes in Mobility, DOI 10.1007/978-3-319-51602-8_12

its evolution, carpooling remained very affordable and is probably the most inexpensive transportation option in France today. It is typically three to four times cheaper than the train [2], and it offers more flexibility than public transit in terms of travel times and geography. These advantages explain in great part carpooling's success in France, although the country is not equipped with high occupancy vehicles (HOV) lanes. HOV lanes are more typical in the United States (U.S.) and enable vehicles with higher occupancy (usually two to three occupants or more) to drive in a lane dedicated to carpooling during peak travel periods.

Information and communication technologies (ICT) have facilitated an evolution in ridesharing services (carpooling and vanpooling) in France and other parts of the world. During the late-1990s, there was a notable increase in the number of ridesharing web sites that offered different forms of carpooling services. In 2007, the number of open-access carpooling web sites in France was estimated at 78 [3]. Chan and Shaheen [4] estimated that there are 638 online ridesharing sites in North America; the majority of those sites support carpooling (612), and many serve both carpooling and vanpooling (127). Growth in the number of organizations made it challenging for any one provider to reach a critical mass, and the operators struggled financially [1].

BlaBlaCar, the focus of our analysis, was founded in 2006 [5]. It rapidly became the most popular carpooling organization in France, and it supports about 90% of the market [6]. The quality and efficiency of the online platform for identifying shared ride opportunities is a key success factor. In 2011, the organization transitioned from a free platform to a fee-based service. The company started to charge users a percentage of the trip fees (between 7.9 and 12.5%), as well as a fixed amount (between €0.7 and €1.2 or $0.90 and $1.50) for each trip, depending on when the reservation gets made (the earlier the cheaper). Although this transition was risky and contested by some users [7–9] it was successful, and the number of users increased exponentially to reach 10 million across 13 different countries by 2014. In April 2015, BlaBlaCar bought competitors carpooling.com and AutoHop, increasing their user base to 20 million members across 18 markets [10].

The BlaBlaCar system is mainly based on occasional long distance or city-to-city trips, with an average trip distance of 300 km (186.4 miles). Passengers and drivers are connected through a web site that is designed to combine social media with a reservations platform. These two associated elements enable a feeling of trust and safety that have contributed to the company's growth. Given its notable success, BlaBlaCar provides an interesting case study for ridesharing.

This paper includes results from an online survey of BlaBlaCar of 618 members in France, conducted in May 2013. In this paper, we explore how this model of carpooling is used and by whom from an equal access perspective. The survey enables an analysis of the respondent profiles and insights into differences in their carpooling practices. We hypothesize that online-enabled carpooling attracts a relatively diverse population segment, and use of the service is reflected in socio-demographic characteristics of its customers.

The first section of this chapter reviews the ridesharing/carpooling literature. Next, we describe the study methodology and its limitations. The results are presented in the following section, including an analysis of respondent socio-demographic characteristics and how they relate to carpooling usage. Cross-tabulations and an ordinal regression model are the methods we used in the analysis. Key findings are summarized and discussed in the conclusion.

2 Background

The available literature on ridesharing is rather limited [4, 11]. In the U.S., the literature is considered the most robust, as the "homeland of carpooling" [12]. Indeed, large-scale organized ridesharing services first appeared in the United States in 1942 as part of the war effort [4], and scientific publications on ridesharing practices and its users started to become available in the 1970s [13].

Some of these publications document the ongoing evolution in ridesharing [4, 11]. In 1987, Roger Teal defined the use of ridesharing, in its broader sense, as "anyone who shares transportation to work in a private vehicle with another worker" [11]. Today, this definition seems rather restrictive, as new ICT technologies have enabled the organization of occasional ridesharing among unrelated individuals for purposes other than work, and the definition of ridesharing has progressively evolved to integrate these new practices. Chan and Shaheen [4] provide the following definition of ridesharing: "It is the grouping of travelers into common trips by car or van. Ridesharing differs from for-profit taxis and jitneys in its financial motivation. When a ridesharing payment is collected, it partially covers the driver's cost. It is not intended to result in financial gain. Moreover, the driver has a common origin and/or destination with the passengers." The diversity of ridesharing systems and practice is also illustrated by the analysis of Furuhata et al. [14], which highlights the existence of six classes of ridesharing matching systems: dynamic real-time ridesharing, ridesharing, long-distance ride match, one-shot ride match, bulletin board, and flexible carpooling.

Given ridesharing's evolution, one can assume that its users have also changed. Teal conducted one of the most in-depth studies on ridesharing users in 1987. He analyzed the factors determining the propensity to rideshare to and from work, using data from 1977 to 1978 Nationwide Personal Transportation Survey. His results show that ridesharing participants were more likely to have a lower-income and be the "second worker" (typically women) of a household that had more workers than vehicles. Moreover, ridesharing users typically commuted longer distances and had a higher commute cost burden (with these two factors being correlated). Teal admits that these variables alone do not accurately predict ridesharing choice. He also recognized the importance of attitudinal factors to explain travel behavior [11]. More recent data on ridesharing users comes from the 2000 US Census, the 2001 National Household Travel Survey (NHTS), and the 2005–2007 American Community Survey (ACS). These data show that carpoolers still tended to have a lower level of income in the 2000s. It also emphasizes that Hispanic Americans and African Americans carpool more than Caucasians and other populations (respectively 23, 16 and 10% of their modal share). Some researchers also have focused on more specific forms of ridesharing. Burris and Winn [15], for instance, analyzed the profile of casual carpoolers (slugging users) and compared it with the profile of classical HOV lane users in Houston, Texas. Their results show that these two categories of carpoolers are quite distinct. Casual carpoolers are more likely to make commute trips versus noncommute trips (96 vs. 80%), are between the ages of 25 and 34, and are more likely to be single or married without children, while other HOV lane users tend to belong to larger households.

Even if these data do not enable a complete overview of the evolution of ridesharing users over the last decades, it provides insights into the socio-demographic factors that are likely to influence different forms of ridesharing use as a modal choice. In contrast, in France, there is very little information available on the profile of ridesharing users, and the recent literature on this topic is limited. There are no data on the modal share of ridesharing at a national level. However, in 2008, the average vehicle occupancy in France was 1.22 [16]. The literature on ridesharing in France is mainly comprised of reports and articles that are a few years old and may not be up to date [17]. These sources provide a discussion of the different ridesharing services and their evolution, as well as opportunities for improvement and development, such as real-time ridesharing systems. While different forms of organized ridesharing exist in France, the key models are Internet based, with either restricted access (in the case of employer-based carpooling) or open-access Web sites [3]. The focus of this chapter is on the second of these forms: Internet and app-enabled carpooling through an open-access web site.

The literature on Internet-enabled carpooling is scarce, but there is one interesting source of data. In 2009, an insurance company [18] conducted a survey among BlaBlaCar users to analyze user behavior and motivations. Their results provide an overview of the user profile and how it relates to their usage. It also provides early insights into the possibility of varying usage practices through a distinction between passengers and drivers. Nevertheless, some elements are missing to enable a deeper analysis, such as level of income or cross-tabulations between socio-demographic characteristics and carpooling practices.

Some researchers in France also studied carpooling as means to broaden the transportation options of disadvantaged populations and to foster accessibility. Carpooling is recognized in France as a mobility solution for disadvantaged populations (e.g., elderly, disabled, or unemployed individuals) [3]. Yet, there is little evidence that these individuals are using organized carpooling [18]. Some other forms of carpooling have been described as more adaptive to the needs of lower-income populations in France. One example is informal carpooling, which takes place among relatives or coworkers, usually for regular trips. Informal carpooling does not include slugging or casual carpooling, as this phenomenon is practically nonexistent in France. These informal practices are considered more common than organized carpooling services among lower-income populations, although there is no data to support this assertion. Some experts believe that real-time carpooling (i.e., carpooling organized on very short notice using smartphones and GPS technologies) has a higher potential than more traditional forms of organized carpooling services that are planned in advance [19]. Real-time carpooling is closer to informal and spontaneous carpooling and implies a high level of flexibility, as people are matched instantaneously a few minutes before the trip [17]. In contrast, the potential of classical carpooling is perceived to be lower for disadvantaged populations (mostly lower-income workers in this case), as it does not offer enough flexibility for individuals that work variable hours or in remote places, for instance. Thus, classical carpooling is perceived as a solution for those with more regular schedules [20].

In this chapter, we focus on a more dynamic carpooling service, BlaBlaCar, which serves longer distances or city-to-city trips. We examine the equity impacts of this service based on socio-demographic characteristics, particularly income level and service use. Equity is an important concept in transportation, as mobility impacts one's ability to access economic and social opportunities [21] including rights, liberties, powers, opportunities, income, wealth, and self-respect, or "primary goods" as defined by Rawls [22]. Our BlaBlaCar survey was designed to help us understand how the Internet and app-enabled carpooling provides mobility and accessibility across a wide variety of users.

3 Methodology

This chapter is based on an online survey that we launched in May 2013 among the users of BlaBlaCar, the primary carpooling service in France, which represents 90% of the market [6]. The objective of the survey was to understand the socio-demographic profile of BlaBlaCar carpooling users in France and their overall usage patterns. The survey was composed of two main sections: one focused on an individual's carpooling practices and transportation behavior and the other on socio-demographic information. Most of the questions listed a range of specified responses; however, certain questions were open ended.

BlaBlaCar circulated the survey through automatic emails that they sent to their users. There was no obligation to answer, and the responses were anonymous. All the respondents were subscribers of BlaBlaCar and included both regular (daily or weekly) and less frequent (monthly or less) users. BlaBlaCar ceased recruitment after approximately 500 surveys were completed. The final survey population included 618 responses, including 471 completed surveys. Both complete and incomplete responses were used when possible.

3.1 Study Limitations

The main limitation of this research is that the representativeness of the results cannot be validated. Also, no survey incentives were provided. Thus, there is likely a self-selection bias among respondents. We noticed a gap between our socio-demographic results and the gender data reported by BlaBlaCar, as the company provides these statistics. While our results reflect a respondent population of 54% women, they only represent 46% of the BlaBlaCar user population in France. Another survey limitation is associated with question phrasing regarding income level. We chose not to ask for a precise figure regarding income level, as we thought this might discourage a response. This made it more challenging to compare the survey data with corresponding national data.

4 Results

As part of our results discussion, we highlight three key areas of our analysis including socio-demographic characteristics, socio-demographics and carpooling usage, and BlaBlaCar use frequency and ordinal regression analysis.

4.1 Socio-demographic Characteristics

In our analysis, we employed socio-demographic data to understand the user profile and how this profile impacts carpooling use. The respondents were asked to provide their individual level of income per year among different categories. Monetary values used in this paper are in May 2013 Euros, as is the EUR/USD exchange rate used for comparison purposes. The lower category (<€10,000 per year or <$13,449) was designed to correspond approximately to the first quartile of the income distribution in 2010 in France, which was €9370 (or $12,602). As a consequence, this category represents about 25% of the French population, and 23% of the respondents. The results for the median income are also quite similar for the sample and the French population. For the sample, the median is situated between €15,000 and €20,000 ($20,173 and $26,896); for France, the median was €17,510 ($23,546) in 2010. Thus, there are no major differences between the sample and the French population with respect to income. Nevertheless, it is important to note that the income levels are very diverse, spreading from very low to very high. Later, we analyze the impact of this diversity on carpooling usage.

As shown in Table 1, with respect to education, the survey reveals an important distinction from the French population. We used the same categories as the French National Institute of Statistics (INSEE) to make comparisons and see an over-representation of individuals with a high diploma level from our survey. Most of the respondents have a university diploma (72% of the respondents in contrast to only 30% for France), and overall, their level of education is greater than the French population in general, no matter which age category is examined [23]. Interestingly, the previous BlaBlaCar survey launched in 2009 demonstrated the same education trends. While Maif [18] does not mention the education level of respondents, the report gives the respondents' socio-professional categories with Executives and

Table 1 Socio-demographic variables for respondents and France

Individual income per Year	France (%)[a]	Survey (%)
Less than €10,000	–	23
Between €10,001 and €15,000	–	11
Between €15,001 and €20,000	–	16
Between €20,001 and €25,000	–	16
Between €25,001 and €30,000	–	9
Between €30,001 and €35,000	–	8
Between €35,001 and €40,000	–	6
Between €40,001 and €45,000	–	3
Between €45,001 and €50,000	–	2
More than €50,001	–	5
Education	**France (2011) (%)[a]**	**Survey (%)**
Primary school certificate or less	21	2
Secondary school certificate	7	1
Certificate of vocational proficiency	25	10
Baccalaureate (high school certificate)	17	15
2 years of college	13	25
Superior diploma (Undergrad to PhD)	17	47
Age	**France (2013) (%)[a]**	**Survey (%)**
18–24	11	22
25–34	16	23
35–44	17	12
45–54	17	20
55–64	16	18
65 years or older	23	4
Gender	**France (2014) (%)[a]**	**Survey (%)**
Male	48	46
Female	52	54
Socio-professional category	**France (2012) (%)[a]**	**Survey (%)**
Tenant farmer	1	0
Craftsman, merchant, entrepreneur	3	4
Executive, superior intellectual profession	10	27
Intermediate profession	13	11
Employee	16	23
Worker	12	2
Retired	27	10
Student	9	18
Other with no professional activity	9	4
Location	**France (2012) (%)[a]**	**Survey (%)**
Rural	14	23
Urban	86	77

[a]Data not applicable; however, other means were used to enable a comparison, such as quartile and median

Superior Intellectual Professions and Intermediate Professions (professions that usually require a high level of education), representing 46% of the respondents. If we add Students to that grouping, 73% of respondents have higher education levels.

With respect to participant age, the results show that carpooling respondents are on average much younger than the French population. Young people seem to be very well represented among carpooling users, with 45% of respondents under 35. However, carpooling gathers a larger age distribution than one might expect, as these results show that the practice is not only reserved to young people. Those aged 45 years and older are well represented in the survey (42% of the respondents). Not surprisingly, the category that is least represented is 65 years and older. Interestingly, the category 35–44 is under-represented in the survey, perhaps due to child care responsibilities and needs associated with younger children, as supported by the literature [11, 20, 21, 24]. Parents, nevertheless, are well represented in our survey, but their children are perhaps older in age.

Our survey results contain 54% women in contrast to 46% men, while the actual proportions of Blablacar users are the opposite with 54% men and 46% women. Finally, respondents were asked their postal codes, to determine if they come from rural or urban areas. The results show that 77% of respondents are from urban areas in contrast to 86% overall for France. To summarize, the survey results suggest that BlaBlaCar users have an average level of income similar to the French population, but they tend to be younger, more educated, and more frequently reside in rural municipalities.

4.2 *Socio-demographic Characteristics and Carpooling Usage*

In this section, we focus mainly on the question of financial resources, although other socio-demographic characteristics are studied to present a more complete analysis. As noted earlier, the level of respondent income is rather diverse, so it is interesting to understand how this diversity relates to usage. To analyze carpooling use relative to income, we divided the respondents into three categories, those who have an income: (1) less than €10,000 ($13,449), (2) between €10,000 and €25,000 ($33,620), and (3) above €25,000 ($33,620). This enables a comparison of the use of low-income, middle-income, and high-income users. The limit of €10,000 corresponds approximately to the first quartile of income in France. The limit of €25,000 is designed to match the third income quartile (that is €24,590 or $33,067) and also corresponds to the upper bound of the middle class, as defined by the Inequalities Observatory [25]. The first category represents 23% of survey respondents, the second 43%, and the third 34%. It is important to note that 74% of the low-income users are students. Students have specific mobility needs and options. The existing literature documents that despite their relatively low financial resources (which often depend on their parents' resources), their mobility patterns are diverse and often reflect experimentation [1]. From our survey, carpooling appears to be an important transportation option for students.

Key sections of this analysis include frequency of carpooling use, trip purpose, passenger versus driver role, motivations, and transportation modes that would have been used in the absence of carpooling.

4.3 Frequency of Carpooling Use

The survey results show that respondents with a relatively low-income level (< €10,000) are more likely to use BlaBlaCar on a regular basis (once a day or once a week): 19% in contrast with 8% for the high-income users. Please note that while some association between income and frequency is evident, income was not found to be a statistically significant predictor of usage frequency in the ordinal regression analysis, discussed later.

BlaBlaCar is more associated with occasional use as it is more focused on long distance and city-to-city travel. Thus, it is not surprising that most of the respondents use it only occasionally. More regular use, however, can be interpreted in several ways. This could mean that carpooling is a convenient and reliable transportation solution for some users, such as students, as they must travel further or they do not have many other options available. Our analysis of trip purpose and alternative transportation modes below helps to explain how carpooling use corresponds to a choice or necessity.

4.4 Trip Purpose

Trip purpose results show that people use carpooling for different purposes, depending on income. Respondents were asked if they carpool for leisure, work, or both. We used this difference between work and leisure to represent roughly the distinction between mandatory and nonmandatory mobility. The data reveal that people with a lower-income (<€10,000 or <\$13,449 a year) are more inclined to use carpooling for work/school than people with a higher-income level (>€25,000 or > \$33,620 a year). As shown in Table 2, 17% of respondents with a lower-income use carpooling for work only (both regularly or occasionally) in contrast to just 11% of respondents with a higher-income. The low-income users who carpool for work/studies only are all under the age of 24.

Given BlaBlaCar's focus on serving occasional and long-distance trips, it is not surprising that the majority of respondents use BlaBlaCar for leisure purposes. Nevertheless, those that use BlaBlaCar for work/study trips are on average younger and less wealthy than leisure users. This confirms a dual practice among different user groups.

Table 2 Carpooling use and car ownership as a function of income level (individual net income per year)

Frequency of use	Less than €10,000 (n = 98) (%)	Between €10,000 and €25,000 (n = 185) (%)	Above €25,000 (n = 147) (%)
Every day or two	2	1	1
About once a week	17	7	7
About once a month	34	24	25
Occasionally	46	65	64
Never	1	3	3
Role	**Less than €10,000 (n = 98) (%)**	**Between €10,000 and €25,000 (n = 183) (%)**	**Above €25,000 (n = 147) (%)**
Passenger	54	26	12
Driver	17	36	51
Both	29	38	37
Purpose of trips	**Less than €10,000 (n = 95) (%)**	**Between €10,000 and €25,000 (n = 180) (%)**	**Above €25,000 (n = 143) (%)**
Work only	17	8	11
Leisure only	58	81	71
Both	25	12	18
Number of cars in the household	**Less than €10,000 (n = 98) (%)**	**Between €10,000 and €25,000 (n = 185) (%)**	**Above €25,000 (n = 147) (%)**
No car	13	9	6
1 car	31	46	37
2 cars	26	31	44
3 cars or more	31	14	14
Alternative to carpooling	**Less than €10,000 (n = 98) (%)**	**Between €10,000 and €25,000 (n = 185) (%)**	**Above €25,000 (n = 143) (%)**
Public transportation	48	29	24
Personal car	41	63	69
Rental car	1	0	2
No alternative	8	4	3
Other	2	4	2

4.5 *Passenger Versus Driver Role*

We also observe a difference in the role taken in a carpool, as users with a lower-income level tend to be passengers, while carpooling users with a higher-income are mainly drivers (51% are drivers only). This is related to car ownership. In our survey, the propensity to have no automobile increases with lower-income levels (13% have no car among lower-income respondents in contrast 9% among middle-income and 6% among higher-income users).

Naturally, the roles of passenger and driver do not imply the same level of comfort and autonomy. The passenger faces the first mile and last mile problem in getting to the carpooling meeting point and his/her final destination. Moreover, a passenger is typically perceived as a guest in the driver's car [26]. A closer look at our results show that the majority of drivers belong to the socio-professional categories of "Executives and Superior Intellectual Profession" (34%) or "Employees" (21%). Passengers on the other hand belong primarily to the category of students (32%), followed by employees (26%).

4.6 Motivations

In our results, the motivation ranking does not vary notably from one income category to another, but there are still some interesting differences to be emphasized. To assess participant motivations, the respondents were asked to rank a series of statements on a scale from 0 to 100. These statements varied depending on the role held in the carpool formation: driver or passenger. See Table 3.

The results show that online carpooling is economically attractive to respondents. Even individuals who can afford other transportation modes seem to prefer carpooling to reduce their overall mobility costs. As a consequence, saving money is the more important motivation across income categories. Indeed, among people who earn more than €25,000 ($33,620) a year, "saving money" is still ranked as the most important motivation, with a rating of 87 in contrast to 92 (still on a scale of

Table 3 Motivations to carpool on a scale from 0 to 100 as a function of income level (individual net income per year)

Motivations to carpool (0–100)	Less than €10,000	Between €10,000 and €25,000	Above €25,000
As a passenger			
Save money	92	89	87
Make my trips more pleasant and social	65	66	59
Help the environment	64	69	66
Save time	47	39	24
Make my trips less tiring	49	44	31
Because I have no choice	46	29	19
As a driver			
Save money	97	83	76
Make my trips more pleasant and social	71	71	64
Help the environment	68	73	65
Help others to get around	61	75	66
Make my trips safer	45	40	35

0–100) for people earning less than €10,000 ($13,449) a year. However, the results also show that users with a low income rank the motivation "Because I have no other choice" higher than users with higher-incomes, with a rating of 46 in contrast to 29 and 19 for the middle- and high-income categories. This is evidence of a constrained practice, which is examined further with respect to other available transportation options in the next section.

4.7 Alternative Mode in the Absence of Carpooling

Respondents were asked what kind transportation mode they would use in general, for the trips they now make using carpooling. The results show that people earning less than €10,000 ($13,449) would use public transportation the most (48%), while the majority of respondents with a higher-income level would drive alone (69%). This is also associated with carpooling role; 88% of the drivers would use a personal car, while 65% of the passengers would use public transportation.

The results also indicate that 38% of the lower-income users consider carpooling as a good alternative to car ownership in contrast to only 20 and 18% of middle- and high-income users. These results show that for some of the users, including those with low financial resources, carpooling is an effective mobility option and not necessarily one chosen by default. However, we also observe that 8% of users belonging to the lower-income category would not have made the trip in most cases (in contrast to 4% of middle-income and 3% of higher-income respondents). To summarize, for most respondents carpooling is a solution among other available options that are more or less convenient. However, the survey highlights the existence of more constrained users who do not have other options.

4.8 BlaBlaCar Use Frequency and Ordinal Regression Analysis

To better understand which respondent factors directly influence BlaBlaCar usage frequency, an ordered logistic regression was performed with usage frequency as the dependent variable. The categorical usage frequency responses are: (1) Every day or two, (2) About once a week, (3) About once a month, and (4) Occasionally (irregular trips). The roughly 3% of completed survey respondents answering "Never" to this question were omitted from the regression analysis, amounting to 460 respondents used in the analysis. The majority of these respondents (62%) reported only using BlaBlaCar occasionally, while 28% reported about once a month, 9% once a week, and 1% every day or two. It should be noted that the lack of spread in frequency led certain variables to not be statistically significant, as discussed later in this section. The results of the regression are shown in Table 4.

Table 4 BlaBlaCar usage frequency ordinal regression

	Value	Std. error	*t*-value
Intercepts (frequencies)			
Every day or two|about once a week	5.207	0.801	6.499
About once a week|about once a month	2.918	0.703	4.149
About once a month|occasionally[a]	0.982	0.684	1.436
Coefficients			
Trip purpose—work	1.623	0.233	6.965
Alternative mode—public transit	−0.629	0.271	−2.323
Role—driver[b]	−0.465	0.281	−1.655
Mobile application use—yes	0.818	0.219	3.742
Student	0.774	0.345	2.240
Age[a]	−0.0947	0.0911	−1.040
Income[a]	0.0393	0.0566	0.694
Education[a]	0.0604	0.102	0.591

[a]Not statistically significant at the 90% level
[b]Statistically significant at the 90% level, but not 95% level

The intercept values between frequency levels indicate cutoff values that dictate what frequency category the model output falls into. For example, if the model outputs a value of 4, it estimates that this respondent would most likely use BlaBlaCar "About once a week," since 4 falls between cutoff values 2.918 and 5.207. The coefficients are the independent predictor variables of the model, some of which are factor (categorical) variables, and others are covariates (continuous variables). Factor variables include Trip Purpose—Work, Alternative Mode—Transit, Role—Driver, Mobile Application Use—Yes, and Student. These variables are all binary and correspond to work trip purposes (vs. leisure trips), public transit as an alternative mode if BlaBlaCar were not available (vs. car and other modes), role as a driver (vs. passenger), use of BlaBlaCar's mobile application (vs. no app use), and student (vs. other professions), respectively. Continuous variables used are Age, Income, and Education. These variables all have an increasing scale and were treated as continuous instead of factor variables. A goodness of fit test on the model using the chi-square test of the residual deviance produced a very small *p*-value (<0.01), indicating a poor overall model fit. This is due to the large number of respondents (62%) that fell into the occasional use frequency category. Even though the goodness of fit is low, the model coefficients are useful to analyze since their magnitudes and directions provide additional insight into user behavior.

The regression results uncover some points we could not discern from socio-demographic statistics alone. Since the intercepts of the model increase with increasing frequency, positive coefficient values lead to higher predicted BlaBlaCar usage, and negative coefficients lead to lower predicted usage frequency. Following this logic, we see that work trip purpose, mobile application use, and student status are all factors influencing higher usage frequency in differing degrees. Work trip purpose has the highest value coefficient of the three, indicating that respondents that use BlaBlaCar for work trips tend to use it more often than those who use it for leisure trips only. Members who have used BlaBlaCar's mobile application tend to take more trips, since downloading and using a smartphone app generally warrants a higher level of commitment to the service. Finally, students take more frequent trips than nonstudents, which may be due to generally higher comfortability with new transportation options and lower student vehicle access.

Factors leading to lower usage frequency include choosing to use public transit in the absence of BlaBlaCar and having the role of driver. The model predicts that those who would have taken public transportation in the absence of BlaBlaCar use the service less frequently than those who would have used a car or other mode to make the trip. This is likely due to public transit accessibility or inaccessibility of an individual's trip. If good public transit options exist from a respondent's origin to destination, it is much easier for them to replace a BlaBlaCar trip with public transit, thus explaining the lower usage frequency seen among these members. If the respondent's origin or destination does not have good public transit options, they are more likely to rely on a car to make the trip. It then follows that these members use BlaBlaCar more frequently because they would make the trip with their private vehicle anyway, and if they are a passenger they may have limited alternatives to carpooling. The other factor that predicts lower usage frequency is being a driver. This an interesting finding because BlaBlaCar operations depend on drivers posting rides, yet drivers are less likely than passengers to be frequent users. This finding makes sense considering that passengers (who may not own a car) are more dependent on the service and therefore use it with more frequency than drivers. This also relates to the finding that higher-income individuals are more often drivers, highlighting that these individuals may not be as financially incentivized to use BlaBlaCar with high frequency.

It should be noted that age, income, and education are not statistically significant at the 90% level in the model. Inclusion of these variables is important, however, since they are key socio-demographic factors that are included in most modeling efforts. They are not statistically significant here mostly due to the homogeneity in BlaBlaCar trip frequency of the respondents. Since 62% of respondents indicated that they use BlaBlaCar occasionally even though our sample contained a generally younger, well-educated population, these factors did not significantly influence usage frequency because the majority of these users fell into this frequency category. Income did not seem to affect frequency either, as both low- and high-income individuals use the service with similar enough frequencies that income was not statistically significant as a predictor variable. Although we found some association in the previous section that low-income members use BlaBlaCar with slightly higher frequency than middle- and high-income members, the regression model shows that income alone is not a significant predictor of usage frequency.

5 Conclusion

Carpooling appeared in France in 1958 and grew to become an important transportation mode. Despite its importance in the French transportation landscape (over three million users), there is limited research on this subject. For this reason, we implemented a survey with BlaBlaCar users to better understand this form of carpooling. Our research is focused on the analysis of the user's socio-demographic characteristics to evaluate what segments of the French population use carpooling and how their usage varies relative to these characteristics.

The survey showed that carpooling users are on average more educated and younger than the French population (87% of the respondents have a baccalaureate level or more in contrast to only 47% for France, and 45% are under 34 years of age in contrast to 27% for France), although their level of income is relatively similar to the general population. People with a low-income level (defined here as <€10,000 or <$13,449 a year) are represented among carpooling users, along with individuals with an intermediate income level (between €10,000 and €25,000 or $33,620) and a high-income level (>€25,000 or >$33,620). Further analysis reveals some differences in their usage practices. Although not found to be a statistically significant predator of usage frequency, carpoolers with a lower-income level tend to carpool slightly more frequently than carpoolers in the higher-income groups. They also have, on average, fewer alternatives available to them than wealthier users: 8% would not have taken the trip, if carpooling did not exist in contrast to 4% for middle-income and 3% for high-income users. Moreover, respondents with a low-income level are most frequently passengers (54% of them carpool only as passengers), while people with a high-income level are more often drivers (51% carpool only as drivers) and rarely only passengers (12%). The ordinal regression analysis reveals that factors such as work trip purpose, mobile application use, and student status lead to higher BlaBlaCar usage frequency. Meanwhile, respondents in the driver role and those choosing public transit as an alternative mode experience a lower usage frequency. Socio-demographic factors, such as age, income, and education, are not statistically significant predictors of frequency, emphasizing that even though many BlaBlaCar members are younger and well educated, these factors alone have little to do with how often individuals use the service.

Because of these differences in carpooling use, we suggest that long-distance carpooling in France reflects a dual practice relative to different population and user segments. Even though it seems safe to say that lower-income users as well as higher-income users gain mobility and accessibility due to carpooling, all categories of income do not have the same range of choice, and for some users, carpooling appears to be the only option. Future research is needed to better understand carpooling in France and how it could be extended to more socio-demographic groups, along with its social and environmental impacts.

References

1. Vincent, S.: Les 'Altermobilités': Analyse Sociologique Des Usages de Déplacements Alternatifs À La Voiture Individuelle. Des Pratiques En Émergence ? Paris 5. http://transports.blog.lemonde.fr/2013/04/28/comment-le-covoiturage-fait-baisser-le-prix-du-train (2008)
2. Razemon, O.: Comment Le Covoiturage Fait Baisser Le Prix Du Train|L'interconnexion N'est plus Assurée (2013)
3. CERTU: Le Covoiturage En France et En Europe, Etat Des Lieux et Perspectives (2007)
4. Chan, N., Shaheen, S.: Ridesharing in North America past present and future. Transp Rev (2012)
5. BlaBlaCar: BlaBlaCar - Le Site Du Covoiturage En Europe. http://www.covoiturage.fr/ (2013)
6. L'Expansion.com: Pourquoi Covoiturage.fr Est Devenu Payant -. http://lexpansion.lexpress.fr/high-tech/pourquoi-covoiturage-fr-est-devenu-payant_323510.html (2012)
7. Brouze, E.: Fin de L'utopie? Le Premier Site de Covoiturage Devient Payant - Le Nouvel (2011)
8. Eve, P.: Covoiturage : J'ai Testé, Mais Les Sites Payants M'ont Dégoûté. Le Nouvel Observateur, Le Plus. http://leplus.nouvelobs.com/contribution/891698-covoiturage-j-ai-teste-mais-les-sites-payants-m-ont-degoute.html (2013)
9. Evenstrood: Blablacar, Le Covoiturage Tué Par La Finance et L'appât Du Gain. http://blogs.mediapart.fr/blog/evenstrood/200614/blablacar-le-covoiturage-tue-par-la-finance-et-l-appat-du-gain (2014)
10. Dillet, R.: BlaBlaCar acquires its biggest competitor Carpooling.com to dominate European market. TechCrunch (2015)
11. Teal, R.: Carpooling: who, how and why. Transp Res **12** (1987)
12. ICARO: Increasing car occupancy through innovative measures and technical instruments (1999)
13. Ferguson, E.: The rise and fall of the American carpool: 1970–1990. Transportation (1997)
14. Furuhata, M., Dessouky, M., Ordóñez, F., Brunet, M., Wang, X., Koenig, S.: Ridesharing: the state-of-the-art and future directions. Transp Res Part B Methodol **57**, 28–46 (2013). doi:10.1016/j.trb.2013.08.012
15. Burris, M., Winn, J.: Slugging in Houston-Casual carpool passenger characteristics. J Publ Transp **9**(5), 23 (2006)
16. Commissariat Général au Developpement Durable. La Mobilité Des Français (2010)
17. CERTU. Covoiturage Dynamique. Etude Préalable Avant Experimentation (2009)
18. Maif. Etude Sur Le Covoiturage (2009)
19. Meissonnier, J.: Pour Un Accès Aux Ressources Urbaines plus Équitables. La Piste Du Covoiturage Dynamique Sur Le Territoire D'un PRU. In: Mobilités Spatiales et Ressources Métropolitaines: L'accessibilité En Question. Grenoble (2011)
20. Chevallier, M.: Usage et Accès À L'automobile Pour Les Ménages À Faible Ressources. Rapport de la Documentation Française. http://portail.documentation.developpement-durable.gouv.fr/documents/dri/PREDIT0059.pdf (2002)
21. Litman T.: Evaluating Transportation Equity. Victoria Transport Policy Institute (2013)
22. Rawls, J.: A Theory of Justice. Belknap Press of Harvard University Press, Cambridge (1971)
23. INSEE: Basse Normandie, Déplacement Domicile-Travail, Des Perspectives Avec Le Covoiturage (2011)
24. Rosenbloom, S., Burns, E.: Why Working Women Drive Alone: Implications for Travel Reduction Programs. University of California Transportation Center (1994)
25. Observateur: Rue 89, 20 Dec 2011. http://rue89.nouvelobs.com/rue89-eco/2011/12/20/fin-de-lutopie-le-premier-site-de-covoiturage-devient-payant-227664 (2011)
26. Laurier, E., Brown, B., Lorimer, H.: Driving and 'passengering' : notes on the ordinary organization of car travel 3(1) (2008)

A Framework for Understanding the Impacts of Ridesourcing on Transportation

Alejandro Henao and Wesley Marshall

Abstract The transportation sector is currently experiencing a monumental disruption with the introduction and evolution of technology and transportation services such as bikesharing, carsharing, on-demand ridesourcing, and microtransit. As these new layers of technology-based transportation options begin to flourish, it is important to understand how they compete and interact with more traditional modes. For example, ridesourcing theoretically takes an underutilized existing resource—empty seats in single-occupancy vehicles—and fills them with passengers. In reality, it is difficult to disentangle the interrelated short- and long-term outcomes and self-selection issues that arise from simply asking whether ridesourcing takes cars off the road or if we are siphoning from walking, bicycling, and transit modes. Beyond travel behavior, these evolving transportation services can also significantly impact our transportation systems, society, and the environment. Due to such complications, these outcomes have yet to be adequately studied. Accordingly, this book chapter provides a framework to investigate ridesourcing impacts. This, in turn, will help cities better account for the impact of technology and evolving transportation services in their planning processes.

Keywords Evolving transportation services · Ridesourcing · Uber · Lyft · Travel times · VMT · Parking demand · Transportation equity · Travel behavior · Mode choice

A. Henao (✉) · W. Marshall
Department of Civil Engineering, University of Colorado Denver, 1200 Larimer Street, Campus Box 113, North Classroom, Room 2012-a, Denver, CO 80217, USA
e-mail: alejandro.henao@ucdenver.edu

W. Marshall
e-mail: wesley.marshall@ucdenver.edu

G. Meyer and S. Shaheen (eds.), *Disrupting Mobility*,
Lecture Notes in Mobility, DOI 10.1007/978-3-319-51602-8_13

1 Introduction

Evolving transportation services such as bikesharing, carsharing, ride sharing, on-demand ridesourcing (e.g., Lyft, Uber), and microtransit (e.g., Bridj) are proliferating all over the world. Many factors—including social networks, real-time information, and mobile technology—allow passengers and drivers to connect through mobile smartphone applications (i.e., apps). In turn, this led to the creation and popularization of technology companies offering app-based on-demand transportation platforms. As these new layers of technology-based smartphone apps and transportation options begin to flourish, it is important to understand how they compete and interact with more traditional modes. Beyond travel behavior, these tools and evolving transportation services can also significantly impact our transportation systems, society, and the environment.

For instance, providing a more diverse array of travel options should theoretically reduce car dependence and lower parking demand; however, there remain unresolved questions about what cities actually gain (or lose) in terms of sustainability-related outcomes such as mode choice, VMT, carbon emissions, not to mention transportation equity issues. Even when replacing single-occupancy vehicle (SOV) trips, there is a potential for negative effects. For instance, if a person shifts from driving to ridesourcing, the ride-source driver may travel additional mileage—before picking them up or after dropping them off—as compared to the mileage that would have been driven had the person originally elected to drive him or herself. There is also a theoretical saturation point where higher ridesourcing supply than demand leaves many drivers circulating without riders, which can cause unnecessary VMT, congestion, environmental issues, and other problems that are not yet documented with these new technology-based modal options.

While there is widespread information online regarding companies such as Uber and Lyft, the academic literature on ridesourcing is extremely limited due to the lack of open data on these services. Without such data, investigating short-term and long-term impacts of ridesourcing on travel behavior—such as the travel modes replaced by ridesourcing and why people shifted from a previous mode—remains difficult. Even with such data, there are still limitations with regard to measuring new trips that may not have occurred before (i.e., induced travel) as well as multimodality (i.e., availability of several modes) and intermodality (i.e., combination of various modes for a single trip or mixed-modes). This combination of problems makes analyzing the impact of these services on the overall transportation system in terms of vehicle miles traveled (VMT), parking demand, transportation equity, and travel behavior exceedingly difficult.

Due to the complexity of this topic, this book chapter proposes a comprehensive framework aimed at starting the conversation on the type of questions that researchers need to be asking and pointing out issues that might arise with conventional research methods. For example, if we ask someone that does not own a car what they would have done without Lyft/Uber for a specific trip, they might answer transit. In theory, the ridesourcing trip gets classified as a negative environmental impact. However, a more comprehensive research framework might reveal that the decision not to own a car in the first place was made in part due to the availability of Lyft/Uber. Taking such long-term car ownerships decisions into account would now expose the ridesourcing trip as a positive environmental benefit.

Beyond looking at the travel modes replaced by ridesourcing, the framework also attempts to include insights from individuals on the process of why a specific mode was selected over the alternatives. For example, what is the role of travel time, travel cost, parking, and other factors in the decision making process? Such insight would help provide researchers with the ability to investigate the impact of ridesourcing on a region or city in terms of VMT and parking demand. It may also facilitate studies across different geographical areas (e.g., urban vs. suburban, city size, density, etc.) where we could find differing impacts in different contexts. In other words, could ridesourcing have, for example, positive impacts in more suburban areas and negative impacts in more urban areas? Or could the contrary be true? The intent is to provide a framework that will allow for such questions to be explored.

The next section overviews the topic of evolving transportation services and covers the limited research in this area. Since we also intend for the proposed framework to help current travel models move past their simplistic focus on traditional modes of transportation (i.e., car, transit, walk, and bike), we then discuss the related literature and the methodological advantages of accounting for what is called "modality style." This is followed by an overview of the proposed framework and preliminary findings. The overall goal is to help researchers study the effects of evolving services and fill this newly found gap in the academic literature so that cities and regional transportation organizations can better account for these issues in their transportation planning processes.

2 Literature Review

This literature review section focuses on travel behavior and transportation impacts of evolving transportation services (e.g., carsharing, ridesourcing) as well as travel demand models and methodological approaches for measuring travel behavior.

2.1 Evolving Transportation Services

Evolving transportation services such as bikesharing, carsharing, ride sharing, on-demand ridesourcing (e.g., Lyft, Uber), and microtransit are disrupting urban transportation systems by influencing travel behavior and competing with more traditional modes. Bikesharing and carsharing systems are continuously evolving and exponentially increasing in cities across the globe [1, 2]. The introduction of these services has implications for travel behavior and mode shift, as well as impacts on the overall transportation system.

The terminology of new and evolving transportation services can be confusing and sometimes inconsistently defined by the transportation sector. Intentionally or unintentionally, many accredited people and organizations use the terminology incorrectly, which can mislead public perception and general use of the services. A recent example is the misused word "ride-sharing" when referring to ridesourcing companies [3]. The Associated Press Stylebook in January 2015 presented an update on the topic: "*Ride-hauling services such as Uber or Lyft let people use smartphone apps to book and pay for a private car service or in some cases, a taxi. They may also be called ride-booking services. Do not use ride-sharing*" [4]. While there seems to be a general consensus that these services are not ride-sharing, there is still no clearly a defined term. Some of the names include: "Transportation Network Companies (TNCs)," "ride-hauling," "ride-booking," "ride-matching," "on-demand-rides," "app-based rides." In an attempt to be consistent with previous academic research [5] and to allow for possible future variations of such schemes to be housed under the same header, this book chapter uses the term "ridesourcing." The definition of ridesourcing is the sourcing of rides from a for-fare driver pool accessible through an app-based platform.

While there is abundant information online regarding companies such as Uber and Lyft, the academic literature on ridesourcing is limited due to the lack of open data on these services. Rayle et al. [5] compared user characteristics of taxi and ridesourcing trips in San Francisco. The findings from this study indicated that ridesourcing users tend to be a lot younger and have higher-incomes compared to the overall San Francisco population. Users also have lower car ownership and frequently travel with companions. Results also show that ridesourcing has an induced travel effect of approximately 8% and both substitute and complement public transit, walking, and biking. Compared to taxis, ridesourcing customers experienced shorter waiting times [5]. These results were similar to carsharing studies suggesting that carsharing users do not usually represent the overall population with regard to socioeconomics, demographics, and travel behavior characteristics. Compared to the representative population, recent studies suggest that carsharing users tend to be younger, with higher levels of education and income, and live in denser areas with better access to public transportation. Carsharing members also have different mobility resources with fewer cars per households, higher levels of bike ownership and public transportation passes, as well as higher transit, walking, and biking mode shares compared to the general population [6–10].

The current carsharing and ridesh aring literature offers a general idea at the aggregate level, but there is no clear understanding at the individual level on the actual motivations why a carsharing user chooses this mode over the alternatives. For example from the previous studies, there is no investigation on the role of travel time, travel cost, or convenience (e.g., parking) on the utility and mode choice of travel demand models. The changes cannot clearly be attributed to carsharing or ridesourcing without knowing the members behavior prior to joining a new service (e.g., car-oriented or multimodal) and controlling for the factors that influence travel behavior over time such as individual and household characteristics, location choice, or transportation resources. This research framework aims to address these problems by implementing a methodology that focuses on a more comprehensive examination of ridesourcing effects on individual travel behavior and overall impacts on the transportation system.

2.2 Travel Demand Models and Transportation Styles

Understanding travel behavior and decisions is a very complex area of study. One of the most common ways to rationalize and forecast travel behavior and decisions is through travel demand models but the issue with travel demand models is that the outcomes are dependent on the assumptions used in the model; and traditionally, travel demand models assume that individuals are aware of the full range of transportation options and choose based on utility maximization theory. Utility maximization derives from economic theory that assigns a utility value for each transportation alternative under the assumption that the mode with the highest utility value is chosen. While this theory is well established in economics and should hold true in any field, it contains some limitations. Calculating the actual utility of any good can be quite complex and typically contains several attributes that are not realistic to measure in many scenarios such as inhibited values, attitudes, perception, and beliefs that could relate to ingrained lifestyles and deeply established habits for certain modes. Individuals adopt different patterns of consumption behavior, not only based on utilitarian needs, but also because they express self-identity, leading to a person making a choice as a result of how to act and who to be [11].

Transportation researchers have studied links between lifestyles and travel behavior since early 1980s by classifying groups, depending on the time spent at different activities, weekly travel-activity pattern, or travel expenditures [12–14]. More recently, the literature evaluates the effects on transportation choice by the influence of a person's mobility decisions including land use [15–17] and mobility resources (e.g., automobile ownership, bicycle ownership, public transportation pass, and carsharing membership) [18, 19]. With respect to our topic, we seek to investigate the diffusion of ridesourcing innovation. For example, we should know how and why an individual decided to download one of the ridesourcing apps and start using the service, how it is being used (e.g., acceptance, use frequency, trip purpose), and how it has changed the mobility resources (e.g., sell or forego buying a car, buy or stop buying a transit pass, join or drop a carsharing membership). These answers will help us unveil the short and long-term influences on travel behavior and mode choice.

Vij et al. [20] introduced the concept of modality style, which can be defined as "a certain travel mode or set of travel modes that an individual habitually uses." This idea counters the conventional assumption that people choose a mode independently for every trip and instead investigates short- and long-term decisions as a function of higher level behavior orientations. In an attempt to capture modality styles and their influence on mode choice decisions, Vij et al. [20] developed a model within the context of travel demand models using the Latent Class Choice Model (LCCMs). LCCM models were first developed in the field of marketing sciences [21] and include two components: a class membership model and a class-specific choice model. LCCM is seen as a solution to the black box of correlation structure from the continuous mixture distribution in most widely known travel demand models [22]. The framework proposed in this book chapter uses the work by these researches as a starting point to incorporate ridesourcing into modality styles and mode choice. Our framework facilitates the analyses needed to disentangle the effects of joining ridesourcing and travel behavior by controlling for factors such as modality resources and modality style prior to using a new service as well as how one's modality resources or style might shift with the introduction of ridesourcing. The framework is explained in detail on the next section.

3 A Framework to Account for Ridesourcing

The ridesourcing research framework should employ a combination of travel attributes (e.g., travel times), revealed-behavior data, and stated-response data structures. Within the framework, we proposed to collect travel data records, travel diaries, and individual surveys focusing on current and previous travel behavior using an innovative approach that combines information gathered from ridesourcing drivers and their passengers. The research survey should ask passengers questions regarding current travel behaviors, available modal options, modality resources (e.g., car ownership, transit pass, etc.) for both typical transportation patterns and for specific trip characteristics to the ridesourcing service. The survey should cover typical travel behaviors prior to ridesourcing options becoming available as well as sociodemographic and socioeconomic status. Data collection from ridesourcing drivers should include times, locations, and total VMT broken down by "cruising for a ride," "driving to passenger," and "ride distance." This data collection effort will facilitate a travel behavior comparison that quantifies, for example, the VMT and parking demand impacts of ridesourcing services.

Differentiating distances driven by type of travel is a very important step in measuring VMT, as it takes into account the full ridesourcing impact compared to driving and other modes. Analogous to the ridesourcing distance breakdown is the travel times used in a door-to-door approach [23] as shown in Fig. 1. Travel times by driving mode include "walking time from the point of origin to the parking space," time spent "cruising for parking," and the time spent "walking from the parking space to the final destination" (Fig. 1a). The inclusion of the additional travel times in the door-to-door approach is crucial, especially when driving to dense areas where parking might be an issue and add to the total travel time. For example, in a review by Donald Shoup on studies from 1927 to 2001 of cruising for parking in congested downtowns, it typically took between 3.5 and 14 min to find a curb space [24]. While ridesourcing travel times do not need to include cruising for parking and walking times, we need to account for "waiting for a ride" time (Fig. 1b).

Our proposed framework disaggregates travel distances and times. Not doing so could lead to bad results. For example, an app like Google Maps provide information on transportation options available and travel mode attributes for a specific origin-destination pair. Google Maps shows the estimated travel times for driving and ridesourcing (Uber in this case) as equal, misrepresenting many trip scenarios. In order to account for travel time differences, we decided to evaluate the accuracy of travel time information provided by Google Maps (e.g., comparing time estimates against actual travel times) using a door-to-door approach and measured ridesourcing travel times and driving travel times for the same origin-destination trip. In 2015, we gathered exploratory data sets for ridesourcing services including Lyft and Uber across the Denver metropolitan region. The preliminary data sets provided insights into the impacts of ridesourcing on our transportation system and travel behavior. The preliminary results for the door-to-door travel times are presented in Figs. 2 and 3. For driving times (Fig. 2), Google Maps estimated times are very accurate for the "driving" only portion of the total driving time but differ significantly (−3 min to +18 min) from actual total travel time. This difference is mainly explained by the "cruising for parking" and "walking time" at the end of each trip. The difference between ridesourcing and actual driving times ranged from −13 to +19 min, with an average difference of 5 extra minutes for ridesourcing. Waiting times for ridesourcing averaged 7.83 min and were as high as 20 min.

As our preliminary results show, parking is a very important area of analysis for travel times. The additional times due to cruising, parking, and walking to final destination creates an incentive for travelers to use ridesourcing instead of driving themselves. The value of time is another factor comparing travel times, as passenger can be doing other activities while waiting for a ridesourcing ride to arrive and during the ride. The information provided by smartphone apps with travel

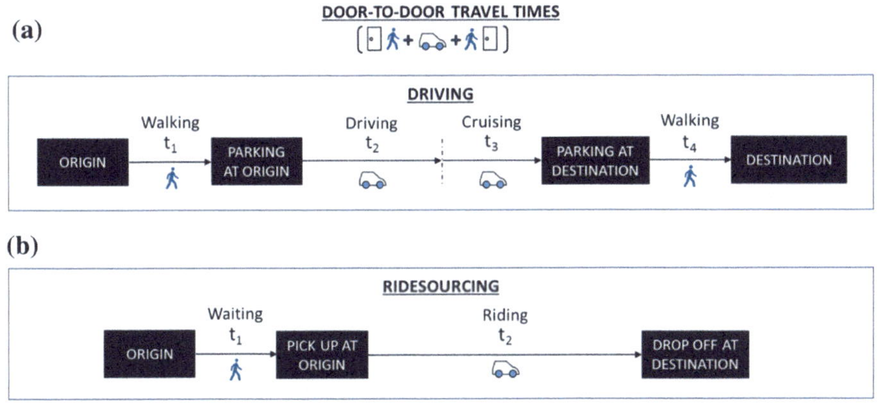

Fig. 1 Door-to-door travel times. **a** Driving. **b** Ridesourcing

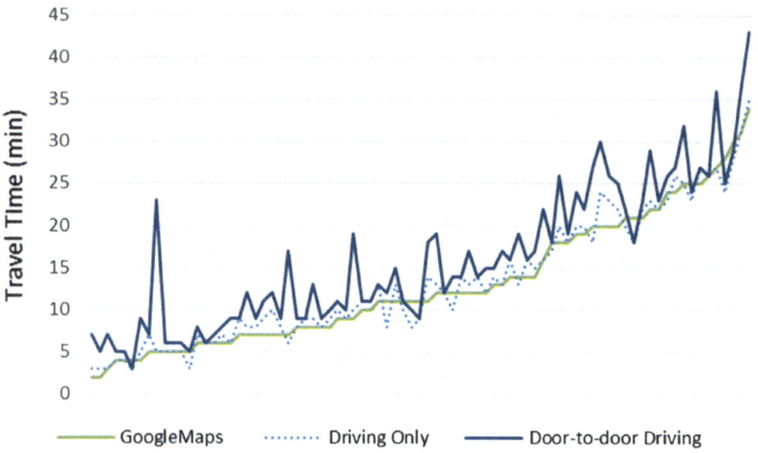

Fig. 2 Google maps estimates and driving travel times

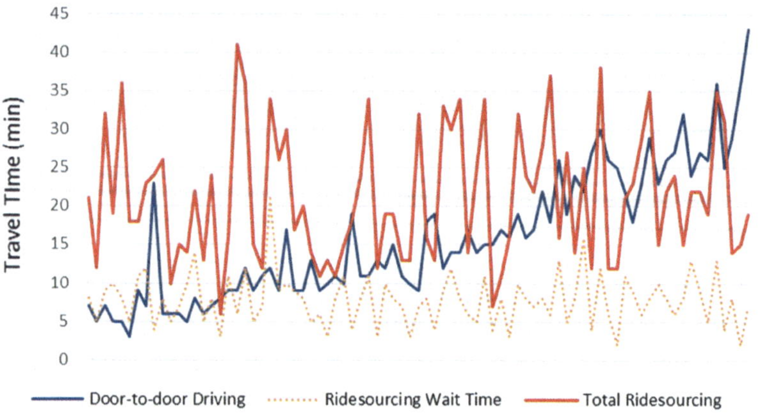

Fig. 3 Driving and ridesourcing travel times

directions such as Google Maps should use a door-to-door approach in travel time calculations for all modes to be truly comparable in absolute terms, as Salonen and Toivonen [23] suggested in their research. If travel information is not accurate or truly comparable between modes, the technology tools might be misguiding and steering certain behaviors due to potential driving bias with shorter travel times. These preliminary results demonstrate the need to disaggregate travel distances and travel times in the research framework, allowing to better account for the impacts of ridesourcing in travel behavior, mode choice, and transportation in general.

If we also sought to evaluate transportation equity issues, the proposed framework could use data collected from the ridesourcing drivers (e.g., passenger waiting time and passenger/trip characteristics) combined with spatially located socioeconomic and sociodemographic data. This would facilitate a transportation equity analysis across different populations and geographic areas to see where these services are more or less accessible and by whom. Such an analysis could also revisit outcomes such as VMT and parking demand from a spatial perspective, which could facilitate, for instance, a better understanding regarding how the findings shift across the rural to urban spectrum.

Regarding modality styles, we used preliminary data to help identify four different classes of modality styles defined in the framework. This classification is as follows:

1. Car: if most subsets only contain the car as the mode of transportation
2. Multimodal with car: if most subsets contain car and at least one more mode of transportation
3. Non-car or multimodal without car: if most subsets do not contain car as a mode of transportation
4. Bi-style: if the subsets are a mix of the three previous classes.

A bi-style modality classification, novel to our framework, refers to an individual with the potential to adopt two completely different modality styles according to several factors and attributes of travel dependent upon trip purpose. For example, a person could normally act as a multimodal traveler, but for trips transporting a child, the person would only consider the car. In this case, the person is bi-style (first: multimodal with car; then: car). Similarly, a person that normally only considers the car as the mode of transportation might behave differently for leisure trips (e.g., going out to eat or drink). This person's classification is also bi-style (first: car; then: multimodal without car). Previous research on carsharing systems provides indirect insights into the bimodality style topic, with the distribution of journey purposes per mode showing that individuals use personal cars, taxi, and carsharing differently depending on the type of journey [25]. Another insight is the potential effect of ridesourcing on drunk-driving accident reduction [26], which implies that people may be using these services more often for leisure trips but not necessarily for other purposes. The bimodality style classification is an important distinction because it allows travel demand modes to better explain travel behavior. Accordingly, more effective planning and policies can be created to help increase the desired benefits and reduce the negative impacts.

The last part of the proposed framework focuses on understanding the mode choice decision process related to ridesourcing services. This would involve building a statistical model of travel mode choice for ridesourcing based on the survey data collected combined with transportation and land use data. The model within the framework would implement an econometric analysis that accounts for the influence of latent modal preference and logit mixture models on travel mode choice behavior [20, 27, 28]. The proposed model is presented in Fig. 4 and would require the following data:

- Characteristics of the individual and household
- Modality style (e.g., car-oriented, multimodal)

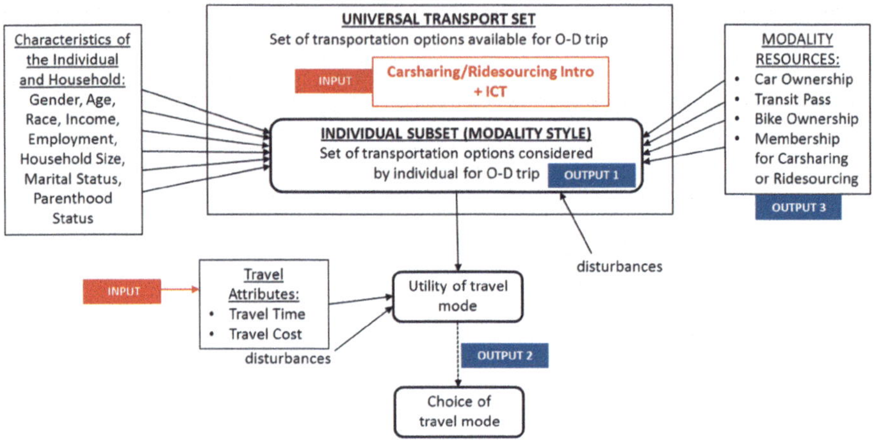

Fig. 4 Travel model framework

- Modality resources (e.g., car ownership, transit pass, etc.)
- Travel attributes (e.g., travel time, travel cost, parking, etc.)
- Land use characteristics (for both origin and destination locations).

This modeling framework would allow us to better understand the motivations behind a ridesourcing user choosing this mode over the alternatives. In turn, this knowledge can be used to create a statistical mode choice with ridesourcing services that could be integrated into regional travel models.

4 Expected Results

The proposed framework can help us guide the ridesourcing research to fill a needed gap in the literature regarding evolving transportation services. The expected outcomes of such research within this framework would include: (i) a developed survey for ridesourcing drivers and passengers that can be utilized for measuring ridesourcing impacts and be implemented in large-scale travel demand models; (ii) a better understanding regarding the role of evolving transportation services in the overall transportation system; (iii) an assessment of the impact of ridesourcing on outcomes such as VMT and parking demand; (iv) an evaluation of transportation equity variations across different socioeconomic and sociodemographic populations as well as different geographical areas due to ridesourcing; and (v) a comprehensive framework for mode choice models that better accounts for the presence of ridesourcing and evolving transportation services.

5 Discussion

We need a research framework to really understand and accurately measure the impacts of ridesourcing in transportation. For example, measuring VMT is not as simple as asking people what they would have done if Lyft or Uber were not present. We need to look at the full scope of individual modality and the impacts on the overall transportation system, including ridesourcing drivers and type of services. We don't know what is happening with ridesourcing and VMT as passengers are switching from other modes of transportation. For example, when the mode being replaced is not driving, the transportation system will automatically experience VMT growth. This is true even when passengers are switching from driving, and the system will likely also experience an increase in VMT due to the drivers' additional travel distances such as "cruising for a ride," "driving to pick up," and "final ride" at the end of the shift. These additional distances from circulating without riders cause unnecessary VMT, congestion, environmental issues, and other problems that are not yet documented with respect to these new technology-based mode options. On the other hand, ridesourcing also likely creates positive effects such as lower parking demand, which would provide an opportunity for mixed land uses and reduced car dependency. Other positives effects include the increase of activities, instead of foregoing nonessential travel, and allowing people to live in places without the need to own a car (although increases in VMT might be unavoidable and should be taken into account). Besides identifying modes being replaced, the framework would capture induced demand trips and intermodal (in connection with transit) travel.

With regards to socioeconomic and demographic attributes of ridesourcing users, as the literature points out, preliminary research suggests that their users do not represent the overall population, sharing similar results with carsharing studies. Looking deeper into transportation equity issues, this proposed framework takes into account the variation across different geographical areas and demographic characteristics. For example, the framework can provide a hypothesis where ridesourcing demand and waiting times are correlated with demographics (e.g., age, income, education, and ethnicity). Similar to equity, research within the framework help assess the magnitude of changes in VMT and parking demand among different geographical areas (e.g., urban vs. suburban, city size, density) and mode share distribution. The results from these research questions could also provide opportunities to incentive the use of ridesourcing to places where driving mode share is very high or better connections to transit are needed. In turn, this could help cities and regions move toward ridesourcing that maximizes the positive effects and minimizes the negative.

With the modality style classification in the framework, we can also identify the type of modality style that most ridesourcing users fall into and formulate hypotheses such as: (i) ridesourcing users are mostly "multimodal with car" for all trip purposes; and (ii) "multimodal without car" for social trips. Finally, the proposed framework could identify the main reasons a passenger chooses a Lyft/Uber ride over other modes of transportation while also identifying the roles that travel time, travel cost, parking, modality style, and modality resources (e.g., whether or not the person has a car available) make in the decision making process.

This book chapter draws upon the existing academic research to identify, measure, and disentangle the impacts of ridesourcing on travel behavior and the transportation system in general. We need to fill a gap in the literature by studying the effects of more recent transportation services and help cities and regional organizations better account for the impacts of technology and evolving transportation services in their transportation travel models and planning processes.

References

1. Shaheen, S., Guzman, S., Zhang, H.: Bikesharing in Europe, the Americas, and Asia. Transp. Res. Rec. J. Transp. Res. Board **2143**, 159–167 (2010)
2. Shaheen, S.A., Cohen, A.P.: Carsharing and personal vehicle services: worldwide market developments and emerging trends. Int. J. Sustain. Transp. **7**(1), 5–34 (2012)
3. Goddin, P.: Redefining Uber: why the term rideshare doesn't fit (2014)
4. Warzel, C.: Let's all join the AP stylebook in killing the term "ride-sharing" (2015)
5. Rayle, L., Shaheen, S., Chan, N., Dai, D., Cervero, R.: App-based, on-demand ride services: comparing taxi and ridesourcing trips and user characteristics in San Francisco, 17, 2015
6. Kopp, J., Gerike, R., Axhausen, K.: Do sharing people behave differently? An empirical evaluation of the distinctive mobility patterns of free-floating car-sharing members. Transportation **42**(3), 449–469 (2015)
7. Sioui, L., Morency, C., Trépanier, M.: How Carsharing Affects the Travel Behavior of Households: A Case Study of Montréal, Canada. Int. J. Sustain. Transp. **7**(1), 52–69 (2012)
8. Cervero, R., Tsai, Y.: City carshare in San Francisco, California: second-year travel demand and car ownership impacts. Transp. Res. Rec. J. Transp. Res. Board **1887**, 117–127 (2004)
9. Martin, E., Shaheen, S., Lidicker, J.: Impact of carsharing on household vehicle holdings. Transp. Res. Rec. J. Transp. Res. Board **2143**, 150–158 (2010)
10. Kopp, J., Gerike, R., Axhausen, K.: Status quo and perspectives for car-sharing systems: the example of drivenow. Strat. Sustain. Mobilities Opportunities Challenges 207–226 (2013)
11. Giddens, A.: Modernity and Self-Identity: Self and Society in the Late Modern Age. Stanford University Press (1991)
12. Salomon, I., Ben-Akiva, M.: The use of the life-style concept in travel demand models. Environ. Plann. A **15**(5), 623–638 (1983)
13. Pas, E.: Weekly travel-activity behavior. Transportation **15**(1–2), 89–109 (1988)
14. Kitamura, R.: Life-style and travel demand. Transportation **36**(6), 679–710 (2009)
15. Kitamura, R., Mokhtarian, P., Laidet, L.: A micro-analysis of land use and travel in five neighborhoods in the San Francisco Bay Area. Transportation **24**(2), 125–158 (1997)
16. Handy, S., Cao, X., Mokhtarian, P.: Correlation or causality between the built environment and travel behavior? Evidence from Northern California. Transp. Res. Part D Transp. Environ. **10**(6), 427–444 (2005)
17. Krizek, K., Waddell, P.: Analysis of lifestyle choices: neighborhood type, travel patterns, and activity participation. Transp. Res. Rec. J. Transp. Res. Board **1807**, 119–128 (2002)
18. Choo, S., Mokhtarian, P.L.: What type of vehicle do people drive? The role of attitude and lifestyle in influencing vehicle type choice. Transp. Res. Part A Policy Pract. **38**(3), 201–222 (2004)
19. Vredin Johansson, M., Heldt, T., Johansson, P.: The effects of attitudes and personality traits on mode choice. Transp. Res. Part A Policy Pract. **40**(6), 507–525 (2006)

20. Vij, A., Carrel, A., Walker, J.L.: Incorporating the influence of latent modal preferences on travel mode choice behavior. Transp. Res. Part A Policy Pract. **54**, 164–178 (2013)
21. Kamakura, W.A., Russell, G.J.: A probabilistic choice model for market segmentation and elasticity structure. J. Mark. Res. **26**(4), 379–390 (1989)
22. Walker, J.L., Ben-Akiva, M.: Advances in discrete choice: mixture models. Handb Transport Econom. **160** (2011)
23. Salonen, M., Toivonen, T.: Modelling travel time in urban networks: comparable measures for private car and public transport. J. Transp. Geogr. **31**, 143–153 (2013)
24. Shoup, D.C.: Cruising for parking. Transp. Policy **13**(6), 479–486 (2006)
25. Le Vine, S., Lee-Gosselin, M., Sivakumar, A., Polak, J.: A new approach to predict the market and impacts of round-trip and point-to-point carsharing systems: case study of London. Transp. Res. Part D Transp. Environ. **32**, 218–229 (2014)
26. Jones, R.G.: Uber can't really prove that it's caused a reduction in drunk driving accidents, in Huffington Post (2015)
27. Vij, A., Walker, J.L.: Preference endogeneity in discrete choice models. Transp. Res. Part B Methodological **64**, 90–105 (2014)
28. Atasoy, B., Glerum, A., Bierlaire, M.: Mode choice with attitudinal latent class: a Swiss case-study. In: 2nd international choice modelling conference, Leeds, UK (2011)

Part III
Innovative Transportation Technologies and City Design

Disrupting Mobility: Decarbonising Transport?

Florian Lennert and Robert Schönduwe

Abstract The transport sector urgently needs to identify decarbonisation pathways. Global demand for mobility is growing. The same applies for emissions from transport, with much of this growth taking place in emerging economies. Numerous scenario studies attempt to determine efficient strategies to decarbonise the transport sector. In this chapter we provide a comprehensive overview of scenario studies and reveal a wide spectrum of options to decarbonisation. Differences in projected GHG emissions, primary energy use and distances travelled are analysed. A typology of scenario studies is elaborated which reveals large differences in possible pathways.

Keywords Decarbonisation · Scenarios · GHG emissions · Primary energy use · Transport · Mobility

1 Introduction

Societies gain enormous benefits from the ability of moving people and goods over space and time. Efficient transportation facilitates interaction of people and the exchange of goods and thus underpins globalisation and human development. However, major challenges are linked to transportation. On the global level climate change is recognised as a major threat to human civilization caused by the extensive use of fossil fuels. The transport sector is uniquely dependent on oil and has grown considerably in the last 50 years. More than a quarter of overall energy use is

F. Lennert (✉) · R. Schönduwe
Innovation Centre for Mobility and Societal Change (InnoZ),
Torgauer Str. 12-15, 10829 Berlin, Germany

F. Lennert
London School of Economics and Political Science (LSE),
Houghton Street, London WC2A 2AE, UK
e-mail: florian.lennert@innoz.de; f.lennert@lse.ac.uk

R. Schönduwe
e-mail: robert.schoenduwe@innoz.de

G. Meyer and S. Shaheen (eds.), *Disrupting Mobility*,
Lecture Notes in Mobility, DOI 10.1007/978-3-319-51602-8_14

allocated to the transportation-sector [1]. As one of the main emitters of CO_2 the transport sector contributes significantly to global warming. Increasing emissions from the transport sector have the potential to undermine efforts to meet economy-wide, long-term emission reduction targets. On the local level, air pollution, noise and motor vehicle accidents pose significant threats to human and ecosystem health [2]. In the context of accelerating urbanisation, existing infrastructures cannot cope with large increases in traffic volume. Congestion is becoming an increasing problem, especially in urban areas. Simultaneously, demand for mobility is growing. The same applies for emissions from transport, with much of this growth taking place in the non-OECD world [3].

Future global transport and mobility will be fundamentally affected by the need to create more resource-efficient, clean transport technologies and to deploy and maintain sustainable transport systems. A long-term transformation of transport infrastructure and services is required to meet climate change mitigation challenges as well as the travel needs and requirements of a rapidly growing global urban population, but also to enable sustainable economic growth with sustainable freight transport links between global agglomeration and periphery.

Significant efforts are under way to advance post-fossil mobility systems deploying alternative propulsion technologies and integrating renewable energy sources with transport infrastructure [4]. New energy and materials technologies are enabling new forms of post-fossil transport. ICT-enabled web and mobile applications are spawning a plethora of new mobility services [5]. Traditional mobility markets are in flux and new players are emerging with disruptive service offerings [6]. These are challenging traditional demarcations between public transport and private mobility and will increasingly necessitate a co-production of mobility services by both traditional public and new private providers. In addition, demographic trends such as aging populations in some key world regions, significant public health implications and the need to maintain economic growth as well as basic equity in mobility provision to all social groups provide for complex transport politics. The politics and governance of land-use provide an additional contested policy arena.

The combined effect of these developments will have far reaching impact on the way public transport, private mobility and logistics will be organised in the future. Shaping this new public space will be a strategic opportunity and challenge for cities, regions and governments globally.

Given long investment cycles for transport capital investment, governments will be increasingly faced with competing claims on future transport infrastructure and long-range investment pathways. Identifying and evaluating cost-effective, equitable and successful policy regimes and switch-over strategies for global transport systems is a central climate policy challenge.

Transport has remained particularly stubborn to mitigating intervention and CO_2 emissions are projected to continue to rise significantly to 2050 even in benign scenario outlooks [7] and more significant yet as a share of overall CO_2 emissions. Given global commitment to decarbonisation agreed at the Paris COP 21 in 2015, accelerating and achieving meaningful decarbonisation of transport systems—

defined as a reduction of transport-related GHG emissions in absolute terms—by 2050 is a necessary condition of meeting the intended 1.5 °C temperature threshold.

Recent scenarios (cf. Sect. 3) offer little confidence that the policy mix currently deployed towards mitigation will have sufficient decarbonisation impact even under assumed benign transport policy regimes to achieve the primary energy substitution and carbon emissions reductions necessary to meet even the intended targets. Looking out to 2050, recent projections appear to offer a stabilisation of current absolute CO_2 emissions from global transport at best and a rather more probable increase of CO_2 emissions, albeit with a reduced rate of increase. This warrants an examination of the efficacy of current mitigation policy design and the decarbonisation levers deployed.

We proceed below with a historic and comparative analysis of transport decarbonisation scenarios and their key parameters. We aim to argue that advancing our understanding of the performance of current transport decarbonisation policy strategies will assist in the identification and heuristic integration of new and more effective levers of decarbonisation into future policy design. Numerous scenario studies attempt to determine efficient pathways to decarbonising the transport sector. In this chapter, we provide a comprehensive overview of scenario studies and reveal a wide spectrum of possible pathways.

2 Transport Scenarios: Overview and Analysis

Scenarios are used to outline future visions of society. As a prerequisite scenarios have to be at least theoretically feasible. Conclusions on future developments are drawn upon a number of assumptions. However, scenarios cannot account for all cause-effect relationships. Inherently, simplifications have to be made. The reduction of complexity can lead to quite different evaluations of assumptions. The key question is therefore in each case which image of the future is guiding specific scenario studies and which policy levers are proposed in attaining decarbonisation projections.

2.1 Scenarios Taken into Account

We reviewed a large number of transport scenario studies compiled within the last 15 years. We selected a sample of studies that:

- concern the future development of the transport sector and take GHG emissions into account
- describe developments further than 2030
- take passenger transport into account
- have at least a national geographic scale.

We comparatively analysed a sample of 59 transport emissions scenarios dating from 2000 to 2015 to identify key input factors and mitigation levers. Table 1 gives an overview on the scenarios we examined and the corresponding source.

A core sample of scenarios with a global outlook was subsequently chosen for more detailed comparison. We selected a sample of 19 scenarios with a global outlook and compared projections across four parameters: CO_2 emissions, primary energy use and fossil fuel share, as well as global travel demand. In Table 1 these studies are marked bold. A detailed comparison of the core sample of global scenario results is carried out in Sect. 3.

Scenario studies are built on different paradigms, modelling principles and world views. We used a typology of four types of scenario studies. This typology is presented and discussed in Sect. 2.2 below.

Table 1 Scenarios examined

	Scenario	Source
1	**World Energy Outlook 2002**	[8]
2	**World Energy Outlook 2004**	[9]
3	**Mobility 2030**	[10]
4	Foresight for transport	[11]
5	**Pathways to 2050**	[12]
6	VIBAT UK	[13]
7	**World Energy Outlook 2005**	[14]
8	Szenarien der Mobilitätsentwicklung unter Berücksichtigung von Siedlungsstrukturen bis 2050	[15]
9	Intelligent infrastructure futures. The scenarios—towards 2055	[16]
10	**World Energy Outlook 2006**	[17]
11	Mobilität 2020. Perspektiven für den Verkehr von morgen	[18, 19]
12	**Climate change 2007: Mitigation of climate change**	[20]
13	**Transport technologies and policy scenarios to 2050**	[21]
14	A sustainable energy system in 2050: promise or possibility?	[22]
15	World Energy Technology Outlook—WETO H2	[23]
16	International passenger transport and climate change	[24]
17	**International Energy Outlook 2007**	[25]
18	**World Energy Outlook 2007**	[26]
19	Backcasting approach for sustainable mobility	[27]
20	Politikszenarien für den Klimaschutz IV. Szenarien bis 2030	[28]
21	VIBAT India and Delhi	[29]
22	**World Energy Outlook 2008**	[30]
23	Modell Deutschland	[31]
24	Renewbility - Stoffstromanalyse nachhaltige Mobilität im Kontext erneuerbarer Energien bis 2030	[32]
25	European Climate Change Policy Beyond 2012	[33]
26	**Roads toward a low carbon future**	[34]

(continued)

Table 1 (continued)

	Scenario	Source
27	Getting into the right lane for 2050	[35]
28	**World Energy Outlook 2009**	[36]
29	Energieszenarien für ein Energiekonzept der Bundesregierung	[37]
30	Energy 2050: Lifestyles subproject	[38]
31	Politikszenarien für den Klimaschutz V	[39]
32	iTREN 2030	[40]
33	ADAM	[41]
34	**World Energy Outlook 2010**	[42]
35	EU transport GHG: routes to 2050?	[43]
36	Langfristszenarien und Strategien für den Ausbau der EE in Deutschland	[44]
37	Renewbility II	[45]
38	The future energy and GHG emissions impact of alternative personal transportation pathways in China	[46]
39	**Global travel within the 2 °C climate target**	[47]
40	**Influence of travel behaviour on global CO2 emissions**	[48]
41	The future of mobility. Scenarios for the United States in 2030	[49]
42	ITF Transport Outlook 2013	[50]
43	Potenziale des Radverkehrs für den Klimaschutz	[51]
44	Treibhausgasneutraler Verkehr 2050	[52]
45	Politikszenarien für den Klimaschutz VI	[53]
46	Economic assessment of low carbon vehicles	[54]
47	eMobil 2050. Szenarien zum Klimaschutzbeitrag des elektrischen Verkehrs	[55]
48	Re-programming mobility	[56]
49	Shell Pkw-Szenarien bis 2040	[57]
50	**World Energy Outlook 2014**	[58]
51	**CECILIA 2050. Optimal EU climate policy**	[59]
52	**IPCC Climate Change 2014: Mitigation to climate change**	[3]
53	**ITF Transport Outlook 2015**	[7]
54	Beyond traffic 2045	[60]
55	Nutzen statt besitzen	[61]
56	Urban mobility system upgrade	[62]
57	World Energy Outlook 2015	[1]
58	Vision Mobilität Schweiz 2050	[63]
59	DEFINE	[64]

2.2 *Typology of Scenario Studies*

Mobility futures and possible reductions in transport-related GHG emissions are evaluated in numerous model-based and explorative studies. These studies can be distinguished by scope, disciplinary background and modelling approaches. These studies also propose different decarbonisation pathways and mitigation options. In our analysis we built on a typology proposed by Creutzig [65]. We expanded this typology and distinguish four types of studies on transportation futures:

- Integrated assessment models (IAM)
- Transportation-sector models
- Culture- and society-based models and studies
- Explorative studies.

Integrated assessment models (IAM) evolve a global, aggregate perspective on transportation futures. While other models focus on specific sectoral strategies, IAMs aim to project impacts of cross-sectoral, economy-wide mitigation strategies. Therefore, IAMs deploy market-based equilibrium concepts and mainly focus on fuel choices and GHG emissions. The narrative focus is on the decarbonization of energy supply, which implies for the transport sector a fuel shift from oil to electricity, hydrogen and/or biofuels.

A serious disadvantage of this approach is the tendency to focus on mitigation strategies in the power sector. This effect is due to the fact that transport-specific mitigation options are usually more cost-intensive than options in other sectors [66]. GHG emission reductions are mainly realised by fuel shift and generic efficiency improvement. Corresponding measures can easily be operationalized in monetary costs. Other options like modal shift, behavioural change or infrastructural options are not sufficiently captured by the modelling algorithms used in IAMs [65]. While IAMs delineate economy-wide decision pathways and proffer systemic levers such as carbon-pricing, they proceed from models and assumption spaces rather than observed data. Some scholars have hence and perhaps unfairly criticised IAMs for creating "a perception of knowledge and precision that is illusory, and can fool policy-makers into thinking that the forecasts the models generate have some kind of scientific legitimacy" [67: 1].

Transportation-sector-specific models evolve a sectoral perspective on mitigation strategies and ignore inter-sectoral equilibrium effects. This means, for example, that a shift in the power sector from fossil fuels towards renewable energies is taken for granted and transport-specific technologies are optimised within this boundary consideration. Optimal mitigation strategies across sectors are ignored [34].

While IAMs often deploy very long time-scales up to 100 years, transportation-sector-specific models often focus on the next decades up to 2050. Compared to IAMs, demand side strategies like infrastructure and modal shift options are better represented. Some direct transport externalities such as road safety or congestion are addressed.

At the core, however, are traditional considerations of incremental efficiency improvements in vehicles, fuel economy and related sectoral technology. The potential for cross-sectoral infrastructure optimisation (or sector coupling), alternative concepts of behavioural change, or place-based shift and avoid strategies is not considered [68]. In this respect there is a broad consistency with IAMs (Fig. 1).

The third type of models and studies is rooted deeper in societal and cultural studies. Starting point of these studies is the assertion that the complex relation between human behaviour, transport and climate change has to be central for elaboration and evaluation of mitigation strategies. It is argued that effective transition strategies rely on a deeper understanding of societal transitions and human–technology interactions. Studies that solely focus on an economic perspective like IAMs and transportation-specific-models wholly ignore complex cultural, societal and political relations. Therefore, culture and society-based studies focus on place-based shift and avoid strategies and underline local best practices as well as the effect of local demand and supply-side policy instruments. Additionally, the set of objectives is further widened in these studies. Beyond road safety and congestion, these studies also consider air quality, spatial and health impacts and deploy urban design and active transport levers. While providing an integrated approach to transport transformation and offering promising decarbonisation potential, these strategies often remain local in ambition and it is yet unclear how their results can be synthesised and scaled beyond local (or regional) best practice solutions. More

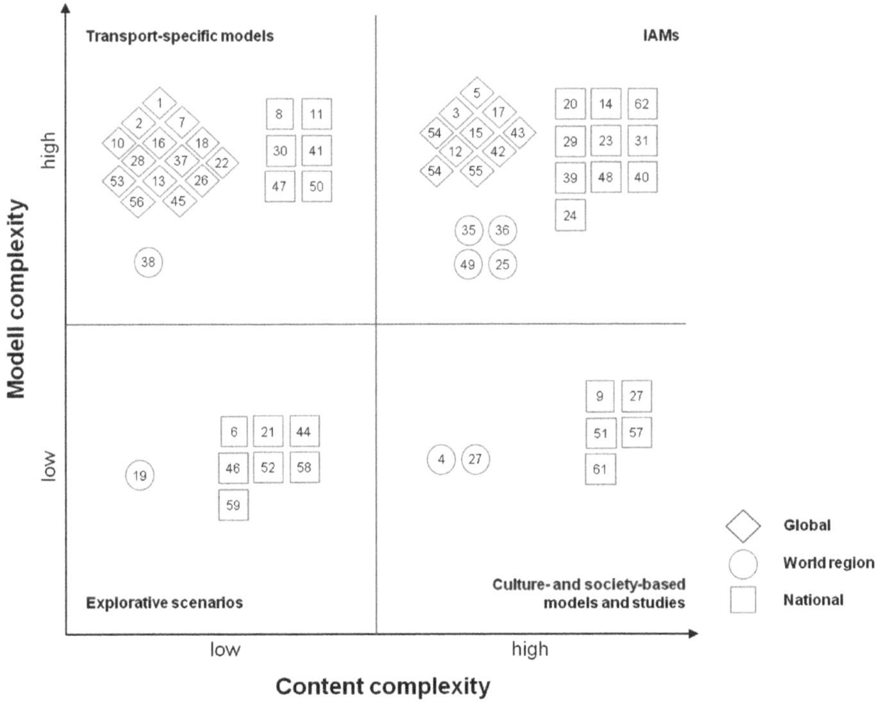

Fig. 1 Classification of observed scenarios

empirical data is also needed on the long-term impact of local transformative transport regimes.

Explorative studies explore trends and describe possible futures in form of narratives and storylines or visualisations rather than numerical estimates. These studies can be especially helpful to explore solution spaces for radical shifts towards low carbon economies [69]. While IAMs and transportations-specific-studies characterise change as being rather slow and incremental over long time periods, explorative studies discuss pathways for rapid change in transport policy or travel patterns and focus on counterfactual scenarios based on disruptive technologies and events [e.g. 6]. In most cases explorative studies remain speculative and are beyond the scope of (global) modelling efforts.

3 Analysis and Comparative Evaluation of Selected Global Scenarios

The EU and other OECD countries have announced very ambitious GHG reduction targets. While even in these world regions increasing emissions from the transport sector have the potential to undermine efforts to meet emission reduction targets, a more critical situation is emerging in other world regions. In non-OECD countries rapid growth of transport volumes is almost unanimously anticipated and threatens to massively accelerate transport-related GHG emissions growth to 2050.

To review the potential performance of transport decarbonisation policies at the global level, we selected 19 global scenario studies and compared BAU and policy projections across four parameters: travel demand, GHG emissions, primary energy and fuel use.

3.1 Distances Travelled

Almost all scenarios share the assumption that global demand for mobility is growing. But only two scenario studies explicitly provided projections for global travel demand. As shown in Fig. 2 Girod et al. estimated travel demand for three [39], respectively five [40] different scenario computations. All eight scenarios show a steep increase in travel demand within the next 85 years.

Projections of travel demand are missing in the other 17 studies; however, all studies expect rapid population growth within the next decades. GDP growth is equally anticipated in all selected studies leading to implicit projections of global travel demand.

Reduction of travel demand is computed in one scenario only. The World Energy Council [21] estimates the potential of a reduction of passenger kilometre travelled of 30% in industrial countries by 2050. While of considerable impact on

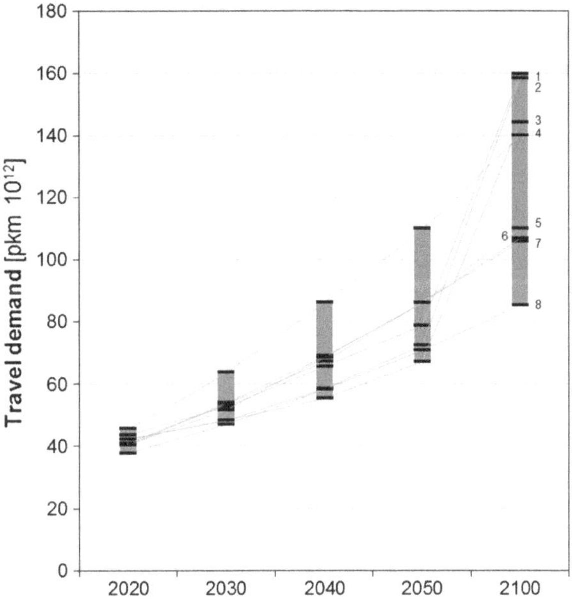

Fig. 2 Global travel demand in different scenarios. *Sources* (*1*) Baseline [39], (*2*) Scenario A [39], (*3*) Scenario B [39], (*4*) POLES [40], (*5*) TIMER [40], (*6*) IEA [40], (*7*) GCAM [40], (*8*) GET [40]

carbon emissions, accelerating travel demand at the global level is projected to neutralise the carbon mitigation effect of even such a significant reduction in transport demand across the OECD. With regard to the policy objectives rapidly growing travel demand marks an extremely challenging starting point.

3.2 *Target Goal: GHG Emissions*

All global BAU scenarios considered in our analysis are projecting increasing and accelerating emission pathways (Fig. 2).

Our review of scenario projections of the last fifteen years reveals that scenario outlooks have had to be adapted over time and that the outlook for GHG emission pathways has progressively worsened. Decarbonisation milestones have also been moved outward in time across a number of policy scenarios, necessitating deeper and faster cuts in emissions in the future then were projected ten years ago. Achievement of these policy milestones appears to become less rather than more probable. As an example, 2030 BAU emission levels projected by the IEA in 2007 are today expected by the IEA [1] to be only achievable under its alternative policy scenario (cf. Figs. 3 and 4).

This illustrates how transport emissions have become of "run-away" concern, as the near-term outlook for even the stabilisation of CO_2 emission has become more negative. More radical path corrections need to be assumed beyond 2030 to achieve any real mitigation effect to 2050.

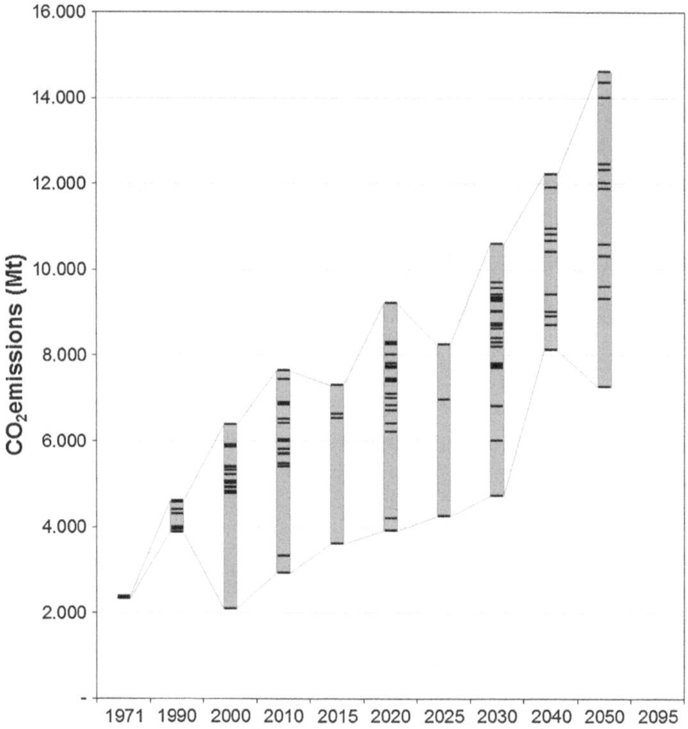

Fig. 3 GHG emissions: BAU scenarios; *grey bars* indicate the range of projected (and historic) CO_2 emissions; *black lines* indicate values of specific scenarios (see Appendix—Tables 2 and 4 for scenarios and values)

3.3 Target Goal: Primary Energy Use

Primary transport energy demand is only projected in a few scenarios. Compared to the results on CO_2-emissions, the variance of the estimates is much smaller and oil retains a dominant share of global transport energy mix in both BAU and policy scenarios. Across all scenarios fossil fuel use in transport is reduced as a share of primary energy source to differing degree; however fossil fuel use remains relatively stable in absolute terms across most policy scenarios. No significant change in fuel use is anticipated in policy scenarios before 2040. Even most benign scenarios of transport decarbonisation project for global transport to consume an equal amount of fossil fuel in 2050 as it does today. This appears a relatively weak ambition of current decarbonisation policy design.

Current demographic, land-use and motorisation trends appear to neutralise projected incremental fuel efficiency and substitution strategies. Some scenarios

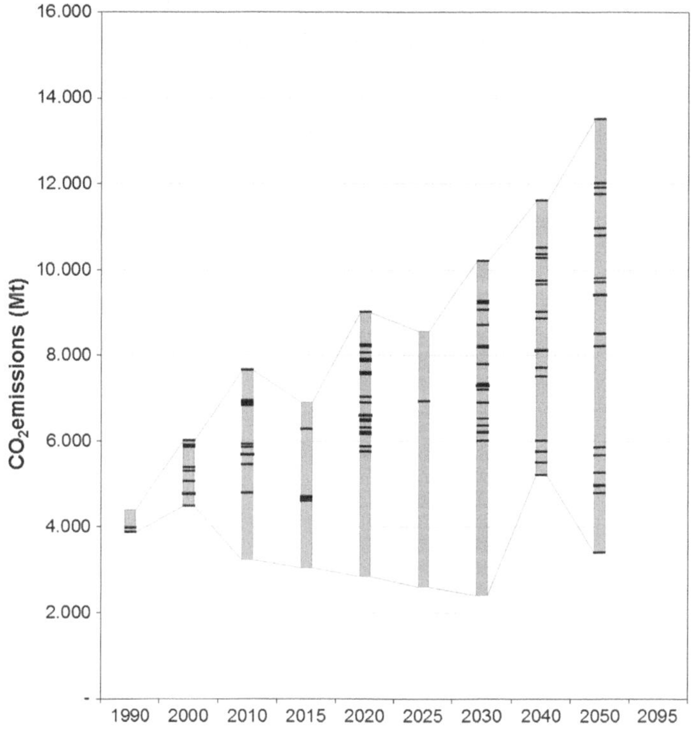

Fig. 4 GHG emissions: policy scenarios; *grey bars* indicate the range of projected CO_2 emissions; *black lines* indicate values of specific scenarios (see Appendix—Tables 3 and 5 for scenarios and values)

deploy assumptions of end-of-pipe CO_2 mitigation such as CCS technology to accomplish required emissions targets to 2050. These leave underlying primary energy and fossil fuel use intact and reveal a potential myopia of type 1 and 2 policy scenarios.

While scenario studies of type 3 (culture- and society-based models and studies) and type 4 (explorative studies) seek to elaborate alternative pathways to individual and fossil motorisation, scenario studies of type 1 and type 2 attach little importance to these option and focus exclusively on combustion-fuel efficiency and incremental fuel substitution along existing mode shares. Underlying patterns of modal share, travel demand and land-use are not analysed or deployed as scenario levers.

It is worthy to note also, that those BAU and policy scenarios extending to 2095/2100 anticipate an extreme increase in overall transport primary energy demand after 2050 (Figs. 5 and 6).

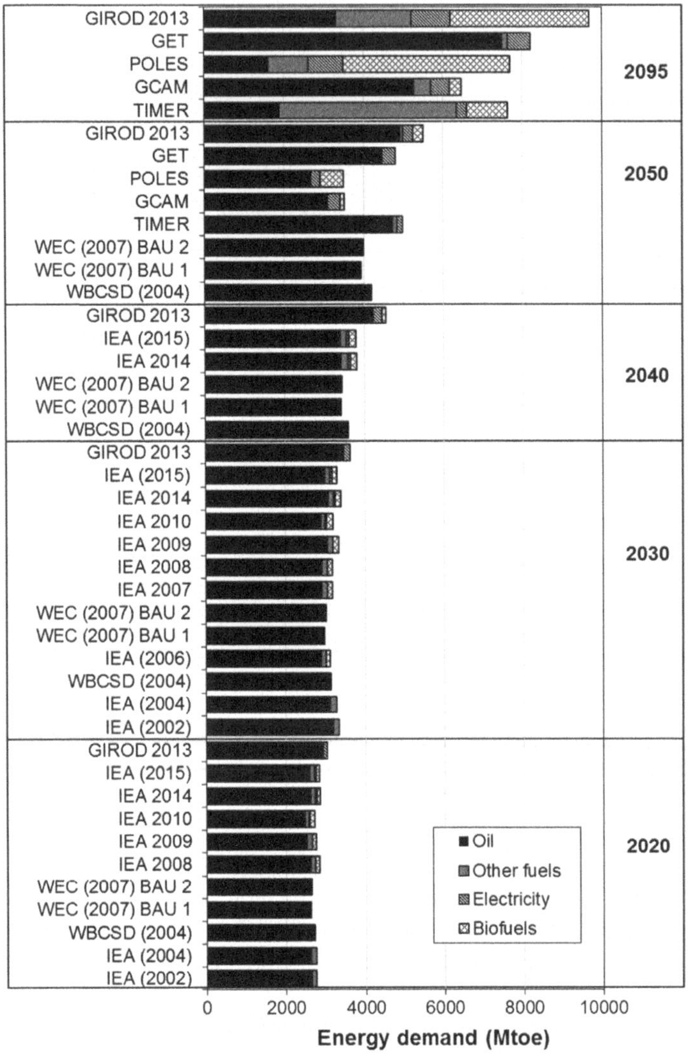

Fig. 5 Primary energy use: projected trends in fuel use in BAU scenarios

4 Conclusion and Research Outlook: Disrupting Mobility?

Based on current scenario projections, a more radical transformation of transport systems appears to be required to achieve decarbonisation targets and is likely to become a significant policy challenge. Global mitigation scenarios currently focus mainly on incremental fuel efficiency and fuel substitution as the key levers for decarbonisation.

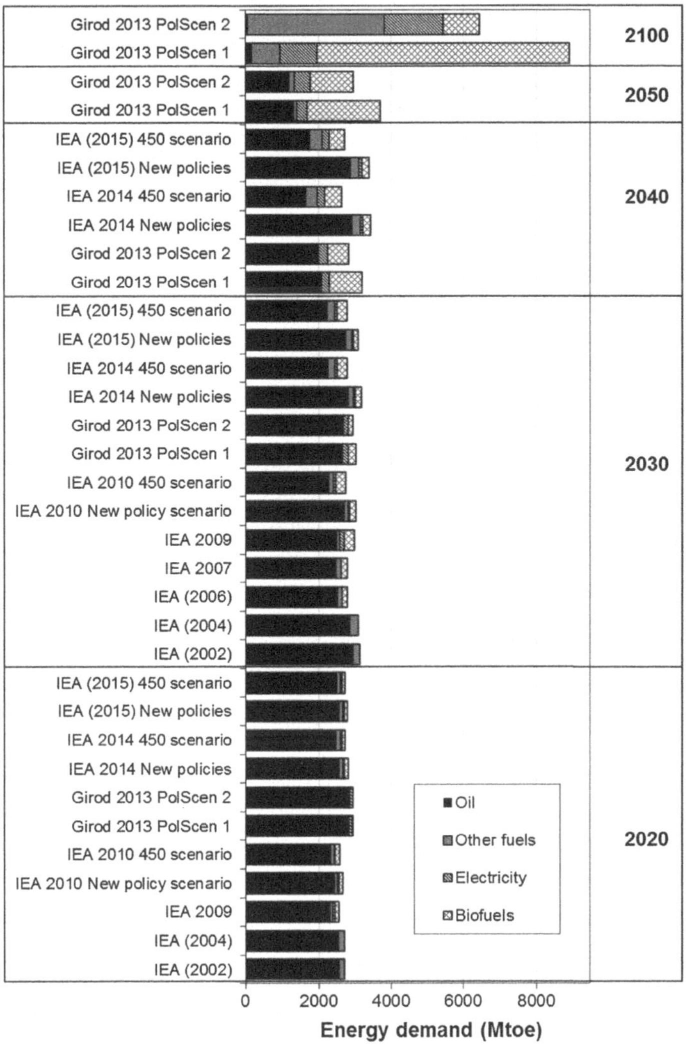

Fig. 6 Primary energy use: projected trends in fuel use in Policy scenarios

Given global demographic and socioeconomic trends and the attendant acceleration of individual motorisation in Asia, South America and Africa, fuel efficiency measures—while impacting on overall carbon efficiency—do not appear to be sufficient to achieve the necessary absolute reductions in fossil primary fuel use and CO_2 emissions. Fuel substitution strategies can offer improved impact but do not address the spatial, resource and economic inefficiency of current individual automobile motorisation. While important other levers such as modal shift and demand and land-use management are included in some scenarios, they are not sufficiently represented, explored or weighted as decarbonisation levers in the core sample of global mitigation scenarios. They do, however, offer significant decarbonisation potential that is under recognised in global policy strategies.

Future research efforts should focus on evaluating the combined and synergetic effects of integrating urban energy, infrastructure and mobility systems and of more resolute modal shift measures, mass transit expansion and sustainable land-use governance.

New Technologies and Service Models

The comparison of global scenarios indicates a path-dependency of current transport policy calculus. Policy remains focused on incrementally optimising existing individual motorisation modes and automobile technologies rather than on leveraging integrated transport strategies and sustainable technologies. Breaking this path-dependency is a key innovation challenge.

Notably, these scenarios do not yet fully incorporate the innovation dynamics of recent years. The potential carbon mitigation performance of emerging new technologies and services such as electric, autonomous and on-demand individual mobility which are currently capturing public attention has not yet been extensively evaluated, in particular in their integrated application. It is yet to be established, whether and which specific new technology systems or service models—either as stand-alone or integrated application—can contribute to meaningful transport decarbonisation to 2050. None of the global scenarios analysed include significant assumptions relating to these proposed technology systems and their potential systemic impact.

Moving forward, and in order to achieve meaningful transport decarbonisation, new mobility systems, technologies and services should be examined with regard to their balanced, simultaneous and effective contribution across the core transport decarbonisation levers available:

- Fuel/vehicle efficiency
- Fuel substitution
- Modal shift
- Demand and land-use management.

Energy Transition as a Driver for Transport Transition

While extensive electrification of existing modes of individual motorisation would provide for significant carbon reduction effects if renewable power is deployed, it would not address issues such as congestion, space consumption and induced CO_2 emissions from expansion of road construction and maintenance. In addition, moving to zero-emission electric mobility requires full fleet conversion and decarbonisation of global power systems within the same time frame, an objective only few nations are expected to meet by 2050. Most scenarios do not yet triangulate between energy ("Energiewende") and mobility system ("Verkehrswende") transformation and the interface between these deserves further research and analysis.

Recent studies have indicated significant systemic decarbonisation potential can be achieved by the coupling of renewable power and electric mobility systems at the local level. Integrating electric mobility as a component of a future renewable energy system appears to be a promising policy vector [70]. To attain maximum decarbonisation impact, however, fuel substitution strategies will need to be integrated with distributed smart grids, energy-efficient urban infrastructures and energy storage systems (i.e. not be solely focused on the electrification of existing fleets and modes).

Modal Shift and Mobility Services

Reducing the share of travel by individual low-ridership combustion engine vehicles can lead to significant reductions in CO_2 emissions. This involves a reduction of use of personal-use and low-occupancy vehicles by promoting the use of more energy-efficient modes such as conventional mass transit (bus, tram, light rail), other shared-ride solutions, as well as cycling and walking. Across a number of European cities daily travel modes have recently shifted away from the automobile and towards public transport or active travel (e.g. London-12%, Berlin-8% from 1998 to 2013, [71]) and individual motorization is a minority mode share in a range of global cities [72]. These are positive trends for decarbonisation.

Conventional car-sharing or short-term rental in principle do not significantly reduce vehicle kilometres travelled and as such do not constitute modal shift. They may have the potential to decrease the overall amount of vehicles required and can potentially enable more efficient first/last mile access to mass transit systems thereby contributing to modal shift. However, even when highly integrated with public transport, a recent study in Germany has indicated conversion to ubiquitous car-sharing would achieve reductions of only 4% of total German transport CO_2 emissions [61]. Better results could be expected from ride-sharing and mass transit services which can significantly increase the number of passenger per vehicle and consequently reduce overall vehicle kilometres travelled.

With regard to new mobility services—as for instance on-demand individual mobility as a stand-alone service innovation for point-to-point transport—early evidence thus indicates these do not necessarily contribute to significant transport decarbonisation. While a reduction in motorisation rates may be achieved with some potential benefit for urban space reclamation, fossil car-based individual mobility services do not in principle reduce vehicle kilometres or related CO_2 emissions. Whether or not new mobility services can avoid or replicate some redundant traffic flows (by avoiding parking spot search traffic or by increasing the number of vehicles in constant circulation seeking riders) remains to be empirically examined and validated at scale and under real-world conditions.

Automation and Vehicle Efficiency

With stable or increasing demand for mobility, automated vehicles will only reduce emissions and energy consumption if they are zero-emission and sustainable from a full system perspective. A priori, automation does not represent a meaningful decarbonisation lever as it does not fundamentally reduce or substitute primary energy used per passenger or vehicle kilometre. While incremental emissions reduction may be achievable through an increase in fuel and circulatory—or traffic—efficiency resulting in an incremental reduction of fuel use and of vehicle kilometres travelled, any meaningful emission reduction could only be achieved by deploying fuel substitution including electrification, as the primary lever. Whether automation is a necessary condition for electrification has not yet been convincingly established. A focus solely on automation could engender complex and expensive vehicles requiring significant investments into public and private digital infrastructure and thus create barriers to rapid fleet decarbonisation and fuel substitution. It may, however, ease the integration of electric fleets with renewable smart grid and charging infrastructure and thus can potentially make a positive indirect contribution to fuel substitution.

Automation also does not in principle contribute to meaningful emissions reduction through vehicle efficiency unless fundamental proportions—i.e. volume, weight, speed—of vehicles become more resource and spatially efficient [73]. A more effective lever in this regard is vehicle miniaturisation. The development of personal electric vehicles capable of blending with low speed active travel in cities and providing first/last miles individual mobility integrated with mass transit, should allow for a significant reductions of emissions from individual mobility. As with electrification, however, automation is not prima facie a necessary condition of miniaturisation or a fundamental decarbonisation lever of vehicle efficiency.

On-demand mobility for individual vehicular point-to-point travel, automated or not, does not appear to offer meaningful decarbonisation potential as it would not in principle and without deploying other levers contribute to significant modal shift, fuel substitution or vehicle efficiency.

A key question to be analysed is thus whether a required “densification” and “electrification” of transport should proceed from the simple conversion of existing motor vehicle systems or from the differentiation of integrated mass transportation systems—both arterial and capillary—offering significant reductions in vehicle kilometres by providing shared mobility at scale and individual first/last mile mobility on demand.

Land-Use Management and Zero-Emissions Mass Transit Systems

The insights from the scenarios analysed suggest strongly that a key lever for successful and deep decarbonisation of transport will be the expansion of zero-emissions mass transit systems and their integration with first/last mile personal electric mobility and active travel. Given demographic and urbanisation trends, spatially and carbon-efficient high through-put public transport has the potential leverage to provide sustainable transport capacity to meet this rising demand. Global scenarios currently offer little intelligence on this core lever which should be a central focus of transport policy. Enabling core mass transportation systems can provide the backbone for integrating decarbonised new mobility services, and energy systems and contribute to sustainable transit-oriented urban development.

Efficient mass and public transport has historically and successfully enabled urban agglomeration and continues to underpin mobility systems in leading global cities such Singapore, Hong Kong, Zurich, Amsterdam, Berlin, London, Barcelona, Paris, Tokyo and New York. Whether automation of individual mobility services can provide an alternative solution to mass transit, in particular in cities with low public transport infrastructure remains to be established. Automation and/or on-demand provision of private vehicular services, however, in principle do not address the energetic and spatial inefficiency of low-occupancy individual motorisation or attendant infrastructure emissions and unsustainable land-use.

While automation, in particular of mass and ride-share transit, has great potential, it does not a priori provide a strategic decarbonisation lever for urban mobility, which will have an increasingly dominant share of global transport emissions. Empirical evidence will yet need to be obtained to evaluate and validate automation and individual mobility services as a decarbonisation strategy. They can strongly support a shift to transport decarbonisation, or further lock in unsustainable travel behaviour and infrastructure design.

In this respect, Mobility-as-a-service and automation strategies should be examined for their balanced contribution to significant modal shift, ubiquitous fuel substitution and sustainable land-use and demand management.

Closing the Gap Between Research Paradigms and New Data Collection Tools

Understanding and differentiating the systemic carbon mitigation performance of emerging and integrated new transport and mobility systems will be fundamental in identifying successful and sustainable transformation paths and to inform long-range policy design. This is a complex and challenging task requiring analytical insights and empirical evidence from across disparate disciplines and domains, relating to complex interactions between technology development, service innovation, user behaviour and preferences, infrastructure and urban design, spatial and economic efficiency and environmental performance.

The comparison of scenario studies reveals a gap between scientific communities. Different paradigms have not been well integrated. Assessing potential new pathways of transport decarbonisation requires a systemic view of the evolving transport transformation across different domains such as energy, transport, ecology, urban design, logistics and dynamic human behaviour. Integrating these multidisciplinary insights will have to build on a critical analysis of transport decarbonisation policy strategies to date and will require policy analysis and decision support systems that can assess and evaluate multidimensional and integrated transport transformation pathways.

Given the combined scenario outlooks discussed above, future policy design will need to focus on optimising the systemic resource efficiency of global transport systems rather than pursuing mono-dimensional levers such as carbon-pricing, fuel efficiency or substitution which only deliver partial success in decarbonisation or suffer self-defeating externalities. A more balanced policy mix is required to account for the real-world complexity of global transport systems and the interdependencies of urban form, energy sufficiency and human and ecological quality of life.

As a priority, transport decarbonisation policy research and development should aim to establish robust policy analysis tools that can integrate the emerging wealth of data from across the scenario domains. This should provide a rigorous basis from which to evaluate and model more sustainable pathways for future mobility systems and to inform public deliberation on transport futures.

However, global transport policy analysis suffers acutely from a lack of reliable and comparable international open data at both the micro and macro level as well as from a lack of multidisciplinary pooling of evidence. New data collection tools can provide real-time empirical data on the movement of people and goods across space and time. These tools enable a much wider range of mobility and transport data to be collated at the individual and spatial level allowing for more precise modelling of the potential impacts of transport policy levers as they interact in real time and space. Integrating this data with existing scenario intelligence should provide for a more balanced configuration of global transport decarbonisation policy. The development of open data research infrastructure to inform public decision-making on sustainable transport pathways should be a core element of a sustainable transport policy agenda.

A collaborative international effort is needed to assimilate and verify the empirical evidence of sustainable transport innovation at the local and regional level and to develop evidence-based global policy strategies for truly intelligent mobility.

Acknowledgments Several informal reviewers and collaborators commented on a number of previous drafts of this paper. The authors would like to thank Axel Volkery, Cathy Marcherais, Lukas Neckermann and Veronique van Acker for their helpful comments and suggestions.

The authors gratefully acknowledge the support of Stiftung Mercator for the preparation of this chapter.

Appendix

See Tables 2, 3, 4 and 5.

Table 2 BAU scenarios used in Fig. 3

#	Source	Scenario
1	IEA [8]	BAU—IEA 2002
2	IEA [9]	BAU—IEA 2004
3	WBCSD [10, 12]	BAU 1—WBCSD 2004
4	WBCSD [10, 12]	BAU 2—WBCSD 2004
5	WBCSD [10, 12]	BAU (IEA 2004)
6	IEA [17]	BAU—IEA 2006
7	IPCC (2007)	WBCSD (2004)
8	WEC [21]	BAU 1—WEC 2007
9	WEC [21]	BAU 2—WEC 2007
10	EC [23]	BAU—EC 2007
11	Meyer et al. [24]	Gompertz constant technology
12	IEA [26]	BAU
13	IEA [30]	BAU
14	McKinsey [34]	No-action scenario
15	IEA [36]	BAU
16	IEA [42]	Current policy scenario
19	Girod et al. [48]	TIMER
20	Girod et al. [48]	GCAM
21	Girod et al. [48]	POLES
22	Girod et al. [48]	GET
23	Girod et al. [48]	IEA
24	IEA [1]	BAU—IEA 2015

Table 3 Policy scenarios used in Fig. 4

#	Source	Scenario
1	IEA [8]	Alternative policy scenario
2	IEA [9]	Alternative policy scenario
3	WBCSD [10]	Diesels potential
4	WBCSD [10]	Hybrids potential
5	WBCSD [10]	Fuel cells (H_2 from NG) potential
6	WBCSD [10]	Advanced biofuels potential
7	WBCSD [10]	Combined technologies scenario
8	WBCSD [10]	Fuel cells (zero-carbon hydrogen) potential
9	WBCSD [10]	Advanced biofuels (also in heavy trucks) potential
10	WBCSD [12]	Pathways 2050
11	WBCSD [12]	Pathways 2025
12	IEA [17]	Alternative policies scenario
13	WEC [21]	BTL 25% in 2050
14	WEC [21]	Diesel 50% in 2050
15	WEC [21]	BTL 25% and diesel 50% in 2050

(continued)

Table 3 (continued)

#	Source	Scenario
16	WEC [21]	Cellulosic 25% in 2050
17	WEC [21]	Hybrid 50%
18	WEC [21]	Cellulosic 25% and hybrid 50% in 2050
19	WEC [21]	FCV 25% in 2050
20	WEC [21]	Pass km reduction 30% in 2050
21	EC [23]	Carbon case
22	EC [23]	H_2 case
23	IEA [26]	Alternative policy scenario
24	IEA [30]	550 policy scenarios
25	IEA [36]	450 policy scenarios
26	McKinsey [34]	Biofuels
27	McKinsey [34]	Traffic flow
28	McKinsey [34]	Driving behaviour
29	McKinsey [34]	Distance driven
30	McKinsey [34]	Optimised ICEs
31	McKinsey [34]	Mixed technology
32	McKinsey [34]	Hybrid + electric
33	IEA [36]	450 policy scenario
34	IEA [42]	New policy scenario
35	IEA (2011)	450 scenario
36	Girod et al. [48]	TIMER
37	Girod et al. [48]	GCAM
38	Girod et al. [48]	POLES
39	Girod et al. [48]	GET
40	IEA [1]	New policies scenario
41	IEA [1]	450 scenario

Table 4 GHG emissions—BAU scenarios (# corresponds to number in Table 2)

#	GHG emissions per year (Mto)									
	1971	1990	2000	2010	2015	2020	2025	2030	2040	2050
1	2320		4814	6010		7449		9024		
2	2351		4914	5977		7375		8739		
3			6370	7640		9200		10,580	12,200	14,350
4			4760	5680		6820		7750	8900	10,300
5			5370				6963			
6		3875	5306		6630			8402		
7	2375	4300	5400	6400		7700		9000	10,400	11,870
8			5850	6850		8250		9250	10,650	12,300

(continued)

Table 4 (continued)

#	GHG emissions per year (Mto)									
	1971	1990	2000	2010	2015	2020	2025	2030	2040	2050
9			5900	6900		8300		9400	10,800	12,450
10		3982	5056	5461		6206		6815		7263
11			2085	2920		4193		5991	8127	10,576
12		3950			6524			8293		
13		4390	5370		7292	7796	8249	8680		
14				3324	3604	3910	4248	4712		
15		4574				7733		9332		
16						7398		8617		
19		4000	5000	5400		6400		7800	9400	12,000
20		4000	5000	5700		6700		7700	8700	9600
21			5000	5800		7000		8200	9000	9300
22		4300	5200	6500		8000		9700	11,900	14,000
23			4900	5800		7100		9300	11,900	14,600
24		4604		7441		8263		9553	10,942	

Table 5 GHG emissions—policy scenarios (# corresponds to number in Table 3)

#	GHG emissions per year (Mto)								
	1990	2000	2010	2015	2020	2025	2030	2040	2050
1			5914		7032		8179		
2			5846		6893		7792		
3		4760	5680		6600		7300	8100	9400
4		4760	5680		6450		6900	7500	8500
5		4760	5680		6500		7200	7700	8200
6		4760	5680		6200		6350	6000	5250
7		4760	5680		6150		6000	5500	4970
8		4760	5680		6300		6200	5750	4950
9		4760	5680		6150		6000	5200	3400
10		5370							4778
11		5370				6926			
12	3875	5289		6265			7336		
13		5850	6820		7850		8700	9720	10,950
14		5850	6850		8200		9200	10,500	12,000
15		5850	6850		7900		8700	9640	10,790
16		5900	6900	4650	7600		8200	9000	9800
17		5900	6950	4600	8250		9250	10,350	11,900
18		5900	6950	4700	7550		8200	8850	9700
19		5850	6850		8050		9050	10,270	11,750

(continued)

Table 5 (continued)

#	GHG emissions per year (Mto)								
	1990	2000	2010	2015	2020	2025	2030	2040	2050
20		6000	7650		9000		10,200	11,600	13,500
21	3982	5056	5439		5861		6184		5850
22	3982	4487	4787		5741		6528		5660
23				6188			7102		
24					7720		8190		
25							7800		
26			3301	3378	3595	3987	4315		
27			3297	3333	3513	3865	4212		
28			3270	3270	3432	3743	4162		
29			3252	3239	3378	3707	4086		
30			3252	3108	2919	2784	2730		
31			3239	3077	2842	2671	2527		
32			3230	3050	2847	2608	2410		
33					7066		7688		
34	4393			6911	7262	7633	8089		
35					6962		6841		
36			5577		6331		7375	7550	8180
37					6000		6216	6391	6234
38					6869		7209	7301	7145
39					7697		8243	7798	6442
40					8150	8558	8898	9660	
41					7969		7789	7107	

References

1. International Energy Agency (IEA), Organisation for Economic Co-Operation and Development (OECD): World energy outlook 2015, Paris (2015)
2. World Health Organization (WHO): Burden of disease from ambient air pollution for 2012. Genf (2014)
3. Sims, R., Schaeffer, R., Creutzig, F., Cruz-Núnez, X., D'Agosto, M., Dimitriu, D., Meza, M. J. F., Fulton, L., Kobayashi, S., Lah, O., McKinnon, A., Newman, P., Ouyang, M., Schauer, J. J., Sperling, D., Tiwari, G.: Transport. In Climate Change 2014: Mitigation of Climate Change. Working Group III Contribution to the Fifth Assessment of the Intergovernmental Panel on Climate Change, pp. 599–670, Cambridge University Press, New York (2014)
4. European Parliament: The world is changing. Transport, too, Brussels (2016)
5. Hinkeldein, D., Schönduwe, R., Graff, A., Hoffmann, C.: Who would use integrated sustainable mobility services—and why? In Sustainable Urban Transport, No. 7, Emerald, pp. 177–203 (2015)
6. Canzler, W., Knie, A.: Mobility in the age of digital modernity: why the private car is losing its significance, intermodal transport is winning and why digitalisation is the key. Appl Mobilities. 1–12 (2016)
7. Organisation for Economic Co-Operation and Development (OECD), International Transport Forum (ITF): ITF transport outlook 2015, Paris (2014)

8. International Energy Agency (IEA), Organisation for Economic Co-Operation and Development (OECD): World energy outlook 2002, Paris (2002)
9. International Energy Agency (IEA), Organisation for Economic Co-Operation and Development (OECD), World energy outlook 2004. Paris (2004)
10. World Business Council for Sustainable Development: Mobility 2030: meeting the challenges to sustainability, Geneva (2004)
11. ICCR, Adelphi Research, University of Cardiff, NESTEAR, ALAMO Online: Foresight for transport. A Foresight Exercise to Help Forward Thinking in Transport and Sectoral Integration, Wien (2004)
12. World Business Council for Sustainable Development: Pathways to energy & climate change. Geneva (2004)
13. Department for Transport: Visioning and Backcasting for UK Transport Policy (VIBAT). London (2006)
14. International Energy Agency (IEA), Organisation for Economic Co-Operation and Development (OECD): World energy outlook 2005. Middle East and North Africa Insights, Paris (2005)
15. TRAMP—Traffic and Mobility Planning GmbH, Deutsches Institut für Urbanistik (Difu), Institut für Wirtschaftsforschung Halle (IWH), Szenarien der Mobilitätsentwicklung unter Berücksichtigung von Siedlungsstrukturen bis 2050, Magdeburg (2006)
16. Curry, A., Hodgson, T., Kelnar, R., Wilson, A.: Intelligent infrastructure futures. The Scenarios—Towards 2055, London (2006)
17. International Energy Agency (IEA), Organisation for Economic Co-Operation and Development (OECD): World energy outlook 2006, Paris (2006)
18. Acatech, Mobilität 2020. Perspektiven für den Verkehr von morgen, München, Berlin (2006)
19. Lemmer, K.: acatech. Handlungsfeld Mobilität. Infrastrukturen sichern. Verkehrseffizienz verbessern. Exportchancen ergreifen, Braunschweig, München, Berlin (2011)
20. Ribeiro, S. K., Kobayashi, S., Beuthe, M., Gasca, J., Greene, D., Lee, D. S., Muromachi, Y., Newton, P. J., Plotkin, S., Sperling, D., Wit, R., Zhou, P. J., Hata, H., Sims, R., Skjolsvik, K. O.: Transport and its infrastructure. In Climate Change 2007, Mitigation of Climate Change, Cambridge, New York (2007)
21. World Energy Council: Transport technologies and policy scenarios to 2050. London (2007)
22. Uyterline, M.A., Ybema, J.R., van den Brink, R.: A sustainable energy system in 2050: promise or possibility? (2007)
23. European Commission: World energy technology outlook 2050: WETO H_2. Brussels (2007)
24. Meyer, I., Leimbach, M., Jaeger, C.C.: International passenger transport and climate change: a sector analysis in car demand and associated emissions from 2000 to 2050. Energy Policy **35** (12), 6332–6345 (2007)
25. Energy Information Administration: International energy outlook 2007. Washington, DC (2007)
26. International Energy Agency (IEA), Organisation for Economic Co-Operation and Development (OECD): World energy outlook 2007. China and India Insights. Paris (2007)
27. European Commission (Joint Research Centre Institute for Environment and Sustainability): Backcasting Approach for Sustainable Mobility. Luxembourg (2008)
28. Öko-Institut, I. f. E.-u. K. Forschungszentrum Jülich, Systemforschung und Technologische Entwicklung (IEK-STE), Deutsches Institut für Wirtschaftsforschung (DIW), Fraunhofer Institut für System- und Innovationsforschung ISI, Politikszenarien für den Klimaschutz IV. Szenarien bis 2030, Dessau-Roßlau (2008)
29. Saxena, S., Banister, D.: Breaking the trend. Visioning and Backcasting for Transport un India & Delhi, Delhi, London (2008)
30. International Energy Agency (IEA), Organisation for Economic Co-Operation and Development (OECD), World Energy Outlook 2008, Paris (2008)
31. WWF Deutschland (WWF), Prognos AG, Öko-Institut e.V., Modell Deutschland. Klimaschutz bis 2050: Vom Ziel her denken, Basel, Berlin (2009)
32. Öko-Institut, DLR Institut für Verkehrsforschung, Institut für Energie- und Umweltforschung Heidelberg (ifeu), Deutsches Biomasseforschungszentrum gGmbH (DBFZ), Technische

Universität Dresden (Professur für Verkehrsströmungslehre), Stoffstromanalyse nachhaltige Mobilität im Kontext erneuerbarer Energien bis 2030. Endbericht Teil 1: Methodik und Datenbasis, Berlin (2009)
33. World Energy Council: European climate change policy beyond 2012, London (2009)
34. McKinsey & Company: Roads Toward a Low-Carbon Future: Reducing Co_2 Emissions from Passenger Vehicles in the Global Road Transportation System. New York City (2009)
35. Netherlands Environmental Assessment Agency (PBL): Getting into the Right Lane for 2050. Bilthoven (2009)
36. International Energy Agency (IEA), Organisation for Economic Co-Operation and Development (OECD), World Energy Outlook 2009, Paris (2009)
37. Prognos AG, EWI - Energiewirtschaftliches Institut an der Universität zu Köln, GWS - Gesellschaft für wirtschaftliche Strukturforschung, Energieszenarien für ein Energiekonzept der Bundesregierung, Basel, Köln, Osnabrück (2010)
38. Anable, J., Brand, C., Tran; M., Eyre; N.: Modelling transport energy demand: a socio-technical approach. Energy Policy. **41**(0), 125–138 (2012)
39. Hansen, P., Matthes, F. C.: Politikszenarien für den Klimaschutz V – auf dem Weg zum Strukturwandel, Treibhausgas-Emissionsszenarien bis zum Jahr 2030, in: Schriften des Forschungszentrums Jülich. Reihe Energie und Umwelt, No. 62, Jülich (2010)
40. Schade, W., Krail, M.: iTREN-2030: Experiences and results for integrated technology, energy and transport policy assessment. Final Report and Deliverable 6 of iTREN-2030 (Integrated transport and energy baseline until 2030). Project co-funded by European Commission 6th RTD Programme. Fraunhofer-ISI, Karlsruhe (2007)
41. Schade, W., Helfrich, N., Peters, A.: A Transport Scenario for Europe Until 2050 in a 2-Degree World. In: 12th World Conference of Transport Research, 11–15.07.2010, Lissabon (2010)
42. International Energy Agency (IEA), Organisation for Economic Co-Operation and Development (OECD), World Energy Outlook 2010, Paris (2010)
43. Skinner, I., Hodgson, I., van Essen, H., Smokers, R., Hill, N.: Towards the decarbonisation of the EU's transport sector by 2050. Final Report, London (2010)
44. Deutsches Zentrum für Luft- und Raumfahrt, I. f. T. T. D., Fraunhofer Institut für Windenergie und Energiesystemtechnik (IWES), Ingenieurbüro für neue Energien (IFNE), Langfristszenarien und Strategien für den Ausbau der erneuerbaren Energien in Deutschland bei Berücksichtigung der Entwicklung in Europa und global, Stuttgart, Kassel (2012)
45. Zimmer, W., Hacker, F., Rausch, L., Fritsche, U., Cyganski, R., Justen, A., Knitschky, G., Lischke, A., Mehlin, M., Müller, S., Schade, W.: Weiterentwicklung des Analyseinstruments Renewbility. Renewbility II - Szenario für einen anspruchsvollen Klimaschutzbeitrag des Verkehrs, Dessau-Roßlau (2013)
46. Kishimoto, P.N., Paltsev, S., Karplus, V.J.: The future energy and GHG emissions impact of alternative personal transportation pathways in China. Cambridge (2012)
47. Girod, B., van Vuuren, D.P., Deetman, S.: Global travel within the 2 °C climate target. Energy Policy **45**, 152–166 (2012)
48. Girod, B., van Vuuren, D.P., de Vries, B.: Influence of travel behavior on global CO_2 emissions. Transp. Res. Part A Policy Pract. **50**, 183–197 (2013)
49. Zmud, J., Ecola, L., Phleps, P., Feige, I.: The future of mobility. Scenarios for the United States in 2030, Santa Monica (2014)
50. Organisation for Economic Co-Operation and Development (OECD), International Transport Forum (ITF): ITF transport outlook 2013. Funding Transport, Paris (2014)
51. Ahrens, G.-A., Becker, U., Böhmer, T., Richter, F., Wittwer, R.: Potenziale des Radverkehrs für den Klimaschutz. Dresden, Dessau-Roßlau (2013)
52. Blanck, R., Kasten, P., Hacker, F., Mottschall, M.: Treibhausgasneutraler Verkehr 2050: Ein Szenario zur zunehmenden Elektrifizierung und dem Einsatz stromerzeugter Kraftstoffe im Verkehr. Abschlussbericht im Auftrag des Umweltbundesamtes zum Forschungsvorhaben "Verkehr 2050 - Entwicklung von Paramtern und Skizzierung eines vereinfachten Energie- und Emissionsszenarios", Berlin (2013)

53. Öko-Institut, I. f. E.-u. K. Forschungszentrum Jülich, Systemforschung und Technologische Entwicklung (IEK-STE),, Deutsches Institut für Wirtschaftsforschung (DIW), Fraunhofer Institut für System- und Innovationsforschung ISI, Politikszenarien für den Klimaschutz VI. Treibhausgas-Emissionsszenarien bis zum Jahr 2030, Freiburg, Dessau-Roßlau (2013)
54. Cambridge Econometrics, Ricard-AEA: An Economic Assessment of Low Carbon Vehicles. Cambridge, London (2013)
55. Hacker, F., Blanck, R., Hülsmann, F., Kasten, P., Loreck, C., Ludig, S., Mottschall, M., Zimmer, W.: eMobil 2050. Szenarien zum möglichen Beitrag des elektrischen Verkehrs zum langfristigen Klimaschutz. Gemeinsamer Endbericht zu den Vorhaben "Wissenschaftliche Unterstützung bei der Erarbeitung von Szenarien zum möglichen Beitrag der Elektromobilität zum langfristigen Klimaschutz" (FKZ: UM 11 96 106) und "Szenarien zum möglichen Beitrag der Elektromobilität im Güter- und öffentlichen Personenverkehr zum langfristigen Klimaschutz" (FKZ: 16 EM 1001), Berlin (2014)
56. Townsend, A.: Re-Programming-Mobility. The Digital Transformation of Transportation in the United States, New York (2014)
57. Shell Deutschland Oil, Prognos. Shell Pkw-Szenarien bis 2040, Fakten, Trends und Perspektiven für Auto-Mobilität, Hamburg (2014)
58. International Energy Agency (IEA), Organisation for Economic Co-Operation and Development (OECD): World energy outlook 2014. Paris (2014)
59. De Koning, A., Huppes, G., Deetman, S.: Scenarios for 2050 for a 2-Degrees World: Using a Four Regions Trade Linked Io-Model With High Sector Detail. CECILIA2050 WP3 Deliverable 3.2, Institute of Environmental Sciences (CML), Leiden University, Leiden (2014)
60. U.S. Department of Transportation: Beyond traffic 2045. Trends and Choices, Washington D. C. (2015)
61. Öko-Institut, INFRAS Forschung und Beratung, Nutzen statt besitzen: Neue Ansätze für eine collaborative economy, Dessau (2015)
62. Organisation for Economic Co-Operation and Development (OECD), International Transport Forum (ITF): Urban mobility system upgrade. How Shared Self-Driving Cars Could Change City Traffic, Paris (2015)
63. Stölzle, W., Weidmann, U., Klaas-Wissing, T., Kupferschmid, J., Riegel, B.: Vision Mobilität Schweiz 2050. Zürich, St. Gallen (2015)
64. IHS Wien, Umweltbundesamt, TU Wien, DIW Berlin, Öko-Institut, CASE, DEFINE - Development of an evaluation framework for the introduction of electromobility. Synthesebericht, Wien, Dessau, Berlin (2015)
65. Creutzig, F.: Evolving narratives of low-carbon futures in transportation. Transp. Rev. **36**(3), 341–360 (2016)
66. Capros, P., Paroussos, L., Fragkos, P., Tsani, S., Boitier, B., Wagner, F., Busch, S., Resch, G., Blesl, M., Bollen, J.: European decarbonisation pathways under alternative technological and policy choices: a multi-model analysis. Energy Strategy Rev. **2**(3–4), 231–245 (2014)
67. Pindyck, R.S.: The Use and Misuse of Models for Climate Policy. National Bureau of Economic Research Working Paper Series, No. 21097 (2015)
68. Avineri, E.: On the use and potential of behavioural economics from the perspective of transport and climate change. J. Transp. Geogr. **24**, 512–521 (2012)
69. Marsden, G., Docherty, I.: Insights on disruptions as opportunities for transport policy change. Transp. Res. Part A Policy Pract. **51**, 46–55 (2013)
70. Canzler, W., Knie, A.: Brave new mobility world? no energy transition without transport transition, Berlin (2016)
71. Kenworthy, J., Newman, P.: The End of Automobile Dependence. How cities are moving beyond car-based planning. Island Press, Washington, Covelo, London (2015)
72. Steiner, J., Wappelhorst, S., Graff, A.: Free-Floating E-Carsharing: Integration in Public Transport Without Range Problems. Presented at European Transport Conference, Frankfurt/Main (2014)
73. Mitchell, W.J., Borroni-Bird, C.E., Burns, L.D.: Reinventing the Automobile: Personal Urban Mobility for the 21st Century. Cambridge (2010)

Accessibility in Cities: Transport and Urban Form

Philipp Rode, Graham Floater, Nikolas Thomopoulos, James Docherty, Peter Schwinger, Anjali Mahendra and Wanli Fang

Abstract This chapter reviews the different pathways which cities are following to become more accessible. By identifying the close link between transport and urban form based on global evidence, it highlights the direct and indirect costs of choices made. It then presents the tipping points which can allow to proceed from sprawling urban development and conventional motorised transport to more compact cities characterised by innovative mobility choices shaped around shared and public transport. The examples used are based on cities worldwide to illustrate emerging trends from both developed and developing countries. Therefore, the recommendations are valuable for a range of stakeholders including local and national policy makers, academics and vehicle manufacturers.

Keywords Accessibility · Compact cities · Tipping points · Transport · Urban density · Urban form · Sprawl

1 Introduction

Access to people, goods, services and information is the basis of economic development in cities. The better and more efficient this access, the greater the economic benefits through economies of scale, agglomeration effects and net-

P. Rode (✉) · G. Floater · J. Docherty · P. Schwinger
LSE Cities, London School of Economics and Political Science, Houghton Street, London WC2A 2AE, UK
e-mail: p.rode@lse.ac.uk

N. Thomopoulos
Systems Management and Strategy, University of Greenwiche, Old Royal Naval College, London SE10 9LS, UK
e-mail: N.Thomopoulos@greenwich.ac.uk

A. Mahendra
World Resources Institute, New Delhi, India

W. Fang
World Bank, Washington D.C, USA

G. Meyer and S. Shaheen (eds.), *Disrupting Mobility*,
Lecture Notes in Mobility, DOI 10.1007/978-3-319-51602-8_15

working advantages. Cities with higher levels of agglomeration tend to have higher GDP per capita and higher levels of productivity. The way in which cities facilitate accessibility through their urban forms and transport systems also impacts directly on other measures of human development and well being. Urban travel currently constitutes more than 60% of all kilometres travelled globally (179]) and, as a result, urban transport is currently the largest single source of global transport-related carbon emissions and the largest local source of urban air pollution.

This chapter presents evidence first on how accessibility in cities is created through the co-dependence of urban form and transport systems and how this relates to urban carbon emissions. It then compares direct and indirect impacts of different urban accessibility pathways. The following section discusses contemporary patterns, trends and tipping points related to the shape of cities, urban mobility and technological innovation to highlight disruptive impacts on mobility. Using this information, the final section presents the enabling conditions for increasing accessibility and low-carbon mobility in cities. This chapter is primarily based on an extensive literature review and aims to assist a further re-framing of the urban transport debate by emphasising accessibility as the underlying objective of mobility and transport in cities. Above all, such a re-framing implies a far greater recognition of urban form characteristics such as land use, the distribution of densities and urban design, in addition to more conventional transport characteristics such as related infrastructure, service levels and travel speeds.

2 Accessibility in Cities and Implications for Carbon Emissions

In any city, patterns of urban development are inseparable from the evolution of urban transport and mobility. Likewise, urban transport cannot be considered independently from urban form [49, 54, 90, 104, 133, 134, 175]. It is a combination of the two that facilitates accessibility[1] within metropolitan regions and thus creates economies of scale, agglomeration effects and networking advantages. Recognition of this interrelationship between transport and urban form is particularly important at a time of unprecedented urban expansion and the emergence of disrupting mobility trends. Some estimates suggest that globally, the total amount of urbanised land could triple between 2000 and 2030 [159] and urban kilometres travelled increase threefold by 2050 [179]. Such unprecedented change would bring with it

[1]Accessibility is frequently contrasted with mobility-based frameworks that dominate urban transport policy (Litman 2009) and draws attention to the interaction of transport conditions, land-use patterns and individual attributes in determining how easily residents of a city can access a range of social and economic opportunities. It has been defined here as: "*the extent to which land-use and transport systems enable (groups of) individuals to reach activities or destinations by means of a (combination of) transport mode(s)*" [114 p. 128].

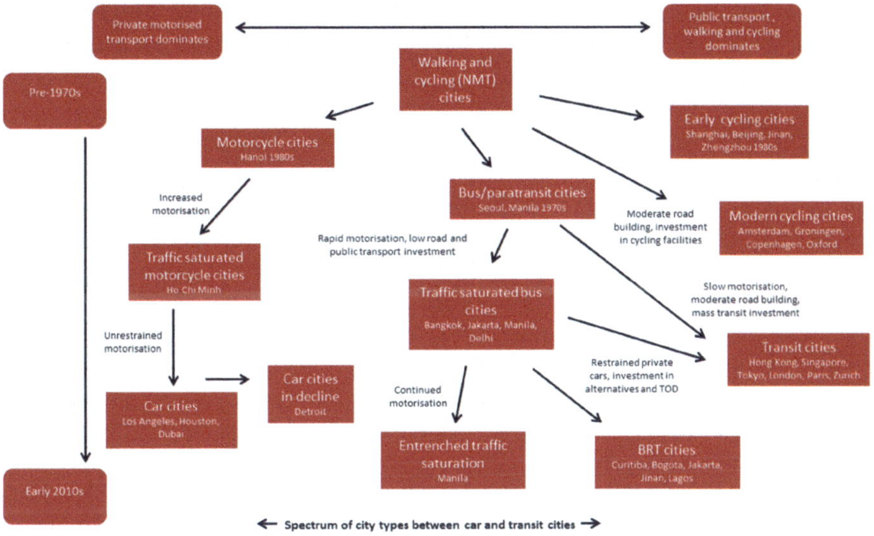

Fig. 1 Urban accessibility pathways based on Barter's city typology and transport development paths [1, 87, 119]

enormous risks associated with locking in energy intense patterns of mobility, accessibility and urban form for decades to come [11].

2.1 *Urban Accessibility Pathways*

Each city has developed its own unique spatial structure and transport system to provide access to people, goods and information. Nonetheless, different principal development patterns have evolved with respect to the most common combinations of urban spatial structures and transport. Given the strong path dependency of these patterns, we call these 'urban accessibility pathways' (Fig. 1). A defining characteristic of these pathways is the degree to which accessibility is based on the physical proximity between origins and destinations or on transport solutions which can overcome spatial separation, and the degree to which these solutions involve private or public motorised transport.

The first principle of achieving accessibility in cities is based on the physical concentration of people, services, economic activities and exchange. Given that scholars have argued that the death of distance—through the digital evolution—is still premature [146], this principle appears to be still valid. In that regard, the most defining characteristics include residential and workplace densities; the distribution of functions and degree of mixed use; the level of centralisation; and local level urban design. More compact and dense cities[2] (also referred to as 'smart growth') are typical examples of facilitating agglomeration economies through greater proximity. Creating accessibility and reducing the need for individual mobility based on physical proximity implies a particular attention to planning, designing, building and managing the specific local condition at a human scale.

Over the last century, the mechanisation of transport and the associated reduction in mobility costs relative to incomes have allowed cities to de-densify and expand horizontally, resulting in the substitution of access by proximity with access by movement. Initially driven by the introduction of streetcars, metro and regional rail systems [68, 85, 104], this process reached a new and entirely different scale with the onset of mass motorisation and the widespread introduction of privately-owned cars [21, 32, 172].

Previously, transit systems allowed for horizontal expansion that both facilitated and required compact, dense urban development and continued to produce human-scale urban environments. Urban design had to acknowledge the fact that at some point in their journey, all public transport passengers remained pedestrians, navigating through public urban space. By contrast, the introduction of the motor car not only facilitated suburban development at far lower density levels, but also introduced a transport mode that needed significantly more space per traveller to operate than any other previous means of transport. In short, public transport requires urban density whilst private car use requires space. In most cities, this has led to extraordinary tensions as a result of the inefficient use of scarce urban space by private vehicles. This provides a particular challenge for dense, developing cities where contemporary motorisation far outpaces the provision of road infrastructure or public transit alternatives.

Today, urban agglomerations can be based on many possible combinations of transport and urban form (Fig. 2), each providing different levels of access. These combinations can range from walkable, public transport-based compact cities to sprawling car-oriented cities [71], and different types can be found in different parts of the world at different levels of development.

More sprawling cities require rapid modes of transport to reduce journey times and often rely on individualised motorised transport modes as the only viable transport for low-density urban areas. In turn, these car-based transport systems require substantially more space than any other urban transport system. For example, at 50 km/h, cars require more than 160 m^2 per person, compared to 4 m^2 for buses (assuming typical occupancy levels) [150]. Space for car parking is an additional need, with cars being idle for most of the time. The average car in the United States is parked for 96% of the time [84] and aggregate parking space in car-oriented CBD's

[2]We define compact urban growth (which can be both new urban development and urban retrofitting) as urban development which is characterised by human-scale built environments with higher density, mixed-use urban form and high quality urban design. Compact urban development typically focuses on urban regeneration, the revitalisation of urban cores, the promotion of public and non-motorised transport, and high standards of urban management [64]. Related concepts include the European City model, smart growth and transit oriented development (TOD).

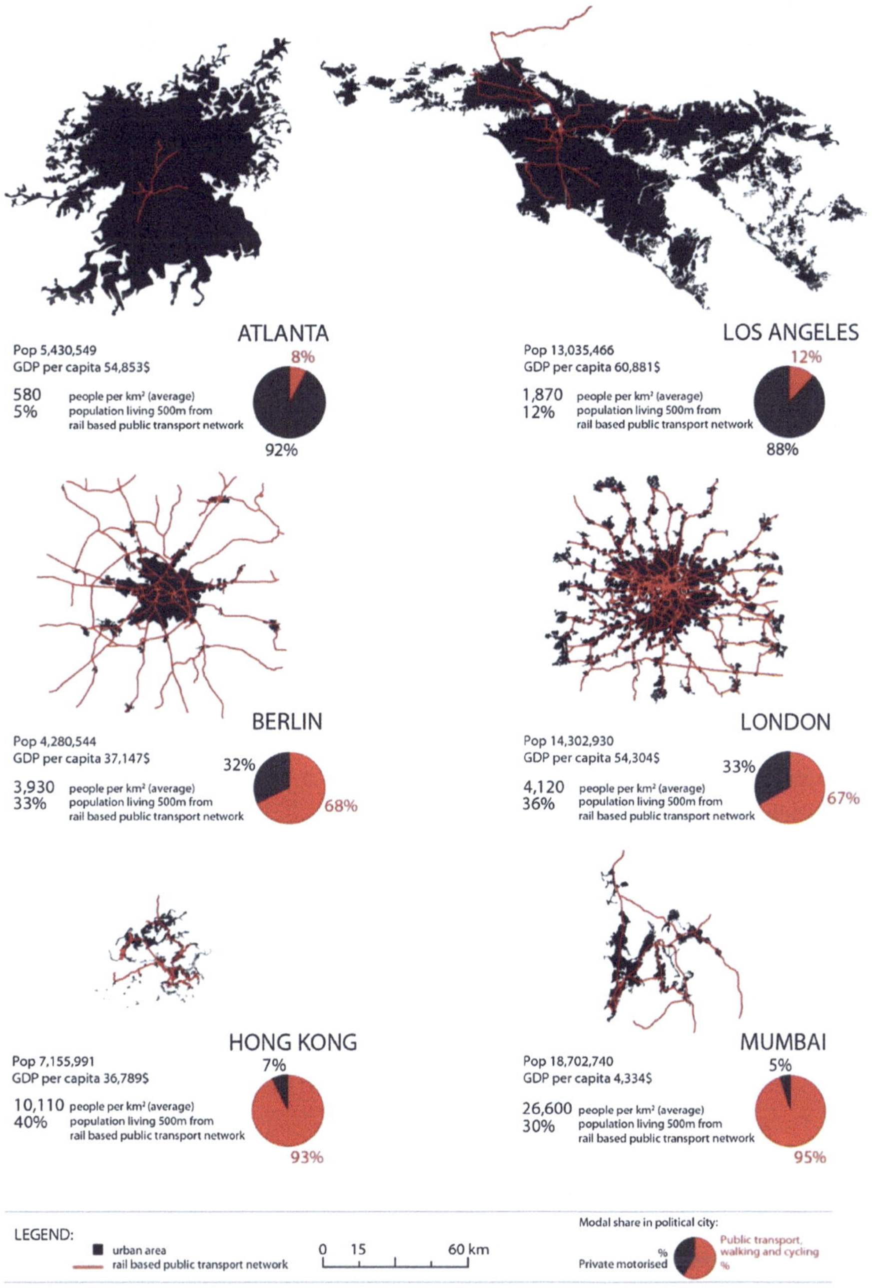

Fig. 2 Urban form and modal share (*black* in pie chart is private motorised) of selected cities [119, 161]

such as in Los Angeles is more than 80% of the CBD land area [121]. Under a business-as-usual scenario, globally an additional 45,000–77,000 km^2 would be required for car parking alone by 2050 [52], a land area equivalent to the size of Denmark. As a result, the space requirements of private vehicular traffic not only imply further de-densification of cities, but they are also a major contributor to congestion and parking pressures on public space, as road infrastructure provision is frequently unable to keep up with rising levels of vehicular traffic [101, 189].

Over recent years, compelling evidence has emerged on the degree to which urban form and transport are interrelated [89, 160]. Controlling for other factors, the difference in transport intensity between high- and low-density areas can be more than 40% in vehicle-miles-travelled per capita [59]. The National Research Council in the US estimates that doubling densities within metropolitan regions can reduce vehicle-kilometres-travelled (VKT) by up to 25% when also concentrating employment [130]. Overall, automobile dependence is negatively associated with higher population and employment density [195]. At the neighbourhood level, density, land-use mix and street design have a significant impact on the likelihood of walking [60].[3]

A particular feature of the transport-urban form relationship is the time lag between spaces and flows: land-use and physical environments change at a far slower pace than activities and related movements [123]. A second feature is the long design life of urban form and transport infrastructure, creating significant 'lock-in' effects. Some of these lock-in effects could be overcome by innovations in transport systems and technology, as discussed later in this chapter. However, where urban form and transport infrastructure is too biased towards sprawling, automobile-dependent patterns of development, it can in turn lead to a change-inhibiting cultural and political equilibrium. It is for these reasons that dealing with urban transport or land-use planning in isolation from their interdependencies can easily lead to adverse effects and unintended consequences. Furthermore, urban transport is more complex than other transport sectors, not just because it involves the integration of different mobility systems, but also because it co-produces accessibility jointly with spatial development.

2.2 *Carbon Emissions from the Provision of Access in Cities*

The co-dependence of urban transport systems with urban form also plays a central role in the global transition to a low-carbon economy [87]. Around ten billion trips are made every day in urban areas around the world. Of these, a significant and

[3]It is important to note that a range of earlier studies were more critical of the potential of reducing travel demand through higher residential densities [36, 39, 72, 76, 81, 97]. However, these studies also tended to look at density in isolation and independent from related changes such as mix use or design quality. Studies supporting the land use transport pattern impacts further include [33, 35, 111].

increasing proportion is undertaken using high carbon and energy-intensive private motorised vehicles which inversely offers a unique opportunity to explore shared mobility options within cities. About 80% of the increase in global transport emissions since 1970 has been due to road vehicles [93].

As a result, transport is one of the major sources of carbon emissions in cities. Overall, the transport sector produces around 23% of global energy-related CO_2 emissions, equivalent to 6.7 gigatonnes of CO_2 in 2010 [93]. While urban car use is the single largest contributor to transport carbon emissions, freight transport—which accounts for up to 20% of urban traffic and up to 50% of urban transport GHG emissions—is often underrepresented [155].[4] In addition, life cycle analysis suggests that carbon emissions embedded in transport infrastructures are substantial,[5] typically adding another 63% for on-road and 155% for rail in addition to emissions from vehicle operations [37]. Emissions are growing more rapidly in the transport sector than in any other sector and are projected to increase by 50% by 2035 and almost double by 2050 under a business-as-usual scenario [52, 93].

Part of this growth is due to rapid urbanisation in emerging economies and developing countries.[6] Over the last decade, significant carbon emission growth was registered for emerging economy megacities. A World Bank study showed that urban transport energy use and carbon emissions were growing by between 4 and 6% a year in the 2000s in Beijing, Guangzhou, Shanghai and Xian [42]. Between 2005 and 2010, carbon emissions from transport in Shanghai even grew by 15% annually, the highest growth rate of any sector [110]. But even within the European Union—an already highly urbanised area with ambitious carbon reduction goals—transport-related CO_2 emissions increased by 36% between 1990 and 2006, while other key sectors achieved modest reductions [58].

While urban transport emissions correlate strongly with income, there are major differences between cities with similar levels of wealth. The carbon intensity of urban accessibility is determined by two main factors: the overall distance of motorised travel required (which is largely informed by urban form characteristics) (Fig. 3), and the carbon intensity of these modes. The latter is informed by the energy intensity of different transport modes and the carbon intensity of their fuels. For example, at similar wealth levels, sprawling Atlanta produced six times more transport-related carbon emissions than relatively compact Barcelona [4, 40, 119]. This finding aligns with analysis conducted for 30 cities in China, which showed that compact cities

[4]Urban freight transport accounts for 31% of energy use and CO_2 emissions throughout Europe [86].

[5]Embedded emissions are upstream CO_2 emissions from energy used for transport, housing or the production of goods and services [51]. These also include emissions that occur as part of constructing or building transport infrastructure or vehicles.

[6]Rapid urbanisation in conjunction with population growth in developing countries and maintenance of existing transport and urban infrastructure in developed countries impose high demand for building materials. 30 billion tonnes of concrete were consumed in 2006 in contrast to 2 billion tonnes in 1950 [186], while cement production accounts for 5% of global anthropogenic CO_2 emissions [181].

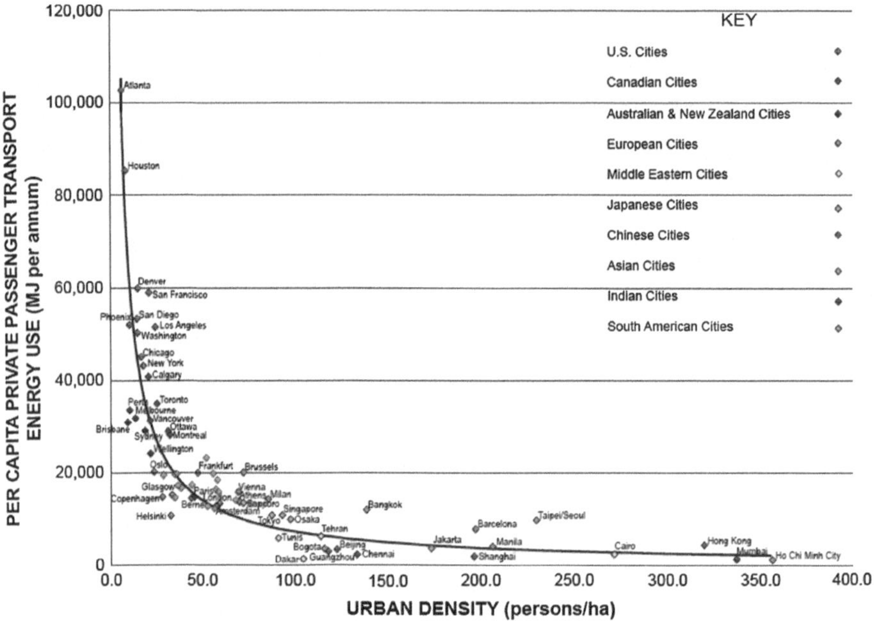

Fig. 3 Population density and transport energy use per capita for selected cities [134]

have higher CO_2 efficiency, particularly as a result of supporting non-motorised transport [171]. The IPCC suggests that over the medium to long-term an urban accessibility pathway consisting of more public transport-oriented compact cities, combined with improved infrastructure for non-motorised transport, could reduce GHG intensities by 20–50% compared to 2010 levels [93].

The substantial impact on carbon emissions of modal choices in cities is illustrated by Fig. 4, which shows the carbon emissions per passenger kilometre for different transport modes. As a result, the share of public transport, walking and cycling is a strong predictor of transport-related carbon emissions at broadly similar wealth levels (Table 1). While all motorised modes have a substantial technical potential to reduce carbon emissions per passenger kilometre—by between 30 and 50% compared to 2010 levels [93]—actual reductions remain highly uncertain.

These relationships matter for the developmental choices which rapidly growing cities face today. A scenario study for US metropolitan areas in cities such as Atlanta and Phoenix suggests a reduction of 7–10% in carbon emissions as a result of a 20–40% reduction in vehicle-miles-travelled due to compact urban development [59]. In the Indian city of Surat, estimates for annual carbon emissions resulting from a projected tenfold increase in the number of trips varies according to the mode choice and trip length—from 1.9 up to 9.5 million tonnes of CO_2. For Mumbai, the same study suggests a range of 10.3–49 million tonnes [144].

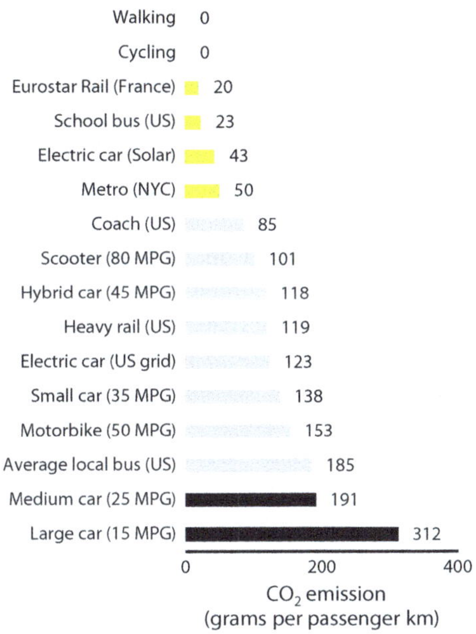

Fig. 4 Emissions per passenger km by urban transport mode [119, 164]

Table 1 Share of green transport modes and carbon emissions per capita per cities [100]

Cities	Share (%) of public transport, walking and cycling	CO_2 emissions (kg per capita per year)
Hong Kong	84	378
Tokyo	68	818
Berlin	61	774
Paris	54	950
London	50	1050
Madrid	49	1050
Montreal	26	1930
Houston	5	5690

3 Assessment: Monetisation of Related Costs and Benefits

This section presents an overview of the economic and social outcomes generated by the different transport urban-form configurations introduced above. The first sub-section will consider the direct economic impacts of different approaches to urban accessibility on urban infrastructure and operational costs, and on transport and associated industries. The second sub-section discusses the implications of different accessibility pathways on broader societal outcomes, including social equity and public health.

3.1 Direct Costs and Benefits

Urban accessibility pathways based on compact and public transport-oriented urban development deliver tangible direct economic benefits compared to sprawling automobile-dependent accessibility models [117]. Developing at higher densities uses land more efficiently and capitalises on economies of scale to reduce operational costs per unit of land.

The rapid urbanisation experienced in recent decades, particularly in the developing world, has placed a huge burden on public administrations to finance and build urban infrastructure including roads, transit systems, and utilities such as water, sewerage and electricity. The OECD has estimated that total global infrastructure investment needs (including reconstruction, upgrade and maintenance costs) could reach US$120 trillion by 2050, which translates to an average of US$3 trillion annually [135]. Under a BAU scenario, India's urban growth alone is projected to require almost 600 billion USD of investment by 2030, including 2.5 billion square metres of roads and 7400 km of metros and subways—20 times the capacity added in the past decade [120]. In China, McKinsey has similarly estimated that BAU urbanisation would require the construction of 5 billion square metres of roads by 2025 [127].

While these estimates imply a massive (and growing) global infrastructure investment deficit [7], they also do not consider alternative, more cost-effective forms of urban development [64] which actively prioritise compact urban growth, affordable mass transit and high levels of non-motorised transport use. Existing evidence suggests that key urban infrastructure (particularly linear and networked infrastructures such as streets, railways, water and sewage systems and other utilities) comes at considerably lower cost per unit at higher levels of urban density [31].

For example, the World Bank suggests that more compact city development in China could save up to US$1.4 trillion in infrastructure spending, equivalent to 15% of the country's 2013 GDP [189]. In the case of developed countries such as the United States, direct cost savings for building road and utility infrastructure in smart growth developments relative to dispersed, car-dependent developments are estimated at between US$5000 and US$75,000 per household unit [116]. Figure 5 illustrates the negative correlation between various infrastructures and urban density.

Within urban transport infrastructure provision, massive capital cost savings can be generated as a result of a shift away from private car infrastructure towards public transport, walking and cycling. Furthermore, innovative urban transport systems such as bus rapid transit (BRT) offer significant cost savings compared to traditional metro and regional railat similar capacity levels.[7] For example, Bogota's TransMilenio BRT infrastructure had a capital cost of US$5.8 million per km (US $0.34 per passenger), compared with estimates for metro rail of US$101 million per

[7]The maximum capacity levels of metro rail systems typically exceeds BRT by a factor of 1.5.

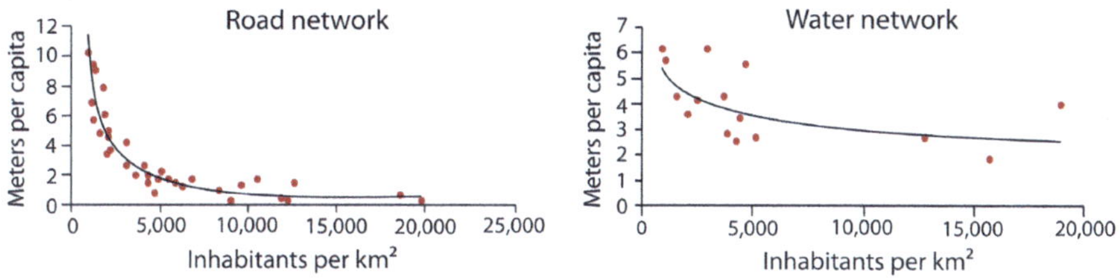

Fig. 5 Impact of urban density on road (L) and water (R) infrastructure requirements (*Source* [156, 131]

km (US$2.36 per passenger) over 3 years [124]. In addition, maintenance costs (which are frequently underrepresented within major infrastructure cost appraisals), are substantially lower on a per capita basis for affordable mass transit and non-motorised transport [30, 169].

The operational costs of urban transport are also directly informed by urban form characteristics, with sprawling urban development leading to higher costs (alongside greater capital requirements) relative to higher density development [25, 26]. As a result, higher density cities have greater opportunities for cost-efficient transport provision.

At significantly lower fuel prices, sprawling Houston spends about 14% of its GDP on transport compared to 4% in relatively compact Copenhagen and about 7% typically in many Western European cities [100, 106]. Research in the United States suggests that transit-oriented urban development (TOD) can reduce per capita use of automobiles by 50%, reducing household transport expenditure by 20% [15]. In New York, it is estimated that density-related cost savings through reduced expenditure on cars and petrol translates to about US$19 billion annually [38]. High operating costs associated with extensive (and spatially inefficient) private vehicle use are further associated with the loss of productivity due to congestion. These financial and welfare costs to cities and citizens can be substantial. For example, the New York City metropolitan region alone is estimated to lose US$13 billion annually as a direct result of traffic congestion, resulting in a notional loss of about 52,000 jobs annually [140]. Investment in improved public transport systems (such as BRT) has been demonstrated to significantly reduce congestion-related costs in numerous cities including Bogota, Lagos, Ahmedabad, Guangzhou and Johannesburg [5, 47, 83, 96, 115, 170].

Figure 6 illustrates the relationship between per capita transport fuel expenditure and density across cities in the US and in Europe. The right-hand graph presents fuel expenditure at local fuel prices, and the one on the left side ecognizedn fuel prices at the 2008 EU average, demonstrating that EU cities tend to be both denser than American cities and significantly more efficient in terms of fuel consumption.

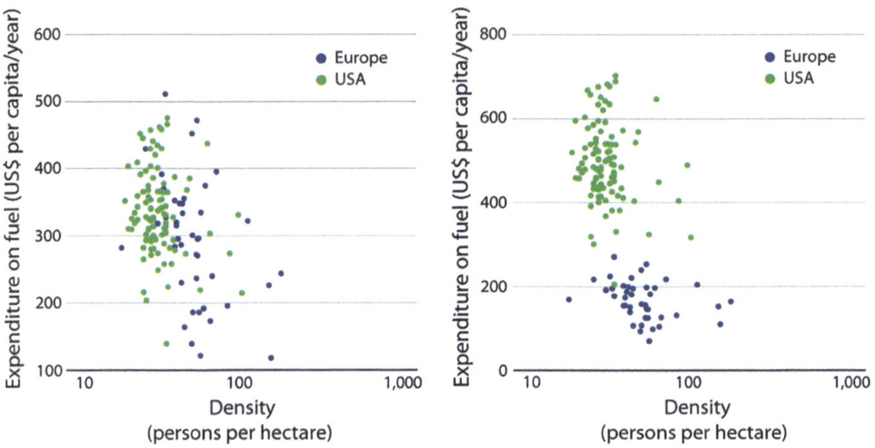

Fig. 6 Fuel expenditure and urban density, 2008 local fuel prices (L), average EU fuel prices throughout I [148]

3.2 Indirect Costs and Benefits

An assessment of the economic impact of increased public transport investment in the United States estimated that US$1 billion dollars of spending on public transportation generated over 36,000 jobs, $3.6 billion dollars of output and $1.8 billion dollars of Gross Domestic Product (GDP) annually in the short term, increasing to $3.5 billion dollars of net annual GDP generated by the 20th year (due to $1.7 billion of additional GDP generated as a result of cost savings) [145].

These findings also extend to non-motorised transport. For example, an extensive evaluation of 58 existing transportation projects in the United States, using an input–output model to assess direct, indirect and induced employment, found that cycling projects and pedestrian-only projects delivered an estimated average of 11.4 jobs and 10 state-level jobs per US$1 million invested, respectively (against an average of 7.8 jobs per US$1 million for road-only projects) [67]; and across 54 capital cities in Europe, North America and Central Asia, the WHO has estimated that an increase in cycling modal share to the levels of Copenhagen (26%) would generate approximately 76,000 jobs in direct activities alone (including bicycle repair, sale and design) [184].

At the same time, growing income and wealth inequalities, particularly in rapidly urbanising developing countries, are exacerbated by the combined effects of urban sprawl, motorisation and spatial segregation [135], with almost one quarter of the global urban population living in informal settlements [174]. Community severance and barriers to sociability provide additional examples of the negative impacts of urban accessibility pathways which incentivise private vehicle use. A comparative study for Mumbai, Istanbul and Sao Paulo emphasises the extent to which socially progressive accessibility in cities depends on compact, socially mixed urban development [151].[8] Despite this, national and city governments across both the developed and developing world currently provide disproportionate levels of investment and institutional support for private vehicle use relative to public and non-motorised transport. In addition, the construction and maintenance of well-designed transport infrastructure can provide large-scale opportunities for the direct employment of the poor and generate high local and national multipliers, particularly where the use of labour-intensive techniques and locally available materials are prioritised.

Along similar lines, different urban accessibility pathways impact significantly across a range of dimensions of public health including road safety, air pollution and activity levels. In 2010, motorised road transport accidents were estimated to account for 1.3 million deaths per year—an increase of 46% over the previous two decades—and a further 78.2 million non-fatal injuries requiring medical care [17]. Global projections continue to show an upward trend in total deaths and injuries, primarily in low- and middle-income countries, with both predicted to double by 2030 [183]. In addition to road traffic accidents, the growth in vehicle-derived urban air pollution in some large emerging economy cities has been particularly rapid: the city of Bangalore, for example, experienced a 34% increase in air pollutants on average between 2002 and 2010 [8], of which 41% of particulate matter (PM10) and 67% of NO_x emissions were emitted by road vehicles [29].

Although the relationship between different urban accessibility pathways and key transport externalities is complex, the negative impacts and high socio-economic costs of urban accessibility pathways that are based on sprawling, car-oriented cities are widely recognised [19, 22, 24, 50, 62, 69, 113, 114, 115, 133, 142, 182]. Higher density urban neighbourhoods and fine-grain street design are related and important predictors for the use of active travel modes such as walking and cycling [34, 102, 112, 118, 194].

4 Patterns, Trends and Tipping Points

Following the discussion of the economic and social effects of different urban accessibility pathways, this section provides an overview of some of the key contemporary trends in urban accessibility and form to identify disruptive practise. Three central elements—urban form, mobility behaviour and technological change—will be discussed separately, each constituting a major factor in determining the accessibility pathways in individual cities.

[8] In Sao Paulo, the most sprawling and socially segregated city of the three, the most disadvantaged groups are on average required to travel twice as long to access basic services as the most privileged—a pattern that does not exist in more compact and socially mixed Istanbul and Mumbai [151].

4.1 Urban Form: Ongoing Urban Sprawl or More Compact Urban Development?

Total global urban land area has grown rapidly in the past century, doubling in OECD countries since the mid-1950s and increasing fivefold outside the OECD [137]. In the United States, the total area of the 100 largest urban areas increased by 82% between 1970 and 1990 [173]. However, despite the continuing trend towards rising global levels of urbanisation, long-run analysis of population densities in cities suggests that there are certain trends towards de-densification.

On the basis of current trends, the world's urban population could double in little more than 40 years but urban land in less than 20 years [12]. Some estimates suggest that, under a business-as-usual urban development scenario, the area of urbanised land will triple between 2000 and 2030 [159]. Assessing a representative sample of 30 cities, Angel [12] concludes that most of these cities reached their peak density more than 100 years ago and declined on average fourfold from their peak to average density levels of 100 persons per hectare around the year 2000—the equivalent of an annual rate of decline of 1.5%.

Annually, the built-up land per person has increased by 2.9% in cities in industrialised nations and 3.6% in developing world [13]. In Angel's study of 120 cities, average built-up density declined between 1990 and 2000, from a mean of 144–112 p/ha. During that period, built-up area densities declined in 75 of the 88 developing country cities and in all 32 developed world cities [12]. In China, population densities in cities have declined by 25% on average over the last ten years [189], whilst built-up land growth rates in Nanjing, Hangzhou and Guangzhou were the highest at over 150% [196]. In contrast, urban development in India has been characterised by far lower rates of horizontal expansion, which is usually attributed to stronger private property rights, weaker local governments and insufficient capacity to develop urban infrastructure [158]. Contemporary policies (e.g. the Delhi Mumbai Industrial Corridor) focusing on smart cities development may challenge this pattern.

Given the significant negative externalities of urban accessibility pathways characterised by sprawling and car-oriented urban agglomerations, many commentators cite major market failures as the cause of sprawl whilst recognising that these are extremely complex and interrelated. Among the most obvious are significant subsidies of related infrastructure and operations, as well as unpriced negative externalities ranging from congestion to health and environmental impacts [23, 187, 191]. Comprehensively planned urban sprawl is most characteristic in China but also common in Korea and Thailand, whereas demand-led sprawl is more characteristic of India, Indonesia and Vietnam.

At the same time, alternative urban development and accessibility pathways are beginning to emerge and re-densification is recorded in many European and some North American cities. Examples of well-planned compact cities include Copenhagen, Stockholm and Hong Kong, whereas other cities such as London, Brussels, Boston, Tokyo, Hamburg and Nagoya have re-densified and moved back towards more concentrated forms [65, 66, 149]. Even cities in China have already started to increase densities with the population density in Beijing's core having already risen by 50% over the past decade [189]. This shift towards public transport investment is anticipated to become more evident since urban rail networks in China will total 3000 km in system length in 2015 and double by 2020, representing over US$645 billion of investment [189]. Bogota, Guangzhou and Ahmedabad are examples of cities that have started to partially redirect their accessibility pathways, with the introduction of mass rapid transit systems.

4.2 *Mobility Behaviour: Conventional Motorisation or New Urban Mobility?*

Cities today show a great variety of travel behaviour patterns—most notably mode choice and trip lengths. This is the case even among cities at similar levels of wealth, indicating that socio-economic factors are only one among several determining factors. This section will look in more detail at past, current and future trends of urban mobility and identifies new trends and emerging tipping points.

Three main categories of urban travel are usually differentiated: public, non-motorised transport and private motorised transport. Globally, public transport —commonly identified as the backbone of urban transport—has surprisingly low shares in terms of actual mobility provision in cities. Estimates suggest that the mode share of all urban public transport trips worldwide was only 16% in 2005 [141]. However, the role of public transport varies significantly between regions and individual cities (Fig. 7), with public transport in Asian and some European cities accounting for up to 51% of trips but fewer than 10% of trips in car-orientated cities in the United States or the Gulf Region [175].[9]

Non-Motorised Transport (NMT) remains the predominant mode of transportation in most African and Asian Cities, particularly in cities where incomes are low and the level of public transport is poor [175]. However, as the figure below demonstrates, NMT also plays a major role in cities in developed countries if pedestrian-friendly environments are provided, with London, Berlin and Copenhagen featuring as prominent examples with NMT shares of around 30% [65]. It is important to note here that data on walking and cycling is often incomplete and as a result non-motorised travel tends to be under-represented in many cities.

[9]While European and American cities have seen the strongest decline of public transport during the post-war decades, negative trends stabilised or even reversed from the 1990s onwards, most notably in larger cities with historically well-established metro and regional rail systems. High income cities in Asia such as Hong Kong, Singapore and most cities in Japan were able to maintain a public transport share of up to 70–90% due to high investment in public transport and a land-use integrated urban planning framework. Developing cities in Asia, Africa and Latin America have registered a greater variation of trends.

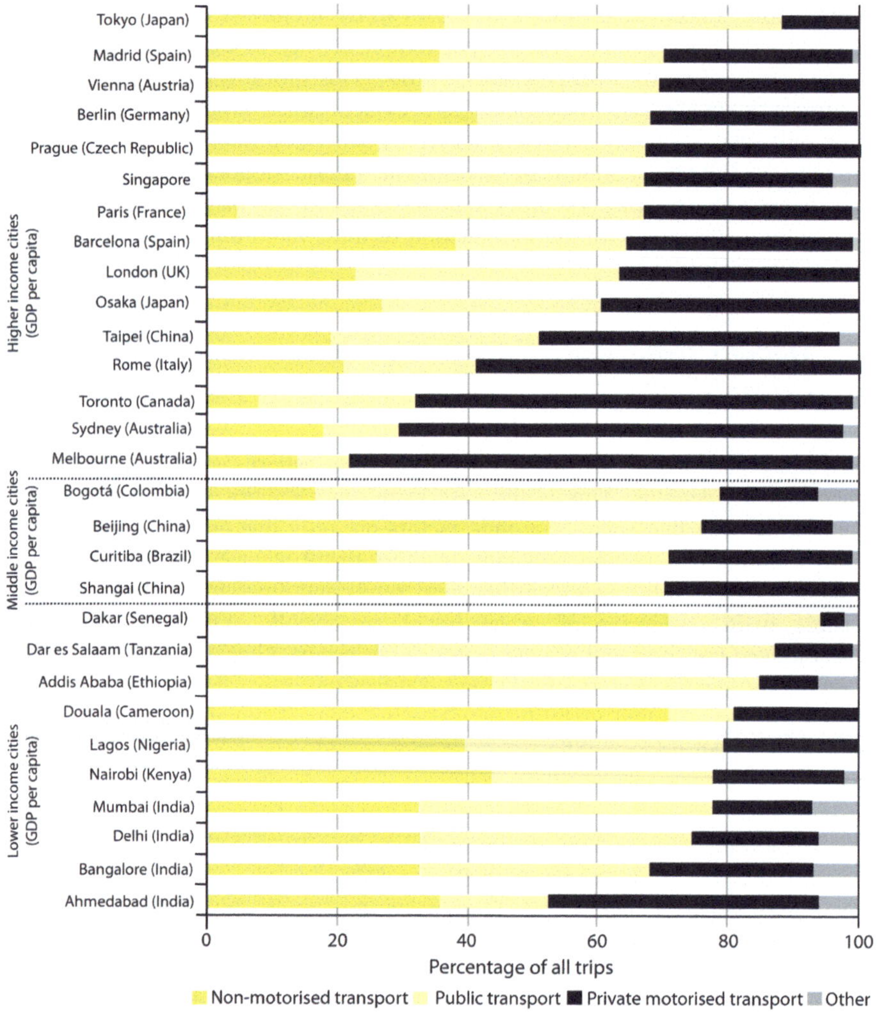

Fig. 7 Modal shares of selected cities [119, 175]

Similarly, the share of private motorised travel varies enormously, even when comparing cities at similar wealth levels. More than 90% of trips in some North American cities are by private vehicles, compared to less than 15 in Tokyo or Hong Kong. Between 1960 and 2010, the number of registered cars worldwide increased more than sevenfold, from nearly 100 million to over 700 million, while the number of registered trucks and buses increased more than tenfold, from nearly 30 million to over 300 million [44, 178]. By 2010 the total number of registered motor vehicles (excluding two-wheelers) in the world stood at over 1 billion. The recent growth in the global vehicle fleet population has been driven by emerging economies, and above all by China.[10]

In developing countries, motorised two-wheelers account for a substantial proportion of the vehicle fleet, and their growth has been at even higher rates then cars. In 2013 alone an estimated 114 million two-wheelers have been added to the global fleet population [175]. Vietnam is a unique example, with 97% of vehicles comprising two wheelers, whilst in India the figure is more than 70% [98]. In absolute terms and in spite of stabilising growth rates in private vehicles, developed countries still have the highest number of passenger cars per 1000 inhabitants.[11]

With regard to current global trends, despite regional variations there is broad consensus regarding a sharply rising motorisation. The International Energy Agency forecasts total growth of the vehicle fleet to almost 1.7 billion by 2035 [91].[12] Others have predicted that the number of passenger vehicles will reach 2.6 billion by 2050 [190]. Once again, this growth of automobiles is highest in rapidly urbanising emerging economies. Figure 8 shows that there will be more vehicles in China by 2025 than in North America or Europe. By 2050 the vehicle stock in China is projected to reach between 486 and 662 million [176].[13]

Nonetheless, it should be noted that there is a considerable risk of overestimating the growth of private vehicle stock, as most growth projections simply extrapolate historic trends without adequately incorporating evidence on changing patterns of mobility and their relationship to income and economic growth [73, 78, 185], in addition to the emerging potential of shared mobility [153]. For example, analysis of traffic forecasting in both the US and UK has indicated that transport planners have consistently overestimated future car traffic growth in the previous two decades (Fig. 9), with significant distortive effects on predict-and-provide based transport planning investments. Furthermore, a growing body of evidence showing a reversal in the previous trend towards consistent traffic growth potentially

---

[10]Over the last decade, the growth of motor vehicles in China has been exponential [105, 163]. In 2000, the number of motor vehicles per thousand people in China was 7, increasing to 44 in 2010 and 54 in 2011 (World Bank, Supporting Reports II - Urban China, 2014.). At the moment there are two hundred times as many motor vehicles (including two-wheelers) in India as there were fifty years ago, with the numbers increasing from 0.7 million in 1961 to 142 million in 2011.

[11]Italy has over 600 cars per 1000 people, Australia 556 and Germany 517. In contrast, Brazil has 179, South Africa 112, China 44 and Indonesia 37 passenger cars per 1000 inhabitants; however these countries have higher growth rates compared to developed countries (World Bank, Supporting Reports II - Urban China, 2014.). At the same time, medium-income megacities have car ownership levels similar or even higher than those in western megacities: 465, 294 and 206 in Sao Paulo, Mexico City and Johannesburg respectively compared to 209 in New York and 331 in London [27].

[12]Less conservative studies estimate the vehicle stock could increase to 2 billion by 2030 if motorisation levels across 45 countries accounting for 75% of the world's population continue to grow on a business-as-usual scenario [41].

[13]India will retain lower vehicle volumes, reflected in the low motorisation index which measures the number of vehicles per 1000 citizens. Still, by 2035, three times as many motor vehicles are expected to be on Indian roads as there were in 2005 [61].

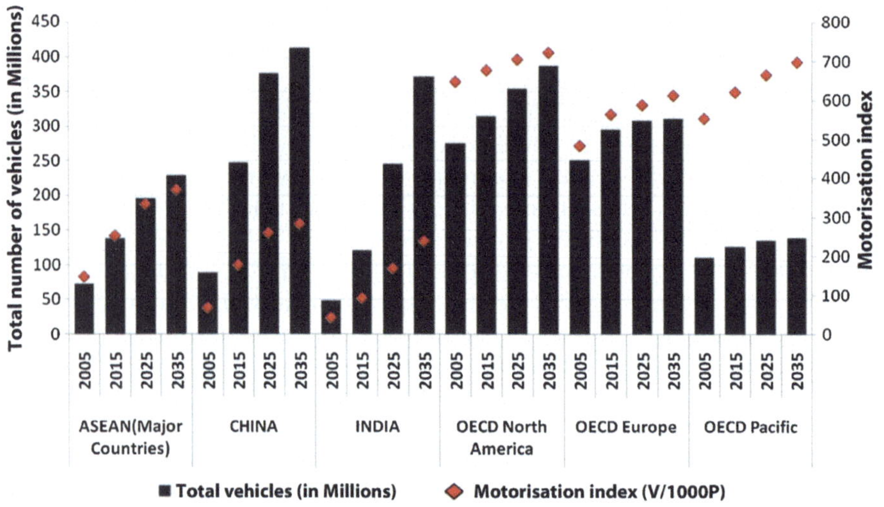

Fig. 8 Number of vehicles worldwide and motorization index by region, 2005 and projections to 2035 [61]

supports the 'peak-car hypothesis', initially formulated in the 1990s and currently linked with autonomous mobility [125, 126, 166].

Despite the global trend towards increasing motorisation, new and alternative patterns of urban form and transport planning have emerged in recent years. For example, between 2000 and 2010 levels of car ownership in New York, London and Berlin has been declining, whereas non-motorised transport, particularly cycling, has been on the rise and public transport passenger numbers have bounced back [28, 149]. Between 1998 and 2013, the share of car and motorcycle use in Berlin and London dropped from 38 to 30% and from 45 to 33% respectively [152].

Figure 10 considers wealth levels and motorisation rates for selected countries (disks) and cities (dots) since significant differences are commonly observed in the relationship between wealth and car ownership at the national and city levels [55]. The correlation between wealth and car ownership is very clear at the country level, but for cities above a wealth level of US$20,000 GDP/capita, there is a less clear relationship between increasing wealth and car use. In fact, there are wealthy cities with relatively low car ownership levels. The tipping points for this shift are complex and interdependent, but may be partially attributed to the propensity of younger urban dwellers to car share or car-pool based on smart technologies given that 87% of 19-year-olds had a driver's licence in 1983, but only 69% of 19-year-olds had a licence in 2010 in the United States [46, 78]. Local opposition in affluent inner city neighbourhoods concerned about quality of life, social capital and historic preservation has initiated a transport policy away from urban motorway programmes. Road safety concerns have led to new street design standards, facilitating safe and enjoyable urban walking and cycling. Furthermore, reducing air pollution in cities around the world has led to restrictive policy measures for vehicles.

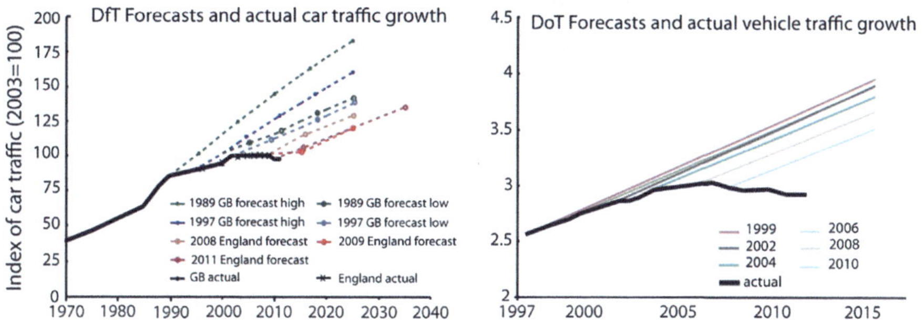

Fig. 9 False projections of car traffic for the United Kingdom (L) and United States I [73, 185]

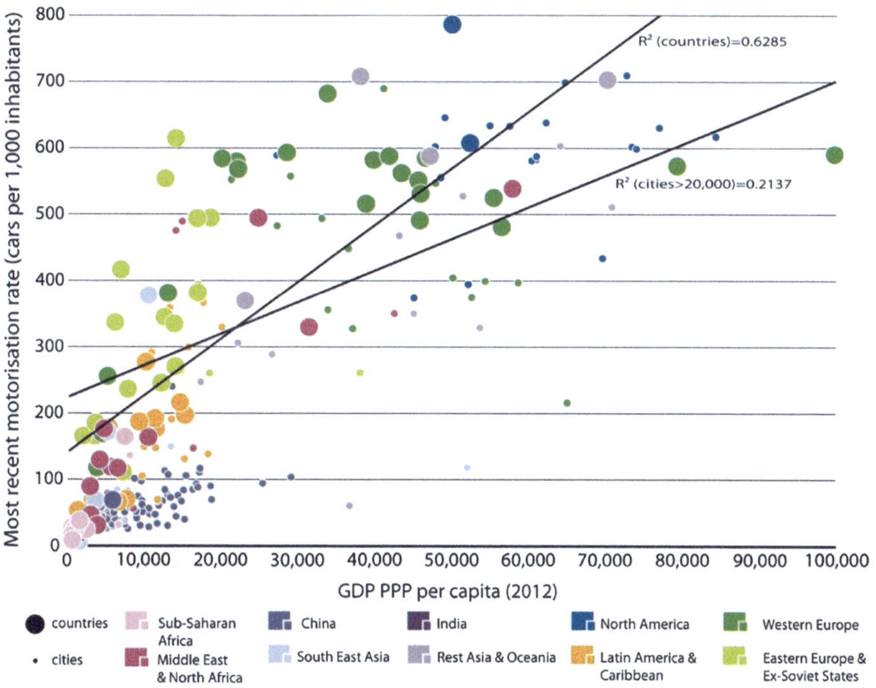

Fig. 10 Wealth and car ownership levels for selected cities and countries [119]

4.3 *Technological Change: Business as Usual or Disruptive Innovation?*

Technology-based mobility innovations in cities may include the introduction of new technology, the innovative use of existing technology and infrastructure or a combination of both. Beyond simply introducing cutting edge technologies to

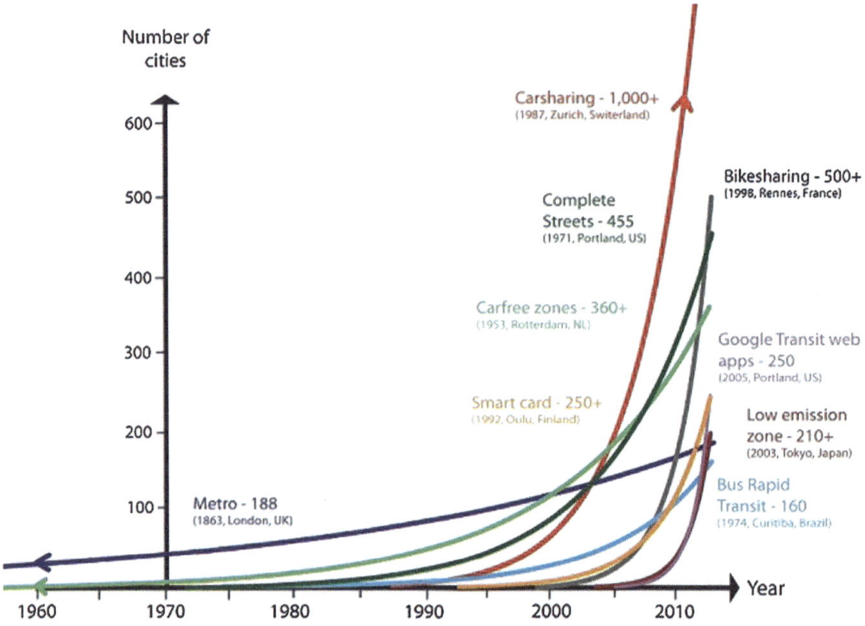

Fig. 11 Global adoption of sustainable transport systems [88]

cities, it is the rate of innovatively adopting and scaling technologies, combined with a broader socio-technical transition, which determines whether or not development pathways can be sustainably transformed. Figure 11 shows the uptake of sustainable mobility concepts, based on the number of cities in which they have been introduced. This figure illustrates that the implementation of several sustainable mobility concepts may be at a tipping point globally, as more and more cities are adopting these solutions to enhance their efficiency, competitiveness, social equity and quality of life. Still, cities will follow diverse paths of new technology adoption depending on their size, developmental stage and income level.

Fundamental technologies for urban transport have not changed substantially over recent decades, so latest developments are highly anticipated to be tested and evaluated by a range of stakeholders worldwide. Among the best examples is Bus Rapid Transit (BRT), which was introduced in Curitiba and then successfully scaled in Bogota, while it is currently operating in more than 166 cities worldwide [57]. Other examples of innovatively using existing technology include transport demand management systems (e.g. congestion charging in London and Stockholm), cycle hire schemes and a range of unconventional urban public transport systems. New urban applications of existing transport technologies include cycle hire schemes, an innovative combination of existing and new technologies that have contributed to the renaissance of urban cycling through shared infrastructure. Similarly, cable car technology has recently been adopted to improve mobility in informal settlements in Latin America. The first so-called metro-cable opened in the

Colombian city of Medellin in 2005, serving about 3000 persons per hour and direction [43] and has since been introduced in Caracas and Rio de Janeiro.

Not surprisingly, Information and communication technologies (ICT) have significantly enhanced existing urban transport systems through more effective and efficient transport management, vehicle use and travel information. In addition, real opportunities for substituting physical travel in cities with digital communication and virtualisation are beginning to emerge [6, 16, 18, 45, 139, 165]. ICT infrastructure, open and big data further allow for crowd-sourced information to update network maps, offer real time transport information and improve service quality [192]. The most important recent enabler for enhancing mobility systems in cities for individual users has been a combination of smart phone technologies and geo-positioning systems, facilitated through high rates of smartphone use penetration.

Equally, digitisation offers major opportunities for shared mobility [103]. Car and bike sharing schemes, for example, have profited enormously from real time information on vehicle/cycle availability. Effective car sharing has already demonstrated a reduction in car ownership levels [122] and potentially even total vehicle kilometres travelled, as users are more open to using public transport when not owning their own vehicle. Furthermore, ride sharing and taxi services have been significantly enhanced through taxi-related smartphone applications, with Uber raising recently US$62.5 billion allowing them to further expand their global operations [132]. Similarly, bike sharing programmes assisted by digitisation have proliferated in both developed and developing cities, with 535 schemes operating in 49 countries in 2013 [108].

However, the most disruptive innovation in this domain has been the combination of shared and electrified mobility services. The successful example of Autolib in Paris has been extended in Bordeaux and Lyon, while agreements have been signed with Indianapolis and London in 2015 and remain to be evaluated as part of the emerging Mobility as a Service approach [99]. High energy conversion rates (around 80% for electric vehicles compared to 25–30% for internal combustion engines [3, 93], elimination of local air pollution and the reduction of noise are among the key advantages of further electrifying urban transport. Storage and charging technologies are therefore the central innovation focus and have led to significant improvement in size, weight, battery capacity and overall cost, with the latter reducing from US$1000/kWh in 2008 to US$400/kWh in 2013 [84]. Consequently, switching propulsion systems of vehicles currently using internal combustion engines (ICE's) to electric drive has been the main focus, while a more holistic re-design and re-application of urban vehicle use and the decarbonisation of electricity are increasingly important components [9].

Based on the 10-year commitment of the US Government to invest US$4 billion to support the development of innovation in autonomous cars [177] and the increasing customer demand for electric vehicles (e.g. Tesla), such developments open up the possibility for disruptive change of motorised transport in cities. Autonomous vehicles can potentially allow for de-privatising vehicles in cities, increasing their utility and offering a one-way door-to-door mobility service.

A modelling scenario has shown that the entire current mobility needs of the city of Singapore could be accommodated through a fleet of shared autonomous vehicles which requires only 1/3 of the current numbers of vehicles in circulation or around 17% in Lisbon [136, 162].

Regardless of the far-reaching technological change that has begun to alter transport systems in cities worldwide, it is highly unlikely in the foreseeable future that technological innovation will fundamentally change the co-dependence of urban form and transport systems and their combined role in providing accessibility in cities.

5 Enabling Accessibility Through Compact Cities and Sustainable Transport

As previously discussed, if the objective is to reduce GHG emissions globally, then efficient urban accessibility pathways based on compact public transport oriented city form must play a central role. City-level policy initiatives are not a substitute for concerted carbon reduction efforts at a national and sectoral level, they can play an important complementary role. Within urban transport policy, three key principles are commonly recognised usually summarised as 'avoid, shift, improve' [77, 176]: first, to reduce the travel intensity in cities through greater physical proximity and co-location of different urban functions; second, to shift from spatially inefficient and energy-intensive private motorised modes to public, shared and non-motorised transport; and third, to improve the efficiency (in terms of energy and space consumption) of road-based vehicles.

Above all, to be more effective in the future, urban policies will have to build on the experiences of those cities already registering a shift towards more compact urban development and sustainable transport, and learn from the related tipping points discussed above. In addition, the promotion of these urban accessibility pathways will have to be driven by sound institutional structures and planning processes as presented in the NCE Cities Paper 02 'Steering urban growth: governance, policy and finance' [63]. Section 5 presents an overview of some of the key barriers and enabling conditions of particular relevance for urban form and transport.

5.1 Barriers

Key barriers may be summarised in resistance from:

- industrial sectors heavily reliant on the business-as-usual urbanisation model
- consumers to alter the predominant car ownership and suburban led lifestyle
- sprawled cities with limited funds available to switch towards a compact city model.

Additionally, a range of institutional and process barriers to a paradigmatic shift in urban and transport planning also exist. Policy integration across urban planning, design and transport is frequently compromised by sectoral and disciplinary silos. Fragmented governance and the lack of coordination between national and local policy frameworks for urban form and transport are widespread [11, 92] and the continued use of narrowly defined cost-benefit analysis (CBA) for transport projects [107, 128, 138, 167, 168] often obstructs more effective and well-coordinated transport-related investments.

5.2 *Planning and Regulation: Shaping Cities Across Temporal and Geographic Scales*

Given the strong interrelationship between urban form and transport, the integration of land-use and transport planning represents a unique policy opportunity. Above all, the provision of strategic infrastructure is one of the most critical public policy instruments informing the long-term shape and character of a city at any stage in its development [80, 109, 129, 175, 193]. Transport infrastructure and services play a key role in determining urban mobility patterns within urban planning, including modal choice [10, 53, 143, 180]. Further instruments include minimum density standards, mixed-use regulation and a density bonus for developers, in order to support compact city development with a hierarchy of higher density, mixed-use clusters around public transport nodes [66].

A further key priority for compact city policy is reforming inappropriate building density limitations (which exist, for example, in many Indian cities) [70, 154]. The World Bank estimates that limitations on building densities in Bangalore leads to urban sprawl which causes welfare losses of 2–3% of household income [188]. Similarly, shifting from minimum to maximum parking requirements for urban development [2, 79, 94] facilitates urban compaction and lower levels of car use. District-level interventions, including the redistribution of road space away from private vehicles and increasing investment in infrastructure for public and non-motorised transport, have proved successful in reducing motorised traffic in cities [74, 147]. For example, as part of Ahmedabad's town planning schemes, land along the expanding BRT corridors is banked by the municipal government for later development into affordable housing. The city's 'Accessible Ahmedabad' plan also fundamentally embraces accessibility planning beyond simply the provision of transport [154].

Regulatory policy instruments also play a key role in shaping urban transport performance. For example, many Chinese cities have started to limit the total number of privately owned vehicles through restrictions on the number of licence plates issued per month, with Shanghai beginning to control the growth of private vehicle registrations as early as 1994 [82].

5.3 Instruments for Shaping Incentives and Mobilising Revenue

Effective fiscal policy represents a key tool for delivering equitable and sustainable urban mobility [63]. Vehicle purchase or registration tax constitute such examples at a national level whereas parking or congestion charging can be applied locally. Yet, such fiscal instruments can play a useful role in increasing the adoption of new, less polluting and more fuel-efficient technologies (such as electric vehicles) if correctly designed, they are often used to support sales of existing vehicle technologies.

Within the urban transport sector, fiscal instruments have several purposes including managing total transport demand, shifting demand to more environmentally and socially beneficial modes, and improving the performance of those modes [75]. Fiscal policy can contribute to these objectives by internalising unpriced externalities; positively shaping incentive structures to promote compact urban development and facilitate increased accessibility; and generating revenues for the purposes of investment in urban infrastructure and services. The single most important fiscal instrument related to vehicle use is fuel pricing. Transport fuel taxation has historically been a key part of government fiscal policy due to its characteristics as a stable, dependable revenue source that is easily administered, and typically has progressive characteristics [56].

However, it is also significant what funds are prioritised for. For example, 70% to 80% of federal funding for urban transport in Mexican cities is typically dedicated to car-based transport, while the share of car use rarely exceeds 30% [14]. By contrast, Bogota's pioneering BRT system was partially financed by redirecting funds away from urban motorway programmes. In addition, private finance can be mobilised through real estate developer charges and fees, property or value capture taxes, loans, green bonds and carbon finance [11, 20]. For example in Hong Kong, the government's 'Rail plus Property' model captures the uplift in property values along new transit routes, ensuring efficient urban form whilst at the same time generating US$27 billion in direct financial benefits for the Hong Kong government since its inception in the 1970s [149].

5.4 Policy for Innovation and Technological Disruption

Automobiles have been the ultimate product of twentieth century industrialisation and have proven to be remarkably resilient. Together with the significant growth rates in car sales seen over the last decades, a certain inertia and lack of desire to innovate and re-invent is inevitable. At the same time, considerable misjudgements about the future of the automobile market are becoming increasingly evident. As previously indicated, a key element of supporting existing urban accessibility pathways that are based on more compact urban development is the application and innovative adaptation of existing technologies, including those for non-motorised transport. Nevertheless, technological innovation and related socio-technical disruption will have to play a significant role in further facilitating the transition to new, more environmentally and socially sustainable urban transport systems. Equally, the role of the international patenting system has to be reviewed whether it acts as a potential barrier to innovation [95, 157].

At the same time, the potential for system-wide technological innovation in the urban transport sector is considerable. Even within the narrower field of vehicle technology, innovations in engine and fuel technologies, digitisation and materials have the potential to significantly improve the efficiency of motorised vehicles in cities [48]. Given the scale of such a socio-technical transformation, the transport industry depends on public policy to provide a framework within which more ground-breaking innovation can take place. Establishing urban test beds may provide the essential policy context in which cities can fulfil their role in assisting urban mobility innovation.

Despite these facts, public policy in many countries frequently stalls or actively impedes the emergence of a 'post-car system' and the evolution of a more viable and resilient urban transport industry. Current policy does not even maximise what is already technologically possible and often incentivises irrelevant or counter-productive forms of innovation. Above all, political leadership coupled to a long-term political commitment will have to set out the principal policy direction and eliminate policy uncertainty as far as possible in order to unlock greater levels of investment in more ground-breaking innovation. Innovation policy would first have to acknowledge that the status quo car system is part of the problem and that future projections of private vehicle use are often linear assumptions, based on the recent past extrapolated into the future as discussed in Sect. 4.2.

6 Conclusion

This chapter introduces the concept of 'urban accessibility pathways' and argues that policy choices leading towards either more sprawling, car-dependent urban development, or alternatively more compact, public transport-oriented cities, have substantial implications, both for the economic and social performance of cities and their role in enhancing or impeding mobility innovation. By focusing on different pathways, the value of public and non-motorised transport has been highlighted in relation to urban compactness. Subsequently, this chapter has pointed out the more than 10-fold difference in transport-related carbon emissions between the most energy-intensive sprawling cities and energy-efficient compact cities linking it with both direct and indirect costs and benefits.

In summary, it has been demonstrated that more compact urban growth, aligned with the increased provision of public transport infrastructure and services and pro-active support for non-motorised transport use, is likely to deliver substantial net economic and social benefits. Despite this, current urban development trajectories across much of the world diverge considerably from more efficient urban accessibility pathways. This chapter provides an overview of current global patterns and trends in the physical development of cities and related urban mobility behaviour. Overall, cities continue to sprawl excessively, with some estimates suggesting that total urban land area could triple between 2000 and 2030. Similarly, in many key emerging economies (such as China and India), private motorised vehicle use and modal share is expanding rapidly, with a range of negative economic and social implications. At the same time, however, evidence on tipping points towards more sustainable development trajectories is emerging in cities across different wealth levels. As a result of socio-demographic change, shifting public opinion, strong political leadership and technological innovation, some cities are moving towards a more compact and public-transport oriented urban development model with increasing shares of non-motorised transport.

The last section then discusses the key policy areas and related instruments relevant to enabling spatially- and energy-efficient urban development. Undertaking strategic spatial planning in relation to key infrastructure developments, managing compact urban growth by identifying areas for intensification rather than simply restricting development, and closely integrating the provision of housing with public transport are among the key planning approaches adopted by leading city governments. In addition, the strategic use of incentives and revenue mobilisation instruments is recognised as essential to ensuring the availability of financing for urban infrastructure investment, respecting the 'polluter pays' principle and creating positive feedback mechanisms between infrastructure investment, transport modal choice and urban form.

And finally, broader policy frameworks may have to be tested more with regards to their impact on innovation and technological disruption. Here, the chapter has identified a substantial number of perverse incentives which act as considerable barriers towards the development of more efficient and effective urban transport. Across all these policy instruments, national and city governments can increasingly learn from existing transformative change in a range of cities and also build on potential tipping points that have emerged in recent years.

Acknowledgements Nick Godfrey, Isher Judge Ahluwalia, Dimitri Zenghelis, Ian de Cruz, Daniele Viappiani, Jeremy Oppenheim, Ricky Burdett, Rachel Lewis.

Disclaimer This chapter is an updated and condensed version of the NCE Paper 03 Accessibility in Cities: Transport and Urban Form which was an output of the New Climate Economy project of the Global Commission on the Economy and Climate (www.newclimateeconomy.net). The latest version of that paper is available by LSE Cities (https://lsecities.net/publications/reports/the-new-climate-economy-report). It builds on the LSE Cities research and publications including the Green Cities and Buildings chapters for UNEP's Green Economy Report, P. Rode's research on integrated planning, design and transport and research by the LSE's Economics of Green Cities programme led by G. Floater, P. Rode and D. Zenghelis.

References

1. ADB: Changing Course: A new paradigm for Sustainable Urban Transport, Urban Development Series. Asian Development Bank, Philippines (2009)
2. ADB: Parking Policy in Asian Cities (final consultant's report). Asian Development Bank, Manila (2011)
3. ADB: e-trikes—Driving Change, 2014, www.adb.org/projects/43207-012/background. Accessed 27 May 2014
4. ATM, Observatori de la mobilitat, Area de Barcelona, Autoritat del Transport Metropolita, Barcelona (2013)
5. Aftabuzzaman, M., Currie, G., Sarvi, M.: Evaluating the congestion relief impacts of public transport in monetary terms. J. Public Transp. **13**(1), 1–24 (2010)
6. Aguiléra, A., Guillot, C., Rallet, A.: Mobile ICTs and physical mobility: review and research agenda. Transp. Res. Part A: Policy Pract. **46**(4), 664–672 (2012)
7. Airoldi, M., Chua, J., Gerbert, P., Justus, J., Rilo, R.: Bridging the gap: meeting the infrastructure challenge with public-private partnerships. Boston Consulting Group, Boston (2013)
8. Alpert, P., Shvainshtein, O., Kishcha, P.: AOD trends over megacities based on space monitoring using MODIS and MISR. Am. J. Clim. Change **1**, 117 (2012)
9. Anderson, R.: Maximizing the potential for metros to reduce energy consumption and deliver low-carbon transportation in cities. Community of Metros, Delhi (2009)
10. Andrade, K., Woods, L., Kagaya, S.: Cycling within urban areas: the cases of England and Japan, European Transport Conference 2011 (2011)
11. Ang, G., Marchal, V.: Mobilising private investment in sustainable transport: the case of land-based passenger transport infrastructure. OECD Environment Working Papers. OECD Publishing, Paris (2013)
12. Angel, S.: Making room for a planet of cities. Lincoln Inst. Land Policy (2011)
13. Angel, S., Sheppard, S., Civco, D.L., Buckley, R., Chabaeva, A., Gitlin, L., Kraley, A., Parent, J., Perlin, M.: The dynamics of global urban expansion, transport and urban development department. The World Bank, UK (2005)
14. Arredondo, J.G.: Invertir para overnos, prioridad inaplazable: Diagnostico de fondos federales para transporte y accesibilidad urbana en Mexico, 2012. Executive Summary. ITDP and British Embassy, Mexico (2013)
15. Arrington, G., Cervero, R.: TCRP Report 128: effects of TOD on housing, parking, and travel. Transportation Research Board of the National Academies, Washington, DC (2008)
16. Banister, D., Hickman, R.: How to design a more sustainable and fairer built environment: transport and communications. IEE Proc. Intell. Transp. Syst. IET **153**, 276–291 (2006)
17. Bhall, K., Shotten, M., Cohen, A., Brauer, M., Shahraz, S., Burnett, R., Leach-Kemon, K., Freedman, G.,Murray, C.J.L.: Transport for health: the global burden of disease from motorized road transport. Global Road Safety Facility, Institute for Health Metrics and Evaluation and World Bank, Washington DC (2014)
18. Black, W., van Geenhuizen, M.: ICT innovation and sustainability of the transport sector. Eur. J. Transp. Infrastruct. Res. **6**(1), 39–60 (2006)
19. Black, W.R.: Sustainable transportation: a US perspective. J. Transp. Geogr. **4**(3), 151–159 (1996)
20. Bongardt, D., Creutzig, F., Hüging, H., Sakamoto, K., Bakker, S., Gota, S., Böhler-Baedeker, S.: Low-carbon land transport: policy handbook, Routledge (2013)
21. Bottles, S.L.: Los Angeles and the automobile: the making of the modern city. University of California Press, Berkeley, California (1987)
22. Bradbury, A., Tomlinson, P., Millington, A.: Understanding the evolution of community severance and its consequences on mobility and social cohesion over the past century. European Transport Conference 2007, Creating a Livable Environment Seminar, Association for European Transport and Contributors (2007)

23. Brueckner, J.K., Mills, E., Kremer, M.: Urban sprawl: lessons from urban economics [with Comments], Brookings-Wharton papers on urban affairs, pp. 65–97 (2001)
24. Bull, F., Armstrong, T., Dixon, T., Ham, S., Neiman, A., Pratt, M.: Physical inactivity. In: Ezzati, M., Lopez, L.A., Rodgers, A., Murray, C.J.L. (eds.) Comparative quantification of health risks. World Health Organization, Geneva (2004)
25. Burchell, R.W., Anglin, R., Beaton, W.P., Brail, R.K., Danielson, M.N., Foxley, S.J., Listokin, D., Newton, L.Q., Schwartz, A., Walker, C.C.: Impact assessment of the New Jersey interim state development and redevelopment plan. New Jersey Office of State Planning, New Jersey (1992)
26. Burchell, R.W., Lowenstein, G., Dolphin, W.R., Galley, C.C., Downs, A., Seskin, S., Still, K.G., Moore, T.: Costs of sprawl–2000, Federal Transit Administration (2002)
27. Burdett, R., Cavusoglu, O., Verdis, S.: City transformations. Urban Age Conference Newspaper, Urban Age Conference, Rio de Janeiro, LSE Cities, London School of Economics and Political Science (2013)
28. Burdett, R., Rode, P.: The electric city. In: Burdett, R., Rode, P. (eds.) Urban age electric city conference. LSE Cities, London School of Economics, London (2012)
29. CPCB: Air quality monitoring, emission inventory and source apportionment study for Indian cities, National Summary Report, Central Pollution Control Board (2010)
30. Campbell, R., Wittgens, M.: The business case for active transportation. Go for Green, Gloucester (2004)
31. Carruthers, J.I., Ulfarsson, G.F.: Urban sprawl and the cost of public services. Environ. Plan. **30**(4), 503–522 (2003)
32. Cervero, R.: The transit metropolis: a global inquiry. Island Press, Washington DC (1998)
33. Cervero, R., Duncan, M.: Which reduces vehicle travel more: Jobs-housing balance or retail-housing mixing? J. Am. plann. Assoc. **72**(4), 475–490 (2006)
34. Cervero, R., Sarmiento, O.L., Jacoby, E., Gomez, L.F., Neiman, A.: Influences of built environments on walking and cycling: lessons from Bogotá. Int. J. Sustain. Transport. **3**(4), 203–226 (2009)
35. Chen, C., Gong, H., Paaswell, R.: Role of the built environment on mode choice decisions: additional evidence on the impact of density. Transportation **35**(3), 285–299 (2008)
36. Cheshire, P., Hilber, C., Kaplanis, I.: Evaluating the effects of planning policies on the retail sector: or do town centre first policies deliver the goods?, SERC Discussion Paper, London School of Economics, vol. 66 (2011)
37. Chester, M.V., Horvath, A.: Environmental assessment of passenger transportation should include infrastructure and supply chains. Environ. Res. Lett. **4**(2), 1–8 (2009)
38. Cortright, J.: New York city's green dividend, CEOs for cities (2010)
39. Crane, R., Crepeau, R.: Does neighborhood design influence travel? A behavioral analysis of travel diary and GIS data. Transport. Res. Part D: Transport Environ. **3**(4), 225–238 (1998)
40. D'Onofrio, D.: Understanding the regulatory environment of climate and the impact of community design on greenhouse gas emissions. Atlanta Regional Commission, Atlanta, GA (2014)
41. Dargay, J., Gatley, D., Sommer, M.: Vehicle ownership and income growth, worldwide: 1960–2030. Energy J. **28**(4), 143–170 (2007)
42. Darido, G., Torres-Montoya, M., Shomik, M.: Urban transport and CO_2 emissions: some evidence from Chinese cities. World Bank, Washington DC (2009)
43. Davila, J.D. (ed.): Urban mobility and poverty: lessons from Medellin and Soacha. UCL & Universidad Nacional de Colombia, Colombia, Development Planning Unit (2013)
44. Davis, S., Diegel, S., Boundy, R.: Transportation energy data book, Oak Ridge National Laboratory. U.S. Department of Energy, Oak Ridge, Tennessee (2011)
45. de Graaff, T., Rietveld, P.: Substitution between working at home and out-of-home: the role of ICT and commuting costs. Transport. Res. Part A: Policy Pract. **41**(2), 142–160 (2007)
46. Deloitte: Global automotive consumer study: exploring consumers' mobility choices and transportation decisions (2014)

47. Deng, T., Nelson, J.D.: Recent developments in bus rapid transit: a review of the literature. Transport Rev. **31**(1), 69–96 (2011)
48. Denis, K., Urry, J.: After the car. Polity Press, Cambridge (2009)
49. Dimitriou, H.T., Gakenheimer, R.: Urban transport in the developing world: perspectives from the first decade of the new millenium, Edward Elgar Pub, Cheltenham (2009)
50. Dora, C.: Health burden of urban transport: the technical challenge. Sādhanā **32**(4), 285–292 (2007)
51. Druckman, A., Jackson, T.: The carbon footprint of UK households 1990–2004: a socio-economically disaggregated, quasi-multi-regional input–output model. Ecol. Econ. **68** (7), 2066–2077 (2009)
52. Dulac, J.: Global land transport infrastructure requirements—estimating road and railway infrastructure capacity and costs to 2050 information paper. OECD/IEA, Paris (2013)
53. Dulal, H.B., Brodnig, G., Onoriose, C.G.: Climate change mitigation in the transport sector through urban planning: a review. Habitat Int. **35**(3), 494–500 (2011)
54. ECOTEC, Reducing transport emissions through planning, HMSO, London (1993)
55. Ecola, L., Rohr, C., Zmud, J., Kuhnimhof, T., Phleps, P.: The future of driving in developing countries, RAND and IfMo (2014)
56. Efroymson, D., Rahman, M.: Transportation policy for poverty reduction and social equity. WBB Trust—Roads for people, Dhaka (2005)
57. Ekosgen, Employment in sustainable transport (2010)
58. Embarq: 166 cities worldwide adopt bus rapid transit—a key tipping point (2013). www.embarq.org/en/news/13/10/30/160-cities-worldwide-adopt-bus-rapid-transit-%E2%80%93-a-key-tipping-point. Accessed 25 April 2014
59. European Commission: EU 27 CO_2 emissions by sector (2007). http://ec.europa.eu/dgs/energy_transport/figures/pocketbook/doc/2007/2007_environment_en.xls
60. Ewing, R., Bartholomew, K., Winkelman, S., Walters, J., Anderson, G.: Urban development and climate change. J. Urbanism **1**(3), 201–216 (2008)
61. Ezzati, M., Lopez, A., Rodgers, A., Murray, C. (eds.): Comparative quantification of helath risks. World Health Organization, Geneva (2004)
62. Fabian, B.: Overview on transport data and MRV potential in Asia. Transport Sector and NAMAs: assessing data readiness for MRV, clean air initiative for Asian cities center (CAI-Asia Center), Pasig, Philippines (2012)
63. Fallah, B.N., Partridge, M.D., Olfert, M.R.: Urban sprawl and productivity: evidence from US metropolitan areas. Papers Reg. Sci. **90**(3), 451–472 (2011)
64. Floater, G., Rode, P., Friedel, B.: Steering urban growth: governance, policy and finance—paper 2. NCE Cities, LSE cities—London School of Economics and Political Science, London (2014)
65. Floater, G., Rode, P., Slavcheva, R., Hoornweg, D., Kennedy, C., Robert, A.: Cities and the new climate economy: the transformative role of global urban growth—paper 1. NCE Cities, LSE cities—London School of Economics and Political Science, London (2014)
66. Floater, G., Rode, P., Zenghelis, D., Montero Carrero, M., Smith, D.A.: Stockholm: green economy leader report, LSE cities—London School of Economics and Political Science, London (2013)
67. Gallagher, R., The rickshaws of Bangladesh (1992)
68. Garrett-Peltier, H.: Pedestrian and bicycle infrastructure: a national study of employment impacts. Political Economy Research Institute, Amherst, MA (2011)
69. Geurs, K.T., van Wee, B.: Accessibility evaluation of land-use and transport strategies: review and research directions. J. Transp. Geogr. **12**(2), 127–140 (2004)
70. Gilbert, R., Irwin, N., Hollingworth, B., Blais, P.: Sustainable transportation performance indicators, TRB 2003 Annual Meeting CD-ROM, Washington DC (2002)
71. Glaeser, E.: Triumph of the city: how our greatest invention makes US richer, smarter, greener, healthier and happier, Pan Macmillan (2011)

72. Glaeser, E.L., Kahn, M.E.: Chapter 56—sprawl and urban growth. In: Henderson, J.V., Jacques-François, T. (eds.) Handbook of regional and urban economics, vol. 4, pp. 2481–2527. Elsevier, New York (2004)
73. Gomez-Ibanez, J.A.: A global view of automobile dependence. J. Am. plann. Assoc. **57**(3), 376–379 (1991)
74. Goodwin, P.: Due diligence, traffic forecasts and pensions. Local Transport Today. **594** (2012)
75. Goodwin, P., Hass-Klau, C., Cairns, S.: Evidence on the effects of road capacity reduction on traffic levels. Traffic Eng + Control **39**(6), 348–354 (1998)
76. Gordon, D.: Fiscal policies for sustainable transportation: international best practices. Energy Found. Hewlett Found. (2005)
77. Gordon, P., Richardson, H.W.: Gasoline consumption and cities: a reply. J. Am. plann. Assoc. **55**(3), 342–346 (1989)
78. Grebert, J.: Renault contribution to the LSE Call for evidence submission (2014)
79. Green, J., Naughton, K.: Woes of megacity driving signal dawn of 'Peak Car' era (2014). www.bloomberg.com/news/2014-02-24/woes-of-megacity-driving-signals-dawn-of-peak-car-era.html. Accessed 9 April 2014
80. Guttikunda, S.K., Mohan, D.: Re-fueling road transport for better air quality in India. Energy Policy **68**, 556–561 (2014)
81. Hall, P.: Forces shaping urban Europe. Urban Stud. **30**(6), 883–898 (1993)
82. Hansjèurgens, B.: Emissions trading for climate policy. Cambridge University Press, Cambridge (2010)
83. Hao, H., Wang, H., Ouyang, M.: Comparison of policies on vehicle ownership and use between Beijing and Shanghai and their impacts on fuel consumption by passenger vehicles. Energy Policy **39**(2), 1016–1021 (2011)
84. Heath, G.W., Brownson, R.C., Kruger, J., Miles, R., Powell, K.E., Ramsey, L.T.: The effectiveness of urban design and land use and transport policies and practices to increase physical activity: a systematic review. J. Phys. Activity Health **3**(Supplement 1), S55–S76 (2006)
85. Heck, S., Rogers, M.: Resource revolution: how to capture the biggest business opportunity in a century. Houghton Mifflin Harcourt (2014)
86. Henderson, V.: Cities and development. J. Reg. Sci. **50**(1), 515–540 (2010)
87. Herzog, B.: Urban freight in developing cities. GTZ, Eschborn (2010)
88. Hickman, R., Banister, D.: Transport. Routledge, Climate Change and the City (2014)
89. Holtzclaw, J.W.: Using residential patterns and transit to decrease auto dependence and costs —executive summary. Natural Resources Defense Council, San Francisco (1994)
90. Holtzclaw, J.W.: Smart Growth—as seen from the air: convenient neighborhood, skip the car. Air and Waste Management Association's 93rd Annual Meeting and Exhibition (2000)
91. IBI Group: The implications of alternative growth patterns on infrastructure costs. Calgary, Canada (2009)
92. IEA World Energy Outlook 2012: OECD/International Energy Agency, Paris (2012)
93. IPCC, Climate Change 2014: Mitigation of climate change—human settlements. Working Group III: Mitigation of Climate Change, Intergovernmental Panel on Climate Change, Potsdam (2014a)
94. ITC: Trade statistics for international business development 8703—cars (incl.station wagon). International Trade Center, Geneva (2014)
95. ITDP: Transforming urban mobility in Mexico: towards accessible cities less reliant on cars. Institute for Transportation and Development Policy, Mexico City (2012)
96. Jaffe, A.B., Lerner, J.: Innovation and its discontents: how our broken patent system is endangering innovation and progress, and what to do about it. Princeton University Press, USA (2011)
97. Johnson, D., Mackie, P., Laird, J., Shires, J., Ercolani, M.: Buses and economic growth, Institute for Transport Studies. University of Leeds, Leeds (2012)

98. Kalthier, R.: Urban transport and poverty in developing countries: analysis and options for transport policy and planning. Deutsche Gesellschaft für Technische Zusammenarbeit, Eschborn, Germany (2002)
99. Kamakate, F., Gordon, D., Managing motorcycles: Opportunities to reduce pollution and fuel use from two-and-three wheeled vehicle. Int. Council Clean Transport. (ICCT) (2009)
100. Kamargianni, M., Matyas, M., Li, W., Schäfer, A.: Feasibility study for mobility as a service concept in London. UCL Energy Institute and DfT, London (2015)
101. Kenworthy, J., Laube, F.: The millennium cities database for sustainable transport, (CDROM database) international union (association) of public transport, (UITP). Perth, Brussels and Institute for Sustainability and Technology Policy (ISTP) (2001)
102. Kersys, A.: Sustainable urban transport system development reducing traffic congestions costs. Eng. Econ. **22**(1), 5–13 (2011)
103. Kim, D.: Blues from the neighborhood? Neighborhood characteristics and depression. Epidemiol Rev. **30**(1), 101–117 (2008)
104. Knie, A.: Neue Beweglichkeit. Internationales Verkehrswesen **63** (2011)
105. Kockelman, K.M.: Travel behavior as function of accessibility, land use mixing, and land use balance: evidence from San Francisco Bay Area, Transportation Research Record. J. Transport. Res. Board **1607**(1), 116–125 (1997)
106. Kutzbach, M.: Megacities and megatraffic. Access **37**, 31–35 (2010)
107. LSE Cities: Transport related carbon emissions in Atlanta and Barcelona: updated comparative calculations. Working paper, LSE Cities, London (2014)
108. Laconte, P.: Urban and transport management—international trends and practices. International Symposium “Sustainable Urban Transport and City”. Tongji University and Nagoya University, Shanghai (2005)
109. Larsen, J.: Bike-sharing programs hit the streets in over 500 cities worldwide (2013). www.earth-policy.org/plan_b_updates/2013/update112. Accessed 11 May 2014
110. Lecocq, F., Shalizi, Z.: The economics of targeted mitigation in infrastructure. Clim. Policy **14**(2), 187–208 (2014)
111. Li, F., Cao, B.: Path and potential of carbon emissions reduction caused by urban energy use: a case study of Shanghai. Urban China Project Report (2012)
112. Limtanakool, N., Dijst, M., Schwanen, T.: The influence of socioeconomic characteristics, land use and travel time considerations on mode choice for medium-and longer-distance trips. J. Transp. Geogr. **14**(5), 327–341 (2006)
113. Lin, J.-J., Yang, A.-T.: Structural analysis of how urban form impacts travel demand: evidence from Taipei. Urban Stud. **46**(9), 1951–1967 (2009)
114. Litman, T.: Transportation cost and benefit analysis. Victoria Transport Policy Institute, vol. 1–19 (2009).
115. Litman, T.: Evaluating public transportation health benefits. Victoria Transport Policy Institute, Victoria (2012)
116. Litman, T.: Evaluating public transit benefits and costs. Victoria Transport Policy Institute, vol. 65 (2014a)
117. Litman, T.: Smart growth savings. Victoria Transport Policy Institute, Victoria, Canada (2014)
118. Liu, Y., Chen, T., Song, X.: Relationship between urban form and urban CO_2 efficiency with policies and recommendations (2012)
119. Lotfi, S., Koohsari, M.J.: Proximity to neighborhood public open space across different socio-economic status areas in metropolitan Tehran. Environ. Justice **4**(3), 179–184 (2011)
120. MGI: Preparing for China’s urban billion. McKinsey Global Institute (2009)
121. Mair, C., Roux, A.D., Galea, S.: Are neighbourhood characteristics associated with depressive symptoms? A review of evidence. J Epidemiol Community Health **62**(11), 940–946 (2008)
122. Manville, M., Shoup, D.: People, parking and cities. ACCESS Mag (2004)
123. McCubbin, D.R., Delucchi, M.A.: The health costs of motor-vehicle-related air pollution. J. Transport Econ. Policy **33**(3), 253–286 (1999)

124. Medley, A.J., Wong, C.-M., Thach, T.Q., Ma, S., Lam, T.-H., Anderson, H.R.: Cardiorespiratory and all-cause mortality after restrictions on sulphur content of fuel in Hong Kong: an intervention study. Lancet **360**(9346), 1646–1652 (2002)
125. Menckhoff, G.: Latin American experience with bus rapid transit. Annual meeting of the Institute of Transportation Engineers, Melbourne (2005)
126. Metz, D.: Demographic determinants of daily travel demand. Transp. Policy **21**, 6 (2012)
127. Metz, D.: Peak car and beyond: the fourth era of travel. Transport Rev. **33**(3), 255–270 (2013)
128. Morichi, S., Raj Acharya, S.: Transport development in Asian Megacities. Springer, Berlin (2013)
129. Mu, Q., Zhang, S.-q.: An evaluation of the economic loss due to the heavy haze during January 2013 in China, China Environ. Sci. **33**(11) (2013)
130. Müller, B., Siedentop, S.: Growth and shrinkage in Germany-trends, perspectives and challenges for spatial planning and development. German J. Urban Stud. **44**(1), 14–32 (2004)
131. Müller, D.B., Liu, G., Løvik, A.N., Modaresi, R., Pauliuk, S., Steinhoff, F.S., Brattebø, H.: Carbon Emissions from Infrastructure Development. Nat. Clim. Change (2013).
132. Nelson, A.C., Moore, T.: Assessing urban growth management: the case of Portland, Oregon, the USA's largest urban growth boundary. Land Use Policy **10**(4), 293–302 (1993)
133. Newcomer, E.: Uber raises funding at $62.5 billion valuation, Bloomberg. December 3, 2015. www.bloomberg.com/news/articles/2015-12-03/uber-raises-funding-at-62-5-valuation . Accessed 6 June 2014
134. Newman, P., Kenworthy, J.R.: The land use—transport connection: an overview. Land Use Policy **13**(1), 1–22 (1996)
135. OECD: Medium-run capacity adjustment in the automobile industry. OECD Economics Department Policy Notes (2013)
136. OECD: The cost of air pollution: Health impacts of road transport (2014)
137. Odgaard, T., Kelly, C., Laird, J.: Current practice in project appraisal in Europe, HEATCO research project (Harmonised European Approaches for Transport Costing and Project Assessment) (2005)
138. O'Connor, K.M., Sauer, S.J.: Recognizing social capital in social networks: experimental results. Johnson School Research Paper Series, vol. 18–06 (2006)
139. Pelletier, M.-P., Trépanier, M., Morency, C.: Smart card data in public transit: a literature review. Transp. Res. Part C **19**, 557–568 (2011)
140. Perego, A., Perotti, S., Mangiaracina, R.: ICT for logistics and freight transportation: a literature review and research agenda. Int. J. Phys. Distrib. Logistics Manage. **41**(5), 457–483 (2011)
141. Pickrell, D.: Transportation and land use. Essays in transportation economics and policy. In: Gomez-Ibanez, J., Tye, W. B., Winston, W. (eds.) A handbook in honor of John R. Meyer. Brookings Institution Press, vol. 403–435, Wasahington DC (1999)
142. Pourbaix, J.: Towards a smart future for cities, Journeys—May 2012 (2012)
143. Prud'homme, R., Lee, C.-W.: Size, sprawl, speed and the efficiency of cities. Urban Stud. **36** (11), 1849–1858 (1999)
144. Putnam, R.D., Leonardi, R., Nanetti, R.: Making democracy work: civic traditions in modern Italy, Princeton University Press (1994)
145. Rayle, L., Pai, M.: Scenarios for future urbanization: carbon dioxide emissions from passenger travel in three Indian cities. Transport. Res. Record: J. Transport. Res. Board **2193** (1), 124–131 (2010)
146. Reno, A., Weisbrod, G.: Economic impact of public transportation investment, transit cooperative research program (2009)

147. Rietveld, P., Vickerman, R.: Transport in regional science: the "death of distance" is premature. Papers Reg. Sci. **83**, 229–248 (2004)
148. Rode, P.: The politics and planning of urban compaction: the case of the London metropolitan region. In: Ruby, A., Ruby, I., Janson, N. (eds.) The economy of sustainable construction. Ruby Press, Berlin (2014)
149. Rode, P., Burdett, R.: Cities: investing in energy and resource efficiency, United Nations Environment Programme (2011)
150. Rode, P., Floater, G., Kandt, J., Baker, K., Carrero, M.M., Heeckt, C., Smith, D., Delfs, M.: Going green: how cities are leading the next economy. LSE Cities, ICLEI and Global Green Growth Institute, London (2013)
151. Rode, P., Gipp, C.: Dynamische Räume: Die Nutzungsflexibilisierung urbaner Mobilitaetsraeume am Beispiel der Berliner Innenstadt, Technical University Berlin (2001)
152. Rode, P., Kandt, J., Baker, K.: Transport equity in Sao Paulo, Istanbul and Mumbai, LSE Cities Working Paper, LSE Cities, London (2013)
153. SGA: Recent lessons from the stimulus: transportation funding and job creation, smart growth America (2011)
154. STF: Shrink your travel footprint (2014). http://shrinkthatfootprint.com/shrink-your-travel-footprint. Accessed 12 June 2014
155. Salon, D., Aligula, E.M.: Urban travel in Nairobi, Kenya: analysis, insights, and opportunities. J. Transp. Geogr. **22**, 65–76 (2012)
156. Salat, S., Bourdic, L.: Urban complexity, efficiency and resilience. In: Morvaj, Z. (eds.) Energy efficiency—A Bridge to Low Carbon Economy (2013).
157. Schubert, J., Wolbring, T., Gill, B.: Settlement structures and carbon emissions in Germany: the effects of social and physical concentration on carbon emissions in rural and urban residential areas. Environ. Policy Governance **23**(1), 13–29 (2013)
158. Scotchmer, S.: Innovation and incentives, MIT press (2004)
159. Sellers, J., Han, S.S., Huang, J., Lu, X.X., Marcotullio, P., Ramachandra, T.V.: Peri-urban development and environmental sustainability: examples from China and India, Asia-Pacific Network for Global Change Research (2009)
160. Smith, K.R., Jerrett, M., Anderson, H.R., Burnett, R.T., Stone, V., Derwent, R., Atkinson, R. W., Cohen, A., Shonkoff, S.B., Krewski, D.: Public health benefits of strategies to reduce greenhouse-gas emissions: health implications of short-lived greenhouse pollutants. Lancet **374**(9707), 2091–2103 (2010)
161. Smith, W.S.: Mass transport for high-rise high-density living. J. Transport. Eng. **110**(6), 521–535 (1984)
162. Sperling, D., Nichols, M.: California's pioneering transportation strategy. Issues Sci. Technol. **28**(2), 59–66 (2012)
163. Spieser, K., Treleaven, K.B., Zhang, R., Frazzoli, E., Morton, D., Pavone, M.: Toward a systematic approach to the design and evaluation of automated mobility-on-demand systems: a case study in Singapore (2014)
164. Staniford, S.: Chinese transportation growth (2010)
165. TfL, Annual: Report and statement of accounts 2012/13. Transport for London, London (2013)
166. Thomopoulos, N., Givoni, M., Rietveld, P. (eds.): ICT for transport: opportunities and threats, NECTAR Series on Transportation and Communications Networks Research. Edward Elgar, Cheltenham (2015)
167. Thomopoulos, N., Givoni, M.: The autonomous car—a blessing or a curse for the future of low carbon mobility? An exploration of likely vs. desirable outcomes. Eur. J. Futures Res. **3** (14) (2015)
168. Thomopoulos, N., Grant-Muller, S.: Incorporating equity as part of the wider impacts in transport infrastructure assessment: an application of the SUMINI approach. Transportation **40**(2), 315–345 (2013)

169. Thomopoulos, N., Grant-Muller, S., Tight, M.: Incorporating equity considerations in transport infrastructure evaluation: current practice and a proposed methodology. Eval Program Plann. **32**(4), 351–359 (2009)
170. Thompson, D.: Suburban sprawl: exposing hidden costs, identifying innovations. University of Ottawa, Ottawa, Sustainable Prosperity (2013)
171. Todd, L.: Analysis of public policies that unintentionally encourage and subsidize urban sprawl. Victoria Transport Policy Institute, Supporting paper commissioned by LSE Cities at the London School of Economics and Political Science, on behalf of the Global Commission on the Economy and Climate for the New Climate Economy Cities Program (2015)
172. Turner, P., Pourbaix, J.: UITP call for evidence response—actions to reduce emissions, urban innovation and smart technologies, UITP (2014)
173. UK Ministry of Transport: The Buchanan Report—traffic in towns: a study of the long term problems of traffic in urban areas. H. M, Stationery Off (1963)
174. Habitat, U.N.: Part three: environmental harmony. Earthscan, London (2008)
175. UN Habitat: State of the world's cities 2010/2011: bridging the urban divide (2010)
176. UNCSD: Sustainable, Low carbon transport in emerging and developing economies, Rio 2012 Issues Briefs, Rio + 20 United Nations Conference on Sustainable Development, 13 (2012)
177. UNEP: Towards a green economy: pathways to sustainable development and poverty eradication, UNEP (2011)
178. UTF: Towards an urban renaissance: the report of the urban task force chaired by Lord Rogers of Riverside; Executive Summary, HMSO, London (1999)
179. Valeur, H.: The horrendous costs of motorized transportation in (Indian) cities, Alog (2013)
180. Van Audenhove, F.-J., Korniichuk, O., Dauby, L., Pourbaix, J.: The future of urban mobility 2.0, Arthur D. Little and UITP (2014)
181. WBCSD: The cement sustainability initiative: recycling concrete—executive summary, WBCSD-CSI (2009)
182. WBCSD: About the cement industry (2012). www.wbcsdcement.org/index.php/about-cement. Accessed 27 May 2014
183. WHO: Mortality: chronic respiratory diseases, deaths per 100,000—data by country (2008). http://apps.who.int/gho/data/node.main.A866?lang=en. Accessed 6 June 2014
184. WHO: Physical activity, February 2014. www.who.int/mediacentre/factsheets/fs385/en, Factsheet. Accessed 12 June 2014
185. WHO: Unlocking new opportunities: jobs in green and healthy transport, World Health Organisation (2014)
186. Wagner, A.: GIZ international fuel prices 2012/2013—data preview April 2013. Eschborn, Deutsche Gesellschaft fur Internationale Zusammenarbeit (GIZ) (2013)
187. West, S.E.: Taxes versus standards, 21st century economics: a reference handbook, vol. 1, p. 247 (2010)
188. World Bank: Guangzhou green trucks pilot project: technology pilot report, World Bank and Clean Air Initiative for Asian Cities Centre (2010)
189. World Bank: Urbanization beyond municipal boundaries: nurturing metropolitan economies and connecting peri-urban areas in India, World Bank (2013)
190. Wright, L.: Bus rapid transit, sustainable transport: a sourcebook for policy-makers in developing cities. Eschborn, Deutsche Gesellschaft für Technische Zusammenarbeit (GTZ) (2002)
191. Wright, L., Fulton, L.: Climate change mitigation and transport in developing nations. Transport Rev. **25**(6), 691–717 (2005)
192. Wu, J.: Environmental amenities, urban sprawl, and community characteristics. J. Environ. Econ. Manage. **52**(2), 527–547 (2006)
193. Yen, I.H., Michael, Y.L., Perdue, L.: Neighborhood environment in studies of health of older adults: a systematic review. Am. J. Prev. Med. **37**(5), 455–463 (2009)

194. Zhang, J., Mauzerall, D.L., Zhu, T., Liang, S., Ezzati, M., Remais, J.V.: Environmental health in China: progress towards clean air and safe water. Lancet **375**(9720), 1110–1119 (2010)
195. Zhang, M.: The role of land use in travel mode choice: evidence from Boston and Hong Kong. J. Am. plann. Assoc. **70**(3), 344–360 (2004)
196. Zhang, M.: Travel choice with no alternative can land use reduce automobile dependence? J. Plann. Educ. Res. **25**(3), 311–326 (2006)

Mobility Patterns in Shared, Autonomous, and Connected Urban Transport

Nicole Ronald, Zahra Navidi, Yaoli Wang, Michael Rigby, Shubham Jain, Ronny Kutadinata, Russell Thompson and Stephan Winter

Abstract A number of recent technological breakthroughs promise disrupting urban mobility as we know it. But anticipating such disruption requires valid predictions: disruption implies that predictions cannot simply be extrapolations from a current state. Predictions have to consider the social, economic, and spatial context of mobility. This paper studies mechanisms to support evidence-based transport planning in disrupting times. It presents various approaches, mostly based on simulation, to estimate the potential or real impact of the introduction of new paradigms on urban mobility, such as ad hoc shared forms of transportation, autonomously driving electrical vehicles, or IT platforms coordinating and integrating modes of transportation.

N. Ronald
Department of Computer Science and Software Engineering,
Swinburne University of Technology, Hawthorn, VIC 3122, Australia
e-mail: nronald@swin.edu.au

Z. Navidi · Y. Wang · M. Rigby · S. Jain · R. Kutadinata (✉)
R. Thompson · S. Winter
Department of Infrastructure Engineering, The University of Melbourne,
Parkville, VIC 3010, Australia
e-mail: ronny.kutadinata@unimelb.edu.au

Z. Navidi
e-mail: z.navidikashani@student.unimelb.edu.au

Y. Wang
e-mail: y.wang205@student.unimelb.edu.au

M. Rigby
e-mail: m.rigby2@pgrad.unimelb.edu.au

S. Jain
e-mail: sjain1@student.unimelb.edu.au

R. Thompson
e-mail: rgthom@unimelb.edu.au

S. Winter
e-mail: winter@unimelb.edu.au

G. Meyer and S. Shaheen (eds.), *Disrupting Mobility*,
Lecture Notes in Mobility, DOI 10.1007/978-3-319-51602-8_16

Keywords Mobility on demand · Demand-responsive transport · Ride sharing · Mobility as a service · Simulated mobility

1 Introduction

The emerging trend of shared economy in transport has been a major talking point in recent years. The success of companies such as Uber and Lyft has brought the attention of academics and industry alike to dig deeper into this trend [1, 2]. A few studies have shown that sharing economy in the forms of carpooling and ridesourcing has been received positively by some [3, 4]. Convenience, time, and monetary savings have been identified as the major motivations for carpooling; some are willing to experience considerable delays to achieve these benefits [4]. Similarly, the users of ridesourcing services receive the same advantages, if not for the more reliable service (in terms of wait time and accessibility) compared to public transport [3]. Despite the potentials, these sharing economy enabling technologies have brought significant disruption in transportation. Therefore, tools that enable impact prediction of such technologies are important to allow a smooth transition into a more sustainable future.

These trends happen in cities, which are complex systems [5, 6], i.e., systems of nonlinear and nondeterministic behavior. Batty characterizes them as "emergent phenomena generated through a combination of hierarchical levels of decision, driven in decentralized fashion" [5, p. 1042]. Then urban transport is a complex (sub-)system itself: It is a phenomenon emerging from the 'decentralized', i.e., uncoordinated mobility of the individuals forming the city's population. The emergent transport patterns are nonlinear because of capacity limits of transport networks, and they are nondeterministic because of human choice that must rely on the information available to them [7]. The latter point, about behavior of people, is evident already from the prevalent preference for ownership and use of a private car, which from the outside is often irrational from an economic, social, and environmental perspective.

How to predict future states of complex urban systems is therefore a nontrivial task, and one that is approached typically by simulation [5, 8, 9]. Since mobility is a derived demand—not a self-contained system, but derived from the needs of people to access and participate in activities [10, 11]—the starting point for investigating (and predicting) mobility patterns cannot be a random distribution of trips within the geographic area of a city: It has to be bound to the distribution of population, of activities, and of the economic and social characteristics at particular locations. The challenge with potentially disruptive technology is, however, that how this demand is expressed in behavioral choices today does not predict how this demand will be expressed in the future. One (hard) question is whether people's current choices or preferences are flexible if the offerings of a mobility system change. For example, how many of them, or which group of them are willing to switch to other, novel modes of traveling in order to pursue their activities? The other (hard) question is whether people, with novel choices and perhaps more flexibility in travelling at hand, will also adapt their activities, or activity locations, in order to satisfy their needs. For example, choices by daily routines—picking up a coffee at a particular place—can easily move with changing routines. But even more significant activities can change, for example, if other work places come into reach that were not accessible before.

The current practice to gain insights into people's choices are surveys. On one hand, *travel and activity surveys* provide insight into current travel behavior. Traditionally, these are paper-based questionnaires, but also tracking-based surveys are trialed or even applied [12]. In future, surveys—i.e., samples of a population—may be replaced by tracking whole populations, and mining the data for activities [13]. The results are used to break down the aggregate behavior that can be observed from traffic monitoring to an individual level. On the other hand, *stated preference surveys* [14] try to find out to what degree people are flexible or would change their behavior in the light of new alternatives. Stated preference surveys have been done for example to find out under what parameters people would choose cycling more often as a mode of travelling [15], or what makes people choose airports or airlines [16]. Despite their elaborate theory of design, these surveys suffer from not being able to predict long-term effects: they capture spontaneous reactions of people on choices, but not their learning over time. This means stated preference surveys are valuable where people know the alternatives already, and are less reliable if people are not familiar with the alternatives yet.

This paper investigates to what extent simulation of the complex system of mobility in the city lends itself to predict future states of choices and behavior. These future states (of demand) have direct impact on costs of systems, and thus, on preferences again. Simulation should allow finding the sweet spots in designs, balancing costs and demand before a novel system is implemented, trialed (and in the past, too often failed) or rejected.

This paper will present four examples of analyses and simulations of novel, ad hoc demand-responsive transportation systems (DRT). A particular focus will be on capturing the social, economic and spatial context of mobility in order to come up with valid, i.e., well-grounded results. First, an analysis is presented of how DRT susceptibility can be predicted for a particular region based on experience in other regions. This is then followed by descriptions of three simulations, focusing on comparing different flexible services, comparing fixed route transit and flexible services, and ridesharing with friends. Optimization and the information provided to users also play a role, however, also needs to be adapted for disruptive services. Finally, we conclude regarding the importance of new analytical and simulation approaches for evaluating these emerging systems.

2 Predictions from Experiences

Transportation systems are an important component of urban development and are often of widespread social impact. Hence, before any policy decision making and implementation of transportation services or facilities, it is devisable to predict travel demand patterns and understand travel behavior of the population. Large number of studies have found that various demographic, geographic, social, and economic factors such as household composition, car ownership, employment status, wealth/incomes, age, gender, ethnicity, lifestyles, habits, and preferences can affect travel demand, behavior, and mode choice [17–20]. The shift to a new mode of transportation from existing modes depends on several factors such as its affordability, travel time, travel cost, convenience, flexibility, technology, and its relative level of service. These factors are not available in advance for an assessment of mode shift as they depend on uptake of the proposed mode in the market.

However, one way to predict the travel demand is by gaining insights from the experience of similar services operating elsewhere in the world. Usability patterns of these services can be studied, and dominant favorable characteristics of users and trips for such transport modes can be identified. Trends of identified characteristics can be found out in the target city using the travel surveys authorities run regularly to collect data for transport planning. These travel surveys record socioeconomic characteristics, demographic characteristics, household attributes and travel details/diaries of a statistically significant and representative sample of the population. In addition to profiling the population they identify the local patterns of travel demand, and thus help to predict the success of a proposed transport mode in the target city.

In order to develop this idea in a formal framework, a review of studies [21–36] on DRT services operating in various places in the world had been conducted, and parameters have been identified that favor the use of DRT, such as a clientele in the age groups of 15–24 and 55 or older, a female clientele, a clientele not in workforce or with no driving license, a clientele of low household vehicle ownership, low household income, single person households, a clientele with lack of near train stations, trips for shopping and social reasons, and a clientele with higher trip waiting times or higher trip walking times.

Variations of these parameters can be determined for any target city, using the mentioned travel surveys. In Melbourne this was the Victorian Integrated Survey of Travel and Activity 2009–10 [37]. Typically the parameters have to be aggregated to larger statistical areas, due to privacy or data availability. In our study for Melbourne the aggregation happened at Statistical Area 3 (SA3) for the population census. Such labeled areas can be used to identify the regions that have distinctly higher likelihood to take up DRT, and hence provide better opportunity of implementation of such novel transport services. This methodology has been validated by applying it on an existing mode, in this case public transport and taxi: The results of predicting the susceptibility of certain regions for these modes of transport were compared with the current, observable demand patterns of these modes.

A high degree of correspondence suggests that the methodology is suited and can be applied to not yet existing modes of traveling in a target city. In principle, the methodology can also predict the nature of competition among the various existing modes of transportation and a proposed DRT service, and will help in decision-making accordingly.

3 Simulation Applications

3.1 Experimenting with Different DRT Systems

An application of disruptive transportation is the replacement of a fixed route bus service by a partially or non-fixed route service. This can occur when the original service has low patronage and operating costs are relatively high.

In 2013, Public Transport Victoria elected to replace a regional bus service by a service operated by taxis, known as Flexiride [38]. The service has some fixed elements, in that the taxi must always start at the same point at set times during the day. If no one is waiting at that point, nor any bookings have been taken, then the taxi returns to regular service. If there are bookings, the driver decides best how to service those passengers and proceeds. Bookings can only be taken before the taxi leaves the first point.

Data from this new service was used as a base case to experiment with different types of services. A model was built using MATSim, and two scenarios were tested: the existing Flexiride service, along with a fully flexible service, where the constraint on start time and start point were removed [39]. This simulation could then be used to compare the two configurations with respect to the number of successful trips, the average waiting time, and the average excess time. More flexibility, of course, delivers better performance for passengers but higher "costs" in terms of longer travel times or more vehicles needed.

One limitation of this model was that only one day can be simulated, which means that the long-term demand for a flexible service cannot be explored. Another study using the same model analyzed how demand could change over many days based on previous experiences with a flexible service [40]. For example, when deciding to make a trip, the estimated travel time could be used to determine whether the trip is worthwhile. This estimate could be based on personal experiences, for example, how long it usually takes to get from home to the destination, or on global experiences, for example, if others are experiencing delays then you might too.

Another limitation was that only flexible services could be included in the model. This means operation could differ in congested scenarios. However, the use of a multimodal simulation tool such as MATSim means future integration is possible.

The advantage of using simulation in this case is to obtain an idea of how different service configurations will perform with different demand patterns. The ability to build in passenger choice in the future will provide more realistic results.

3.2 Comparing Buses and on-Demand Systems

Public transport operator companies face financial challenges when it comes to providing transit in suburban areas. The main problem is that the demand density is so low that makes it difficult to financially maintain a high frequency public transport and the low frequency is most unattractive to users. Demand-responsive transport (DRT) has promising features to solve this problem. However, there are high financial and operational risks associated with the implementation of a DRT system, which could be foreseen and prevented by the help of simulation. Following preliminary work [41], we use simulation to test our hypothesis that replacing conventional public transport (CPT) with DRT would improve people's mobility in a small area and in turn solve the problem of underutilized CPT. To this end, the two modes have been compared in term of user performance and cost to the operator.

A limited number of studies have been focused on comparing these modes or finding the demand switch point, which is the highest demand that makes DRT outperform CPT [42–45]. These studies concluded that DRT outperform CPT in areas with low demand density. Other studies tried to define the demand switch point by analytical methods [46, 47]; however, their studies were focused on DRT operating as feeders.

Two important issues were missing in all these studies: the study of a dynamic ad hoc DRT and considering user preferences. To bridge this gap, this work utilizes an agent-based traffic simulation, MATSim, with an embedded dynamic routing algorithm. This model has the capability of modeling all the passengers as individual agents and offers them ad hoc DRT service to study the impact of replacing CPT with DRT on an individual level. Moreover, the diversity of investigated scenarios helps find the demand switch point.

To test the hypothesis, an extensive comparison was conducted investigating several variations in network shape, demand density and transit frequency. The combination of four transit system (DRT and CPT with three frequencies: 7.5, 15, and 30 min) and five demand density (1, 2, 3, 4, 5, 10 and 15 requests per minute) were investigated in two different network shape: grid and star shape (X) with 16 km^2 area.

The scenarios were evaluated in terms of the quality of passengers' mobility and the operator cost. The quality of the mobility has been defined in this work as the passengers' perceived travel time and presented as Virtual in-vehicle time (VIVT). The importance of people's perception of travel time in their decision about their mode of transport has been verified previously [48, 49]. It is calculated according to passengers' walking time, waiting time, in-vehicle ride time, and number of transfers. The cost to the operator was calculated according to the size of the fleet, the operating hours, and the kilometers driven by each vehicle. In this study, the uptake for both services and their ticket price is assumed to be equal.

The results demonstrate that replacing CPT with DRT results in a significant improvement in people's mobility. The VIVT is the lowest for DRT users in all scenarios, which shows the superiority of this mode in terms of user performance. Moreover, the percentage of people waiting less than 10 min has been calculated for all scenarios. The average percentage for CPT users is 99% in both networks, while it goes up to 70 and 80% in star shape and grid network respectively.

After verifying that replacing DRT with CPT results in a better situation for users, it is necessary to evaluate their cost. DRT's cost is less than any frequency of CPT in grid network and the demand switch point can be defined according to the frequency. However, the highest demand for which DRT can outperform CPT (in terms of cost) is almost 7 requests per minute.

As a conclusion, replacing CPT with DRT results in improvement in people's mobility in small areas with low demand, mostly without any extra cost to the operator. This means that this new system can solve the problem of uneconomical transit services for the operators and provide a better mobility option for inhabitants. A high quality, door-to-door public transport service also improves the social equity of a suburb for its habitants and increase the standard of living. The fact that the operator cost depends on the demand in DRT operation demonstrate its flexibility, which is very useful in adapting the supply to demand in different time periods. This work demonstrated the power of simulation tools to evaluate the performance of a new system and compare it to old ones on a limited time scenario. However, it is expected that by developing more complex models, it is also possible to study the DRT performance on a longer run (for instance over a week or a month).

3.3 Motivating Behavioral Change

Human mobility and travel decision-making are complex behaviors that are affected not only by physical conditions, e.g., space-time limits, but also by sociopsychological factors. Especially for such collaborative behaviors with multiple participants as ridesharing, decisions are made in many aspects. Trajectory analysis has indicated that many people could do ridesharing according to their space-time concurrence [50]. But despite the environmental and economic benefits, there is still a low rate in participation of ridesharing [51–53]. The contradiction is partially explained by some surveys showing that the willingness to share rides with social contacts (a first or second degree socially connected person, hereafter called "friend") is significantly higher than with strangers [52, 53]. Therefore, friendship can be a good drive for ridesharing. While the low willingness to share rides with strangers signifies that many of the existing rides are in fact invalid for a certain person, ridesharing exclusively with friends and declining offers from strangers nearby may lead to higher detour cost and even less opportunities to get a ride within given space-time budgets.

To prove the benefit of ridesharing with friends, two null hypotheses against the objective are to be rejected. The first one is that sharing rides only with friends significantly increases detour cost. The second is that the number of successfully matched rides is significantly lower with friends than with anyone. Detour cost and matching rate are influenced by social similarity and spatial distribution of friends. Social similarity contributes to the willingness to share rides with and detour tolerance for a person, while spatial distribution decides detour cost.

Agent-based microsimulation provides a way to cope with the complex behavior process. An agent-based transport simulation is set up using NetLogo assuming that everyone in the study area has a car. The simulation measures the reduction of the total amount of cars by changing the behaviors of private car owners. Based on a regular gridline road network with artificially generated origin/destination points of trips, the model systematically tests different parameters to reach the general conclusions applicable to different urban contexts. The parameters include social network structure (average degree of friendship), spatial distribution of friends (spatially clustered vs. random), and varied tolerance and willingness for different social levels (i.e., direct friends, indirect friends, and strangers). Two populations with small world social network structures are simulated, with 2000 and 5000 agents respectively. There are three matching patterns: (1) any driver and passenger must be direct friends, (2) any driver and passenger must be either direct or indirect friends, and (3) no one has to be friends.

The detour costs of the cross categories of social network structures and matching patterns are calculated for statistical analysis. The analysis contributes three major findings. First, detour cost with friends is not necessarily significantly higher than collaboration with anyone. Especially when friendships are spatially clustered, sharing with friends saves more. Second, even if not excluding strangers, giving priority to friends drastically increases successful matching between friends. Finally, a successful matching rate is not positively associated with the size of choice set. It also depends on the spatial distribution of friendship.

3.4 Using Simulations to Test Optimization Strategies

In order for a DRT service to operate efficiently, an important aspect to consider is the optimization of the vehicle routes. By carefully planning the vehicle routes, the operating cost can be minimized while still guaranteeing a certain level of service quality. Thus, the main use of the optimization is to determine the "sweet spot" of the trade-off between service quality and cost, which eventually leads to an economically justifiable DRT service framework.

The routing problem encountered in a demand-responsive transport is categorized as a Pick-up and Drop-off Problem with Time Windows (PDPTW), where each customer has a load to be picked up and dropped off within the nominated time windows. Occasionally, late arrivals (outside of the time windows) are allowed with some penalties being accrued. The PDPTW is typically considered to be a static problem with customer requests being fixed in advance. A number of solvers have been developed for the PDPTW, including exact solution methods [54–58] and heuristic algorithms [59–70]. When the scale of the problems increased and more layers of complexity are added into the problem, most exact solution methods are computationally expensive and, thus, implementation usually relies on heuristic algorithms [71].

The inevitable challenge when implementing most kinds of heuristic algorithms is the fine-tuning of their parameters. Arguably, the performance of these algorithms is heavily dependent on the selection of these parameters, and a set of parameters can be good for one scenario but not the others. One of the benefits of using a simulation platform is the ability to fine-tune the selection of these parameters prior implementation. In most cases, the methods used to fine-tune these parameters require iterations of specific conditions, which are impossible to perform in practice.

To illustrate this point, an example will be drawn from the algorithm discussed in [72]. The considered algorithm is a two-layered neighborhood search. At the bottom layer, the neighborhood algorithm is used to optimize the route of a vehicle given a certain set of trip requests to be served. At the top layer, the passenger allocation is optimized using another neighborhood search algorithm, requiring multiple call of the lower layer optimization.

In the proposed algorithm, there are two parameters at each layer, defining the neighborhood size and the number of iterations. At each iteration, the algorithm first randomly chooses and evaluates the costs of a number of solutions that are considered as the "neighbors" of the solution at the current iteration. Then, the best neighbor is defined as the solution for the next iteration (this is done even though the best neighbor is worse than the solution of current iteration, which is useful to avoid being trapped at a local extremum) and the process is repeated multiple times according to the predefined number of iterations. The detail of the algorithm is outlined in [72].

The optimization result when using this algorithm heavily depends on these parameters. With higher neighborhood size and number of iterations, it is expected that the algorithm would asymptotically approach the best possible result, at the cost of an increasingly heavy computational burden. Therefore, it is of the operator's interest to select a set of parameters that provides an excellent optimization result with a minimal computational time.

In order to do this, a pilot study can be carried out in a simple simulation platform. As an example, the pilot study is carried out by considering a single vehicle case serving a total of 30 trip requests in a certain predefined network. Figure 1 shows, firstly, the optimization result for various parameters values (note that the neighborhood size is obtained by multiplying the neighborhood size factor and the number of stops in the route, i.e., 60) is shown and, secondly, a similar plot for the total wall time is produced. By superimposing the contour maps of both graphs, it is possible to determine a pair of values for these parameters that produces

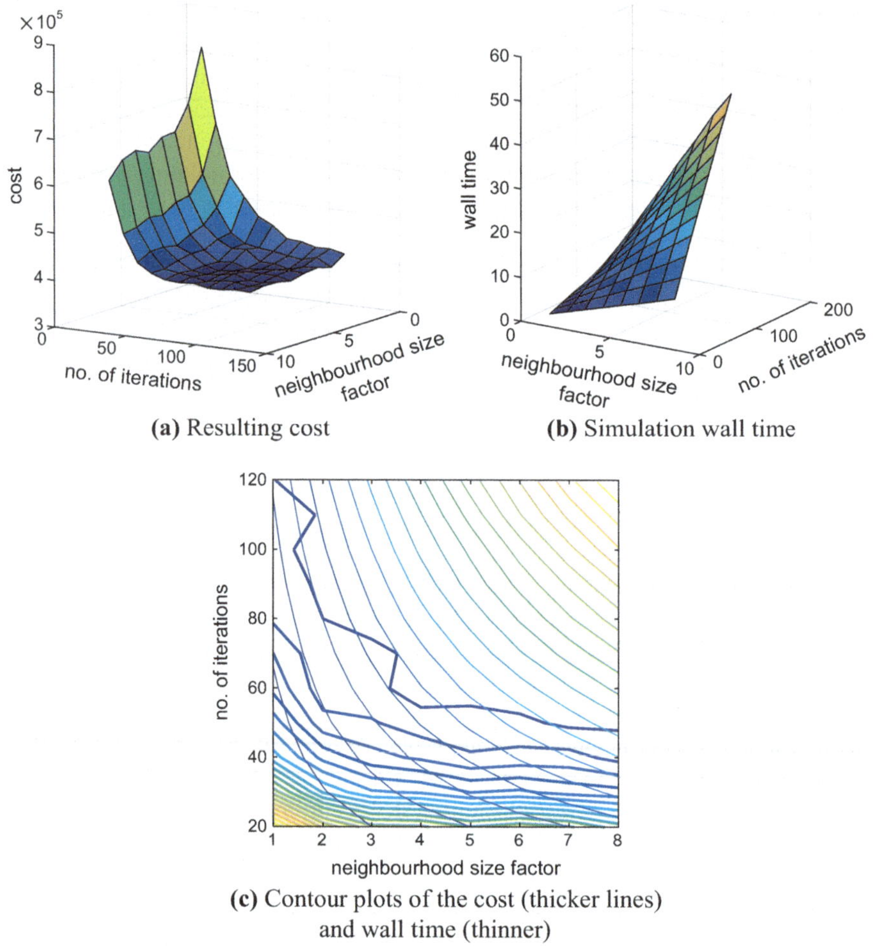

Fig. 1 Results of the fine-tuning of the neighborhood search algorithm

the best result while still minimizing the wall time of the optimization (in this example, via visual observation). Thus, the pilot study enables a convenient way to determine the optimal parameter values, which otherwise is impossible to obtain in practice.

The benefit of using a simulation platform for routing optimization is not limited to the illustrated example. Another important aspect is to decide which algorithms to use. This is especially important in a dynamic vehicle routing problem, where heuristic is the prevalent approach. For instance, in the dynamic case, an optimal solution might become relatively of poor quality when new ad hoc requests emerge. Therefore, many suggest exploiting some known stochastic information about the demand [73–76]. A comparison study can be easily carried out by using a simulation platform. Therefore, this again shows the importance of a simulation platform when deciding the most suitable optimization algorithm in a DRT service.

4 The Role of User Interfaces

Overall, this paper is addressing the impact of a novel mobility service on people's modal choices. But the acceptance in the market does in the end not only depend on the mobility options themselves, but also on the information, and particularly the ease of access to this information about the use of a service. For example, simple payment options can make a service more attractive.

Especially services operating on a demand-responsive basis have to consider the interaction with their users. These users do have to express their demand on the fly, and then to choose from the offerings of the service. This dialog has to happen in the most convenient and intuitive way, which is not trivial from two perspectives. One is the communication about places and times in the world, a known usability challenge for all travel planners. The other is the real-time aspect of this negotiation in ad hoc demand-responsive services.

For example, a service may use predictions of demand in order to balance supply and demand as in Sect. 3.4, but in unexpected situations of poor service, such as during times of high demand or low supply, it may be impossible for a system to satisfy some requests. In such situations, a person would require intuitive information to inform their travel planning alternatives. Here subtle changes to a trip's constraints, that is, flexibility in space or time, may increase the success of getting a ride. For this reason a graphical, map-based user interface that communicates potential pick-up locations by matching a drop-off (Fig. 2) may be used for situation awareness [77]. Such an interface to spatial information may also be used to motivate behavioral change: Compromises in space-time may actually yield greater individual utility due to a range of other affordances like physical exercise or carbon emissions [78].

5 Conclusions and Outlook

This paper collects various approaches to predict human behavior when novel and disruptive forms of transport are added to the complex system of urban mobility. In each case the focus is on long-term impact, not preconceived conceptions. In particular simulation permits the exploration of the detailed operation of disruptive services, and thus also helps understanding the operator emergent patterns prior to the introduction of any new service. As the studied 'disruptive' transportation services are flexible, either in route or time or both, it is more difficult to predict how they will perform compared to scheduled services along given routes. The

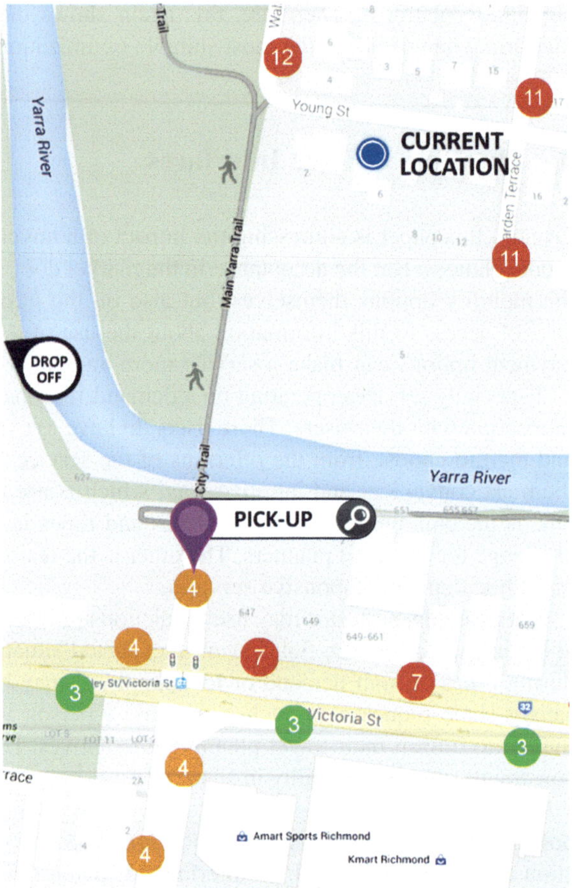

Fig. 2 The ease of use of travel service interfaces can be deciding on market acceptance (base layer © Google Maps 2016)

simulation has to solve the vehicle's scheduling and routing problem, and additionally consider the social, economic and spatial context of people and their mobility demand in order to come to valid conclusions. This paper has put a special effort in explaining how each model has considered this context. It is this context that determines emergent patterns. Without a particular care for this capturing and modeling of the relevant factors of this context the results of simulations would be as random as their assumptions on population, demand or acceptance of pricing. For example, the prediction method based on use patterns (Chapter "Multimodal Transportation Payments Convergence—Key to Mobility") had been successfully validated by applying the same methodology not only to a novel service, but also to an already existing service in a known environment. Since the predicted success of the existing service in the particular area matched the actual usage sufficiently, the methodology can also deliver trustable results for the novel service in the same area.

Furthermore, agent-based microsimulation allows for deep modeling of the individuals' preferences and decision making, which then enables to study this aspect in itself, due to the full control of parameters in such experiments (in contrast to the real world). Expanding our studies in this direction is part of future work. It will include factors such as ad hoc mode choice or modeling at which time in advance individuals make a request for ad hoc transport demand. Again, the challenge will be to identify the relevant parameters for such detailed modeling of behavior and choice.

Simulation as a tool, and context awareness as an indispensable part of a simulation model, is also suited to consider not only single modes for their potential to impact (and potentially disrupt) people's behavior and choices, but integrate these novel modes with other modes and study bigger pictures.

Acknowledgements The authors acknowledge support through the Australian Research Council (LP120200130).

References

1. Nicas, J.: Google takes on uber with new ride-share service. 31 August 2016. www.wsj.com/articles/google-takes-on-uber-with-new-ride-share-service-1472584235. Accessed 1 Sept 2016
2. Shaheen, S.: Mobility and the sharing economy (editorial). Transport Policy (2016)
3. Rayle, L., et al.: Just a better taxi? A survey-based comparison of taxis, transit, and ridesourcing services in San Francisco. Transp. Policy **45**, 168–178 (2016)
4. Shaheen, S., Chan, N., Gaynor, T.: Casual carpooling in the San Francisco Bay area: understanding user characteristics, behaviors, and motivations. Transport Policy (2016)
5. Batty, M.: Cities as complex systems: scaling, interaction, networks, dynamics and urban morphologies. In: Meyers, R.A., (ed.) Encyclopedia of Complexity and Systems Science, vol. 1041–1071. Springer: New York, NY, (2009)
6. Batty, M., et al.: Entropy, complexity, and spatial information. J. Geogr. Syst. **16**(4), 363–385 (2014)
7. Simon, H.A.: Bounded rationality, in utility and probability. In: Eatwell, J., Milgate, M., Newman, P. (eds.) pp. 15–18. The Macmillan Press Ltd, New York, (1990)
8. Benenson, I., Torrens, P.M.: Geosimulation: Automata-based Modeling of Urban Phenomena. Wiley, Chichester, UK (2004)
9. Torrens, P.M.: Geosimulation, automata, and traffic modelling. In: Hensher, D.A., et al. (eds.) Handbook of Transport Geography and Spatial Systems pp. 549–564. Elsevier: Amsterdam (2004)
10. Mokhtarian, P.L., Salomon, I.: How derived is the demand for travel? Some conceptual and measurement considerations. Transp. Res. Part A **35**(8), 695–719 (2001)
11. Axhausen, K.W., Gärling, T.: Activity-based approaches to travel analysis: conceptual frameworks, models, and research problems. Transport Rev. **12**(4), 323–341 (1992)
12. Cottrill, C., et al.: Future Mobility Survey. Transport. Res. Record: J. Transport. Res. Board **2354**, 59–67 (2013)
13. Scholz, R.W., Lu, Y.: Detection of dynamic activity patterns at a collective level from large-volume trajectory data. Int. J. Geogr. Inf. Sci. **28**(5), 946–963 (2014)
14. Louviere, J.J., Hensher, D.A., Swait, J.D.: Stated Choice Methods: Analysis and Applications. Cambridge University Press, Cambridge, UK (2000)
15. Tilahun, N.Y., Levinson, D.M., Krizek, K.J.: Trails, lanes, or traffic: valuing bicycle facilities with an adaptive stated preference survey. Transp. Res. Part A: Policy Pract. **41**(4), 287–301 (2007)

16. Hess, S., Adler, T., Polak, J.W.: Modelling airport and airline choice behaviour with the use of stated preference survey data. Transp. Res. Part E: Logistics Transp. Rev. **43**(3), 221–233 (2007)
17. Contrino, H., McGuckin, N.: Demographics matter travel demand, options, and characteristics among minority populations. Public Works Manage. Policy **13**(4), 361–368 (2009)
18. Kattiyapornpong, U., Miller K.E.: Understanding travel behavior using demographic and socioeconomic variables as travel constraints. In: ANZMAC 2006: Advancing theory, maintaining relevance: Proceedings of the 2006 Australian & New Zealand Marketing Academy Conference: [Queensland University of Technology, School of Advertising, Marketing and Public Relations] (2006)
19. Litman, T.: Understanding Transport Demands and Elasticities (2013)
20. Rasouli, S., Timmermans, H.: Applications of theories and models of choice and decision-making under conditions of uncertainty in travel behavior research. Travel Behav. Soc. **1**(3), 79–90 (2014)
21. ActiveAge, An introduction to Demand Responsive Transport as a Mobility Solution in an Ageing Society, 2008
22. Anspacher, D., Khattak, A.J., Yim, Y.: Demand-responsive transit shuttles: who will use them? Calif. Partners Adv. Transit Highways (PATH) (2004)
23. Bearse, P., et al.: Paratransit demand of disabled people. Transp. Res. Part B: Methodol. **38** (9), 809–831 (2004)
24. Enoch, M.P., et al.: Evaluation study of demand responsive transport services in Wiltshire. Final report, Loughborough University, Loughborough (2006)
25. Häme, L.: Demand-responsive transport: models and algorithms. In: Department of Mathematics and Systems Analysis. Aalto University, Aalto (2013)
26. Koffman, D.: Operational experiences with flexible transit services, vol. 53. Transportation Research Board (2004)
27. Laws, R.: Evaluating Publicly-Funded DRT Schemes in England and Wales, Loughborough University (2009)
28. Lerman, S.R., et al.: A model system for forecasting patronage on demand responsive transportation systems. Transp. Res. Part A: General **14**(1), 13–23 (1980)
29. Maddern, C., Jenner, D.: Telebus mobility and accessibility benefits: final report. In 12th International Conference on Mobility and Transport for Elderly and Disabled transport (TRANSED): Hong Kong (2007)
30. Mageean, J., Nelson, J.D.: The evaluation of demand responsive transport services in Europe. J. Transp. Geogr. **11**(4), 255–270 (2003)
31. Nelson, J.D., Phonphitakchai, T.: An evaluation of the user characteristics of an open access DRT service. Res. Transp. Econ. **34**(1), 54–65 (2012)
32. Rosenbloom, S., Fielding, G.J.: Transit Markets of the Future: The Challenge of Change, vol. 28. Transportation Research Board (1998)
33. Ryley, T.J., et al.: Developing Relevant Tools for Demand Responsive Transport (DRT). In: ATCO Conference, Liverpool (2013)
34. Scott, R.: Demand Responsive Passenger Transport in Low-Demand Situations December 2010 (2010)
35. Spielberg, F., Pratt, R.H.: Demand-Responsive/ADA-Traveler Response to Transportation System Changes (2004)
36. Wang, C., et al.: Multilevel modelling of demand responsive transport (DRT) trips in greater Manchester based on area-wide socio-economic data. Transportation **41**(3), 589–610 (2014)
37. Victorian Integrated Survey of Travel & Activity 2009–10, Survey Procedures and Documentation, The Victorian Department of Transport (2011)
38. Public Transport Victoria, New PTV FlexiRide service for Yarrawonga and Mulwala: Melbourne, Australia, 2 (2013)
39. Ronald, N., Thompson, R.G., Winter, S.: A comparison of constrained and ad-hoc demand-responsive transportation systems. Transp. Res. Rec. **2536**, 44–51 (2015)

40. Ronald, N., Thompson, R.G., Winter, S.: Modelling ad-hoc DRT over many days: a preliminary study. In: 21st International Congress on Modelling and Simulation (MODSIM): Gold Coast, Qld, Australia, pp. 1175–1181 (2015)
41. NavidiKashani, Z., Ronald, N., Winter, S.: Comparing demand responsive and conventional public transport in a low demand context. In: First International Workshop on Context-Aware Smart Cities and Intelligent Transport Systems, Sydney (2016)
42. Daganzo, C.F.: Checkpoint dial-a-ride systems. Transp. Res. Part B: Methodol. **18**(4–5), 315–327 (1984)
43. Diana, M., Quadrifoglio, L., Pronello, C.: Emissions of demand responsive services as an alternative to conventional transit systems. Transp. Res. Part D: Transport Environ. **12**(3), 183–188 (2007)
44. Diana, M., Quadrifoglio, L., Pronello, C.: A methodology for comparing distances traveled by performance-equivalent fixed-route and demand responsive transit services. Transp. Plann. Technol. **32**(4), 377–399 (2009)
45. Edwards, D., Watkins, K.: Comparing fixed-route and demand-responsive feeder transit systems in real-world settings. Transp. Res. Record: J. Transp. Res. Board **2352**, 128–135 (2013)
46. Chang, S., Yu, W.J.: Comparison of subsidized fixed-and flexible-route bus systems. Transp. Res. Record: J. Transp. Res. Board **1557**, 15–20 (1996)
47. Quadrifoglio, L., Li, X.: A methodology to derive the critical demand density for designing and operating feeder transit services. Transp. Res. Part B: Methodol. **43**, 922–935 (2009)
48. Beirão, G., Sarsfield, J.A.: Cabral, Understanding attitudes towards public transport and private car: A qualitative study. Transp. Policy **14**(6), 478–489 (2007)
49. Hensher, D.A., Stopher, P., Bullock, P.: Service quality—developing a service quality index in the provision of commercial bus contracts. Transp. Res. Part A: Policy Pract. **37**(6), 499–517 (2003)
50. Santi, P., et al.: Quantifying the benefits of vehicle pooling with shareability networks. Proc. Natl. Acad. Sci. **111**(37), 13290–13294 (2014)
51. Amey, A.M.: Real-time ridesharing: exploring the opportunities and challenges of designing a technology-based rideshare trial for the MIT community, Massachusetts Institute of Technology (2010)
52. Chaube, V., Kavanaugh, A.L., Perez-Quinones, M.A.: Leveraging social networks to embed trust in rideshare programs. In: System Sciences (HICSS), 2010 43rd Hawaii International Conference on: IEEE (2010)
53. Wessels, R.: Combining Ridesharing and Social Networks, UTwente (2009)
54. Baldacci, R., Bartolini, F., Mingozzi, A.: An exact algorithm for the pickup and delivery problem with time windows. Oper. Res. **59**, 414–426 (2011)
55. Desrosiers, J., Dumas, Y., Soumis, F.: A dynamic programming solution of the large-scale single-vehicle dial-a-ride problem with time windows. Am. J. Math. Manage. Sci. **6**, 301–325 (1986)
56. Psaraftis, H.N.: Scheduling large-scale advance-request dial-a-ride systems. Am. J. Math. Manage. Sci. **6**, 327–367 (1986)
57. Zhou, J.: Routing by mixed set programming. In: Proceedings of the 8th International Symposium on Operations Research and Its Applications, pp. 155–166 (2009)
58. Ropke, S., Cordeau, J.F.: Branch and cut and price for the pickup and delivery problem with time windows. Transp. Sci. **43**, 267–286 (2009)
59. Badaloni, S., et al.: Addressing temporally constrained delivery problems with the swarm intelligence approach. Intell. Auton. Syst. **10**, 264–271 (2008)
60. Bent, R., Hentenryck, P.V.: A two-stage hybrid algorithm for pickup and delivery vehicle routing problems with time windows. Comput. Oper. Res. **33**, 875–893 (2006)
61. Gronalt, M., Hartl, R.F., Reimann, M.: New savings based algorithms for time constrained pickup and delivery of full truckloads. Eur. J. Oper. Res. **151**, 520–535 (2003)

62. Hasle, G., Kloster, O.: Industrial vehicle routing. In: Hasle, G., Lie, K.A., Quak, E. (eds.) Geometric Modelling, Numerical Simulation, and Optimization, pp. 397–435. Springer, Berlin (2007)
63. Hosny, M., Mumford, C.: New solution construction heuristics for the multiple vehicle pickup and delivery problem with time windows. In: Proceedings of the Metaheuristic International Conference (2009)
64. Huang, Y.H., Ting, C.K.: Ant colony optimization for the single vehicle pickup and delivery problem with time window. In: Proceedings of the International Conference on Technologies and Applications of Artificial Intelligence (2010)
65. Koning, D.: Using column generation for the pickup and delivery problem with disturbances, Utrecht University (2011)
66. Lu, Q., Dessouky, M.M.: A new insertion-based construction heuristic for solving the pickup and delivery problem with time windows. Eur. J. Oper. Res. **175**, 672–687 (2006)
67. Nagata, Y., Kobayashi, S.: A memetic algorithm for the pickup and delivery problem with time windows using selective route exchange crossover. In: Proceedings of the 11th International Conference on Parallel Problem Solving from Nature, pp. 536–545 (2010)
68. Pankratz, G.: A grouping genetic algorithm for the pickup and delivery problem with time windows. OR Spectr. **27**, 21–41 (2005)
69. Ropke, S., Pisinger, D.: An adaptive large neighborhood search heuristic for the pickup and delivery problem with time windows. Transp. Sci. **40**, 455–472 (2006)
70. Xu, H., et al.: Solving a practical pickup and delivery problem. Transp. Sci. **37**, 347–364 (2003)
71. Pillac, V., et al.: A review of dynamics vehicle routing problems. Eur. J. Oper. Res. **225**, 1–11 (2013)
72. Kutadinata, R., Thompson, R., Winter, S.: Cost-efficient co-modal ride-sharing scheme through anticipatory dynamic optimisation. In: Submitted to the 23rd World Congress on Intelligent Transport Systems (2016)
73. Bent, R., Van Hentenryck, P.: Scenario-based planning for partially dynamic vehicle routing with stochastic customers. Oper. Res. **52**, 977–987 (2004)
74. Gendreau, M., et al.: Parallel tabu search for real-time vehicle routing and dispatching. Transp. Sci. **33**, 381–390 (1999)
75. Gendreau, M., Laporte, G., Séguin, R.: Stochastic vehicle routing. Eur. J. Oper. Res. **88**, 3–12 (1996)
76. Yang, W., Mathur, K., Ballou, R.: Stochastic vehicle routing problem with restocking. Transp. Sci. **34**, 99–112 (2000)
77. Rigby, M., Winter, S.: Enhancing launch pads for decision-making in intelligent mobility on-demand. J. Location Based Serv. **9**(2), 77–92 (2015)
78. Broll, G., et al.: Tripzoom: an app to improve your mobility behavior. In Proceedings of the 11th International Conference on Mobile and Ubiquitous Multimedia (MUM '12), pp. 57: 1–57:4. ACM, New York, NY, USA (2012)

Transit Leap: A Deployment Path for Shared-Use Autonomous Vehicles that Supports Sustainability

Bern Grush and John Niles

Abstract The concept of Transit Leap is introduced and explained as robotic, shared-use, multi-passenger vehicle applications that start small, expand by demand, merge, and spread. It is an approach to deploying automated vehicles that is meant to blunt the long-established worldwide trend of ongoing increases in the number of private vehicles. Transit Leap is an alternative to year-by-year automotive feature creep, which is currently the most likely path to ubiquitous robotic mobility, absent public policy intervention. Transit Leap helps bypass the interim challenges of semi-autonomous and mixed-autonomy scenarios, and supports equity in mobility, as well as environmental quality and the financial viability of public transit networks.

Keywords Autonomous vehicles · Disruptive innovation · Driverless · Labor disruption · Mobility as a service · Public transit · Public–private partnerships · Self-driving · Service deployment · Technology forecasting · Transit Leap · Transportation as a service · Transportation policy · Urban economics · Vehicle automation

1 Introduction

Two popular and somewhat utopian views of the future of the self-driving automobile are shaped by vehicle ownership. One is the consumer-friendly, extra safe, super convenient, congestion-busting, personally owned household vehicle that requires no attention to operate and smoothes out traffic flows with tight vehicle spacing and no collisions. The other is an on-demand commercial robo-cab that

B. Grush
Grush Niles Strategic, 515 Rosewell Avenue, Suite 504, Toronto, ON M4R 2J3, Canada
e-mail: bgrush@endofdriving.org

J. Niles (✉)
Grush Niles Strategic, 4005 20th Ave West, Suite 111, Seattle, WA 98199-1290, USA
e-mail: jniles@endofdriving.org

G. Meyer and S. Shaheen (eds.), *Disrupting Mobility*,
Lecture Notes in Mobility, DOI 10.1007/978-3-319-51602-8_17

rolls up to wherever you are within a minute of your request via smartphone and zips you to exactly where you told the app you want to go.

How can urban regional governments and nations prepare for one or both of these scenarios as vehicles become more numerous?

1.1 World Vehicle Growth Is High and not Slowing

In 1995 about 625 million vehicles moved about the planet [1]. Twenty years later, this number doubled to 1.2 billion. Despite all the benefits of mobility, the impact of congested streets and highways from growing vehicle counts in an increasingly urbanizing world is widely recognized to be problematic. As *The Economist* magazine states dramatically in a global overview:

> Megacities are seizing up. Surveys of São Paulo suggest that half of all adults spend at least two hours a day traveling. Lagos has such epic traffic jams that an army of street hawkers plies the roads, selling peanuts, Christmas trees and puppies to a captive market of drivers [2].

Traffic measurement in USA and Europe [3] reveals the same growth of grid-lock. Recognition of this issue is hardly new. In 2009 transportation scholars Dan Sperling and Deborah Gordon published *Two Billion Cars*, a book detailing reasons to address the growing populations of motorized vehicles [4]. The book's final chapter, "Driving Toward Sustainability," listed 16 policy initiatives to ensure that the two billion vehicles projected for 2030 might have a lighter footprint on the planet. Thirteen of these were directed at alternative fuels and fuel economy, while three focused on reducing vehicle usage. The automated vehicle had not yet mainstreamed, so the book missed it. But now it has become a factor to be considered.

Yet two billion cars are too conservative. In 2013 Bill Ford speaking at the annual Milken Global Conference projected four billion vehicles by 2050, echoing the 2007 analysis by Joyce Dargay at The University of Leeds [5]:

> By 2050, the population is expected to be around 9 billion people. With most of this growth happening in major cities, some 4 billion cars are expected to be on the road by then. If we continue on the path we're on, Ford said, the result will be what he called "global gridlock [6].

An even more startling projection from Dargay: the global vehicle population by 2100 will finally saturate at close to eight billion vehicles. This may seem like scare mongering, but it is the predictable future given the ongoing, rapid, worldwide growth in the automobile population, despite the temporary plateau in automobile use in recent years in developed countries, such as Australia, United Kingdom, and United States.

This chapter suggests a path for automobility that can satisfy the projected demand for worldwide personal, motorized mobility in 2050 with just one billion vehicles, and by 2100 with fewer than two billion. By adding automation concepts to the ideas for more environmentally friendly fuels and trip reduction from Sperling and Gordon, humans can achieve automotive sustainability.

1.2 Humans Will Continue to Demand Motorized Mobility

A significant, per capita, reduction in car travel around the world is unlikely. There is much to commend in efforts to promote walkable and bicycle-friendly communities as a way to reduce driving. Efforts to configure streets to be more complete by including walking and biking as well as motoring are now visible and encouraging. This effort with motor vehicle controls can result in car-free or at least car-limited local environments. But consumption of motorized automobility is a force in the wider world of intercity travel that cannot be turned back.

Powered automobility—provided first by animals—has a 7500-year history supporting a wired-in socio-biological preference that cannot be extinguished. We consider both horseback riding and electric bicycles as forms of powered automobility, although the four-wheel motorized living-room called a car that keeps the traveler dry in the rain is where technology has led us. Individualized decisions as to trip timing, destinations, and route choice are central to powered automobility. A personal, motorized mode is preferred by most humans and in most travel circumstances [7].

The nominal path to the worldwide future is growth in demand for motorized vehicle mobility continuing until a natural saturation of vehicle ownership and use is reached at a level calibrated to wealth. As human population settles at 11 billion over the next century, the trend of vehicle population, pushed by gradual increase in human wealth that correlates with smaller family sizes, will continue to approach eight billion, where it should finally plateau.

Global experience to date shows that all human populations strive in the long run toward the level of automobility achieved in the developed world. The USA is one of a handful of countries leading the trend to ownership saturation, and the current American level of consumption of vehicle miles traveled is indicative of a human tendency rather than a uniquely American behavior.

Current automobility is provided principally by unnecessarily large, collision-prone, and pollution-emitting vehicles equipped with internal combustion engines, even as electric propulsion is now gradually gaining market share. The overuse and abuse of these vehicles has proven difficult to manage; and the problem has largely resisted proposed remedies to date at a scale that is meaningful to the future of nations and the planet. The sheer number of vehicles is a primary characteristic of the issue.

2 Could Shared Fleets Dominate Future Automobility?

The jury is out on the long-run preference for vehicle ownership as motor vehicle automation grows, even though hope-filled forecasts describe a utopia of wide-spread, electric, crash-proof robo-cab vehicles, kept in constant use, shared by urban customers for all trip types.

Will most automated vehicles be owned as family vehicles are now? Or will the advantages of shared fleets be available and selected by the great majority of travelers so that the population of household vehicles shrinks dramatically as decades pass—by 90% according to the most optimistic projections? Will the car become more of a travel service and less of an accessory—i.e., all about the trip, nothing about status?

Many academics and consultants are on record predicting "few people will own automated vehicles; most will share them," but there are many reasons—rational or otherwise—why most people reveal a preference for ownership, even while a tiny-but-growing few have found ways to avoid owning a vehicle. The backdrop of culture, habit, status, privacy, and convenience of owning can be stacked against the rational, economic notions of sharing and is used very effectively by automotive marketers. That started long ago.

Zipcar founder Robin Chase has said "no sane person would own a car" when they become automated, but what she says about the non-automated cars of today is nuanced with conditions. Says Chase, a champion of the sharing economy, sharing of non-automated vehicles already offers a distinct advantage over traditional car ownership. "If you are financially smart and you are living in the city and you don't need a car to get to work, you are insane to own one," she says, "You will always be saving money by renting them when you need them" [8].

In support of sharing as the rational choice after automated vehicles grow in capabilities, some simulation-based research has been generated for cities such as Austin [9], Lisbon [10], Manhattan [11], Stockholm [12], and others. Consistently, these researchers find that each simulated automated vehicle can replace about ten current, family-owned vehicles (that's where the above "90%" comes from). But these simulations are realistic only in a constrained context. They have been parameterized using the origin–destination data collected in the simulated cities, but in most cases the researchers imply or reviewers conclude that such figures can be extrapolated to the world vehicle population. Ronald Bailey writes [13]:

> Researchers at the University of Texas, devising a realistic simulation of vehicle use in [Austin] that took into account issues like congestion and rush-hour usage, found that each shared autonomous vehicle could replace eleven conventional vehicles. Notionally then, it would take only about 800 million vehicles to supply all the transportation services for 9 billion people. That figure is 200 million vehicles fewer than the current world fleet of 1 billion automobiles.
>
> In the Texas simulations, riders waited an average of 18 s for a driverless vehicle to show up, and each vehicle served 31 to 41 travelers per day. Less than half of one percent of travelers waited more than five minutes for a vehicle. In addition, shared autonomous vehicles would also cut an individual's average cost of travel by as much as 75 percent in

comparison to conventional driver-owned vehicles. This could actually lead to the contraction of the world's vehicle fleet as more people forgo the costs and hassles of ownership.

There are several problems that often arise with these simulations and the conclusions drawn from them. These studies, constrained by the availability of useable origin--destination data, often propose unwarranted generalizations that cannot be reasonably extrapolated to suburbs and rural areas or work-/service-related vehicles. Extrapolations such as echoed by Bailey may also assume an inevitable and general willingness of all or most travelers to use shared vehicles. While there is much good to be said for a sharing economy, there is no evidence that most humans will engage this way. Barriers are easily found, including social reasons related to privacy, health, and status.

We can find ways to overcome some of these barriers, but it will not "just happen." Humans make many non-rational decisions based on personal, contextual or experiential criteria. In the coming decades, the success of the massive shared fleets these researchers simulate will depend more on revealed preferences and behavioral economics than on the capability of the artificial intelligence software controlling the cars.

An analogy to this occurred 110 years ago, when the car was hailed as the solution to the horse problem, characterized by the stink, flies and disease from manure on city streets. Society dove headlong into full-bore automobile-centric planning and automobile user-preference as horses were pushed out of our cities. There was neither understanding nor mitigation of the eventual global effects, as illustrated in Fig. 1. One-hundred and ten years ago people were largely unable to foresee these effects and paid little attention to the few warnings on offer. Of course humanity is free to repeat this error, and the likelihood of doing so is high, especially as there are payoffs from status quo business models and ownership-thinking.

Robotic vehicles have the potential to make our problems worse, especially congestion, sprawl, and a demand for yet more traditional infrastructure such as

1890:
Save us from horses!

2010:
Save us from cars!

Photo source: Washington Museum of History and Industry

Photo source: Wikipedia Commons

2050: Save us from automated vehicles!

Fig. 1 Will History repeat itself?

roads and parking facilities. They may wipe out any residual value in financially stressed public bus systems. As well, they would tend to entice away from transit those people who own cars but choose to ride transit. Sam "Gridlock" Schwartz warns:

> It can become a vicious circle: the more transit becomes dominated by less affluent people, the more it becomes associated with poverty. And the more it gets associated with poverty, the less appealing it becomes for the affluent. Equity declines [14].

3 Municipal and Regional Governments Can Respond

Regional and local governments could start now to develop policy direction that is more likely to make a desirable outcome of more sharing prevail than just waiting and hoping would do.

If local jurisdictions wait-and-see, they risk the consequences of being swept up by exponential innovation. Governments find private sector innovation hard to track, regulate, and manage. Uber is giving regulators headaches in 2013–16, but the disruption to be wrought by robotics in 2035 will reverberate far more dramatically. Picking winners may work temporarily, but public jurisdictions are prone to commit to consumer choices of the moment and stick with them for too long before being swept away by the next unanticipated innovation. The 20-year transportation future that starts now is harder to predict than any prior 20-year future since 1908, when Ford introduced the Model T.

The only way to escape this conundrum is to innovate and integrate to seek a better way through the technology tsunami. Government agencies must complement traditional notions of infrastructure to go far beyond physical facilities to encompass the methods, business models, vehicle access and use models, data, and labor models that create transportation value. Road surface, train tracks, heavy transit vehicles, schedules, and routes will soon explain less and less of the total picture.

Local governments that fight commercial, robotic, shared fleets—like some fight Uber—will lose. Without paid drivers, the cost per passenger kilometer in flexible, driverless vans, minibuses, and robo-taxis will be a fraction of the cost per passenger kilometer in today's municipal buses, compelling a fleet change based on economics.

Some pundits propose that cities or states set up testing grounds to be leaders and promoters for technology development. Why? City governments do not test pharmaceuticals. States and Provinces do not test new nanotechnologies. Why should they test robotic vehicles? Let corporations and existing standards bodies do that. Instead, have technology suppliers prove the technology.

As their contribution to the future, local authorities should begin thinking through how their communities can encourage or orchestrate the building of large shared, robotic fleets using public–private partnership approaches. Local authorities should plan to disrupt their own public transit agencies head-on as a pathway to creating public robotic fleet services in a way that ensures equitable access for every citizen. Equity is a concept dangerously missing from the young, mid-chic, urban middle-class business model of today's transportation network companies based on smartphone apps [15].

Today, at a time when robotics are still not ready to take over, cities could create the preconditions for the equitable future they want to create.

4 Evolution Toward Robo-Cars

Private ownership of automated vehicles will lead to predictable large vehicle counts, a continuation of present trends, just with more and better technology features. Since newer vehicles with automation would eventually not require a licensed operator, a large cohort of young, old or disabled passengers could then utilize a dedicated vehicle without a chauffeur. Hence some families will see owning one or more additional vehicles as a very rational decision. The marketing forces of the automotive industry will always prefer a high-volume, well-featured consumption model stoked by year-over-year improvements rather than a shared-vehicle model, even while responding to whatever opportunities exist to sell into shared fleets.

Will autonomous vehicles be gradually mixed in with human-operated vehicles or will they be somehow isolated to carefully constrained, perceptually safer applications? There are many operational, social, and liability complexities involved in freely mixing driver-out and driver-in vehicles on the same roadway—the biggest elephant in this room being distracted driving [16].

As this myriad of problems becomes solvable, traditional automotive manufacturers will prefer the mixed-driving model, since it justifies many years of new safety, intelligence, infotainment, status, and convenience features, while nurturing an ongoing preference for ownership. Vehicle turnover is king. Car companies are likely to continue using marketing techniques based on behavioral economics in every conceivable way. And they will mine the rich marketing opportunities across the full spectrum of partial-to-complete robotic enablement, making driving in congestion more comfortable, and taking advantage of the cultural predilection for "my car, my style, my way" [17].

Using increasing automation as a generator of new and compelling features for each model year is an example of the consumer product upgrade practice known as "feature creep." Clearly, automation and safety-related features should not be disparaged, yet the same year-over-year business model of incrementalism stokes consumer envy and sustains sales. Traditional manufacturers are not likely to readily abandon this underlying success formula for creating consumer demand and maintaining competitive advantage.

But new players such as Google that promise full, driver-out robotics sooner than the traditional players see evolutionary feature creep as unworkable in the early stages. Executive Astro Teller of Alphabet, the Google business division overseeing its automated vehicle, in a March 2015 keynote address at the South by Southwest Interactive in Austin, Texas, said this best:

> Even though everyone who signed up for our [self-driving car] test swore up and down that they wouldn't do anything other than pay 100 percent attention to the road, and knew that they'd be on camera the entire time...people do really stupid things when they're behind the wheel. They already do stupid things like texting when they're supposed to be 100 percent in control...so imagine what happens when they think "the car's got it covered." It isn't pretty. Expecting a person to be a reliable backup for the system was a fallacy. Once people trust the system, they trust it. Our success was itself a failure. We came quickly to the conclusion that we needed to make it clear to ourselves that the human was not a reliable backup—the car had to always be able to handle the situation. And the best way to make that clear was to design a car with no steering wheel—a car that could drive itself all of the time [18].

This and the recent stories about Tesla Level 2 drivers [19] suggest that feature creep will fail as a path to vehicles becoming fully automated. Well before creeping toward driver-out, a jump to full autonomy will be demanded. But it is obvious that society cannot move to pervasive road robotics quickly; it will most probably have to creep. Teller's comment also predicts problems for mixing autonomous and non-autonomous vehicles. Until February 2016 [20], collisions involving Google's autonomous vehicle operations have been blamed on drivers of non-autonomous vehicles, who mostly rear-ended Google's cars. It may be that autonomous cars conform to speed limits more consistently or tend to stop more frequently or more suddenly than do human-controlled cars [21].

5 Transit Leap

If driver-in/driver-out mixing is going to be problematic, it would make much more sense to put robotic vehicles to work earlier in constrained, less-mixed applications. Instead of waiting for the fortunes of evolution, leaders should promote an intentional revolution.

For this path to the future, we introduce the concept of *Transit Leap*, which initially focuses on lower risk, partially isolated applications with which to begin cautiously. They start out highly constrained and incorporate extreme oversight, before branching out and eventually disrupting traditional transit. The EU's CityMobil2, a small-vehicle (minibus) demonstration in several cities, is an early example, with vehicles such as shown in Fig. 2. Notably, these trials included research into financial, cultural, and behavioral aspects as well as effects on land use policies and how the new systems mesh with existing infrastructure [22].

There are many other smaller scale, spatially constrained applications available for Transit Leap, such as military, university, and employment campuses. Parking lot shuttles at airports could be serviced by 10- or 12-passenger vehicles running at

Fig. 2 Transit Leap in Europe (*Photo source* CityMobil2)

modest speeds on clearly marked lanes and tightly constrained to regular service on regular routes. Retirement communities could use such vehicles for local on-demand trips including for shopping, entertainment, and worship, with the vehicles beginning to determine best routes, rather than being constrained to fixed routes.

Human attendants, initially in place to provide continuity and comfort to early users and to help address changing labor demands, would be eased out gradually. Such applications are numerous, can start almost immediately, and can be gradually expanded to include longer routes, allowances to handle passenger requests by smartphone (more like a jitney than a shuttle), and to increase route flexibility, length, and detail.

Urban areas could begin with short and simple bus routes at low speeds on constrained lanes at grade and without barriers, treated like reserved bicycle lanes. If adjacent lanes carry driver-in motorcars, they would be traffic-calmed. These city systems would benefit from the experience of the earlier parking shuttles, campus applications, and the retirement communities. This would encourage a degree of local government interest in supporting these earliest systems. City transit routes could expand in number, distance, and flexibility until transit is dominated by multi-sized autonomous vehicles and each is tailored—i.e., scaled to purpose [23]. During the latter half of this shift, true robo-taxi services could be phased in and would merge so that robo-taxi and robo-transit offer a continuous service spectrum.

Figure 3 shows five levels of Transit Leap across five stages of spatial reach, each absorbing the prior stage, and eventually blending into spatially continuous, fully pervasive automation over increasingly larger areas until they all interconnect. Level 1 starts with very small, independent local applications and ends at Level 5—

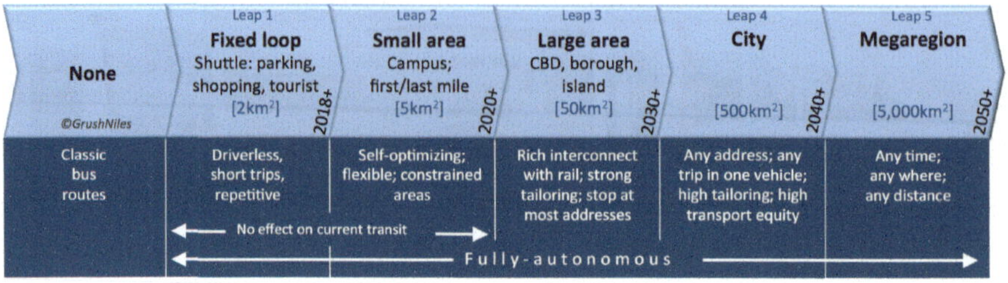

Fig. 3 Five levels of Transit Leap: add by spatial aggregation; encourage transit use, lower ownership and higher density (Copyright, Grush Niles Strategic.)

essentially nationwide. This would take 30–40 years, the same amount of time it took the motorcar to completely displace the horse.

This is distinct from the five Society of Automotive Engineers (SAE) levels of feature creep yielding increasing levels of automation: [L1] driver assistance, [L2] partial automation, [L3] conditional automation, [L4] high automation, and [L5] full automation. The major difference is that the vehicles deployed in the gradually expanding, ridership-growing, spatial applications of Transit Leap would all be SAE level 5 from the outset. The constrained, protected spatial applications allow greater technical autonomy, turning in-vehicle operators into trip assistants and guides from day one.

5.1 Transit Leap and Mobility Digitization

It is important to the workability of Transit Leap that it fits into the larger picture for the direction that society and technology is taking.

As we enter the era of *mobility digitization* the movement of people and goods will experience the effects of digital technologies similar to those we have seen for music, print, broadcast, hotels, entertainment, and hundreds of other aspects of human activity. Part of this will be a move away from ownership toward *usership*—the buying of more trips and fewer vehicles may be expected, or at least hoped for. We can expect an untold number of innovations that will result in new entrepreneurial activity and commercial choices for mobility. These activities and choices can be leveraged to the benefit of cities and urban transit. Transit Leap is one instance of such a lever.

The first stage of mobility digitization was the aggregation of hundreds of thousands of part-time drivers and their underutilized cars by *Transportation Network Companies* (TNCs) such as Uber and Lyft. The second stage, *Mobility as a Service* (MaaS), strengthens that capability by aggregating all forms of transportation—cars, buses, taxis, subway, streetcars, bicycles, carshares, motorbikes—into a single app. MaaS, providing trip coherence with minimal hassle and without car ownership, has already débuted. Maas Global launched *Mobility as a Service* in four cities in Scandinavia in 2016 and expects to launch in more cities soon. MaaS is an instance of the *Mobility Internet*.

These first two app-based mobility digitization technologies tend to reduce the need for car ownership, but the most far-reaching of all digital mobility technologies is vehicle automation. Robotics, far more than just an issue of safety and convenience, is a powerful optimizer of time, space, human attention, and energy. Vehicle robotics will change fundamentally how, why, and how much we travel. It will influence how we sprawl. It will tend to flatten our cities.

Some formats of vehicle automation are expected to reduce vehicle ownership while increasing trip counts and trips lengths, while other formats and circumstances will increase the demand for vehicle ownership. This contradiction alone—driven as much by behavioral economics and choice availability as by technology maturity—will cause more uncertainty in planning and infrastructure over the next few decades than any other single factor of mobility digitization.

Technologies for mobility digitization cannot be stopped, yet surprises in the scope and results of specific developments around the world are already common. Inability to predict the timing, direction, and effects of vehicle automation and mobility digitization now has become the single most troublesome aspect of urban and regional infrastructure planning.

Transit Leap is a deployment system for mobility digitization rather than one of its fundamental technologies. Its value lies in its ability to channel fast-arriving technology developments into *publically available* mobility services. Such services as we described above can preserve and enhance urban transit's roles supporting transportation equity, urban planning, employment skill-mix transitions, congestion abatement, safety, and building livable communities. Critically, Transit Leap can maximize the capability and value of planning in the face of dramatic change.

6 Conclusion: Innovation and Integration

Public jurisdictions can now reasonably begin the process of deciding how robotic mobility technology is to be deployed. They could use vehicle control automation technology to completely transform surface transportation from transit that is cripplingly expensive and used across all trip types for only 5–7% of passenger kilometers in the US and Canada. Shared vehicles (taxicabs, transportation network companies, and carshares), although growing in number now, still produce statistically miniscule passenger kilometers on a North American basis.

Setting and beginning work on a long-run target of 80% of all passenger kilometers to be traveled in shared vehicles—i.e., vehicles that belong to public, private, or co-op fleets and that are busy from 40 to 80 h per week instead of only eight or nine hours—should motivate an urban region to the point where a community of business and government leaders could begin to innovate just how such a fleet could be financed, maintained, managed, and priced. Leaders could begin to

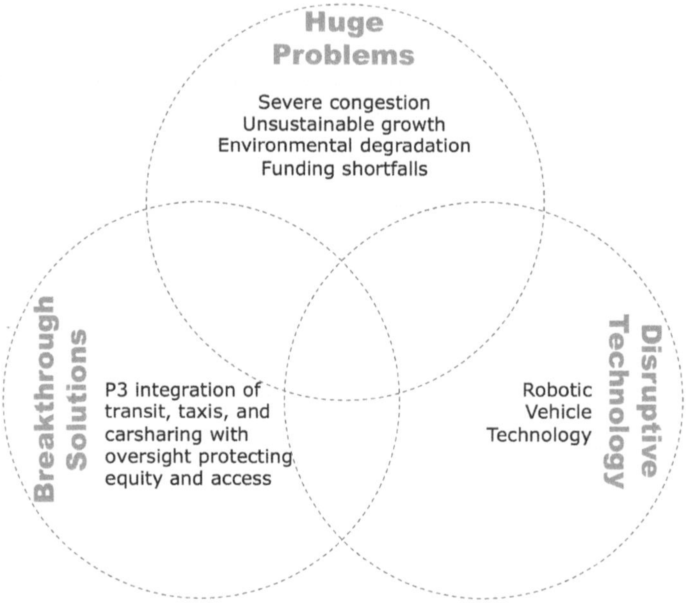

Fig. 4 Public–Private partnerships for innovation

figure out how to park this fleet off peak, how to power it, and how to re-purpose liberated parking areas.

Real estate interests in the community could begin a process to decide how to turn parking garages to other uses or parking lots into parks or building sites. Planners and public works departments might convert street parking into bicycle paths. If community leaders on a market-wide scale do not set such an assertive target and push to implement it, automotive manufacturers will inevitably continue to operate a high, personal-vehicle consumption model for the world's cities.

Public–private partnerships (P3s) for innovation could create opportunities for regions to ensure access and equity to all as well as enormous opportunities for manufacturing and jobs. The approach for moving forward is shown in Fig. 4, derived from the above mentioned online presentation by Goggle (Alphabet) [24].

Urban leaders focused on transportation systems should start thinking now who would be best to deploy such fleets. Which kinds of organizations? With the present pace of technology development, it is not too early for civic leaders to begin forums to discuss the incentive and regulatory structures that would fit community values. Ideas should be considered regarding ownership models. The alternatives of fleets owned and managed by large corporations à la the Wal-Mart company-owned store model or franchised as family-run fleet clusters on the McDonalds model should be put into planning scenarios. Universities and professional groups should be asked to think about a future role in sponsoring affinity fleets run by co-op transportation operators. What kinds of government guides for pricing, service, and response times will be needed to maintain equity, or can the forces of the competitive market include motivations for sustainable social and transportation equity in some unexpected ways? All this, and more, is worth discussing now in government-business forums.

7 Recommendation: Avoiding Ugly Disruption

The difference between the incremental feature-creep model normally pursued by auto manufacturers and the disruptive model of moving directly from driving to not driving as is being pursued by Google, the European Union, and others is the important difference that addresses the vehicle autonomy-mixing. There are numerous problems of mixing robotic-driven and human-driven vehicles at any ratio—whether 1, 50 or 99%. Adopting an incremental, gradual, mixed-traffic model would lead to years of contention regarding traffic rules, overly cautious robotics, insurance liability, ethics conundrums, and new legions of distracted drivers using robotics that operate for 90 or 99 but not 100% of a trip.

If urban leaders emphasized full-solution, Transit-Leap innovation instead of preparing or waiting for household-vehicle feature creep from the auto industry, less contentious, incremental improvements would likely emerge.

If public agencies used innovative business and financing models to replace and grow public transit passenger kilometers in increments safe for passengers and in ways that allow for thoughtful mitigation of inevitable labor disruptions, the subsidization burden of transit could shrink, and ridership could grow. Travelers would be attracted out of household vehicles, expanding the ways TNCs do this now. Rather than resisted, TNCs should be regulated and integrated into a new hybrid solution of privately operated fleets governed for accessibility and equity. However much Lyft and Uber may be good for young, car-less travelers in our cities, TNC services are not designed to be available to low-income, digitally and economically disadvantaged travelers. Cities have a critical role to ensure access and equity, especially as current transit methods and technologies are disrupted.

With multiple service levels keyed to variables such as vehicle age, ride features, number of stops, ride sharing, convenience, comfort, and more, a range of prices can be supported to be affordable for all users. There are ways with very little subsidy to have transportation available to everyone at a level affordable for each.

Transportation leaders in the developed world where private automobile ownership is the highest should lead the way in establishing a clear direction toward a preferred future of massively used, massively shared robotic fleets. Benefits will come from demonstrating sustainable, environmentally friendly models of urban style and fashion for the rest of the world to imitate as vehicle automation emerges in practice.

References

1. Dargay, J., Gately, D., Sommer, M.: Vehicle ownership and income growth, worldwide: 1960–2030. Energy J. **28**, 163–190 (2007)
2. The Economist, Jam today, 27 February 2016. www.economist.com/news/leaders/21693577-get-worlds-biggest-cities-moving-stop-subsidising-driving-jam-today. Accessed 29 Feb 2016
3. INRIX, Urban Mobility Scorecard Annual Report (2015). http://inrix.com/scorecard. Accessed 29 Feb 2016
4. Sperling, D., Gordon, D.: Two Billion Cars. Oxford University Press, New York (2009)
5. Dargay, J., Gately, D., Sommer, M.: Vehicle ownership and income growth, worldwide: 1960–2030. Energy J. **28**, 163–190 (2007)
6. Undercoffler, D.: Bill ford: the future if self-driving cars is closer than you think, Los Angeles Times, 30 April 2013. http://articles.latimes.com/2013/apr/30/autos/la-fi-hy-autos-bill-ford-milken-talk-20130501. Accessed 29 Feb 2016
7. Grush, B.: Social evolution and road pricing, Tolling Review (2014). www.researchgate.net/publication/264041743_Social_Evolution_and_Road_Pricing. Accessed 29 Feb 2016
8. McKenna, P.: Urban transit's uncertain future, Nova Next. 31 Jan 2016. http://www.pbs.org/wgbh/nova/next/tech/cities-autonomous-vehicles. Accessed 29 Feb 2016
9. Fagnant D., Kockleman, K.: Dynamic ride-sharing and optimal fleet sizing for a system of shared autonomous vehicles. Proceedings of the 94th Annual Meeting, Transportation Research Board (2015). www.ce.utexas.edu/prof/kockelman/public_html/TRB15SAVswithDRSinAustin.pdf. Accessed 29 Feb 2016
10. International Transport Forum, Urban Mobility System Upgrade: How shared self-driving cars could change city traffic, Corporate Partnership Board, OECD (2015). www.internationaltransportforum.org/Pub/pdf/15CPB_Self-drivingcars.pdf. Accessed 29 Feb 2016
11. Burns, L., Jordan, W., Scarborough, B.: Transforming personal mobility, The Earth Institute, Columbia University (2013). http://sustainablemobility.ei.columbia.edu/files/2012/12/Transforming-Personal-Mobility-Jan-27-20132.pdf. Accessed 29 Feb 2016
12. Rigole, P.: Study of a Shared Autonomous Vehicles Based Mobility Solution in Stockholm (Master of Science Thesis), Stockholm (2014). http://kth.diva-portal.org/smash/get/diva2:746893/FULLTEXT01.pdf. Accessed 29 Feb 2016
13. Bailey, R.: The End of Doom, Environmental Renewal in the Twenty-First Century. St. Martins Press, New York (2015)
14. Schwartz, S., Rosen, W.: Street Smart: The Rise of Cities and the Fall of Cars. Public Affairs, Philadelphia (2015)
15. Silver, N., Fischer-Baum, R.: Public Transit Should Be Uber's New Best Friend (2015). http://fivethirtyeight.com/features/public-transit-should-be-ubers-new-best-friend. Accessed 29 Feb 2016
16. Richtel, M.: A Deadly Wandering: A Tale of Tragedy and Redemption in the Age of Attention. William Morrow, New York (2014)
17. KPMG, Me, my car, my life … in the ultraconnected age (2014). www.kpmg.com/US/en/IssuesAndInsights/ArticlesPublications/Documents/me-my-life-my-car.pdf. Accessed 29 Feb 2016
18. Teller, A.: How to Make Moonshots (2015). https://backchannel.com/how-to-make-moonshots-65845011a277. Accessed 29 Feb 2016
19. Davies, A.: Obviously Drivers Are Already Abusing Tesla's Autopilot, Wired, 22 Oct 2015. www.wired.com/2015/10/obviously-drivers-are-already-abusing-teslas-autopilot. Accessed 29 Feb 2016
20. Woods, R.: Google's self-driving car hits a bus, and it could be Google's fault, Neowin, 1 Mar 2016. www.neowin.net/news/googles-self-driving-car-hits-a-bus-and-it-could-be-googles-fault. Accessed 29 Feb 2016

21. Richtel, M., Dougherty, C.: Google's Driverless Cars Run Into Problem: Cars With Drivers, New York Times. 1 Sept 2015. www.nytimes.com/2015/09/02/technology/personaltech/google-says-its-not-the-driverless-cars-fault-its-other-drivers.html. Accessed 29 Feb 2016
22. CityMobil2: About CityMobil2. www.citymobil2.eu/en/About-CityMobil2/Overview. Accessed 29 Feb 2016
23. Burns, L., Jordan, W., Scarborough, B.: Transforming personal mobility, The Earth Institute, Columbia University (2013). http://sustainablemobility.ei.columbia.edu/files/2012/12/Transforming-Personal-Mobility-Jan-27-20132.pdf. Accessed 29 Feb 2016
24. Teller, A.: How to Make Moonshots (2015) https://backchannel.com/how-to-make-moonshots-65845011a277. Accessed 29 Feb 2016

Biking and the Connected City

Victoria Adams, Sudeeksha Murari and Christopher Round

Abstract There is a growing consensus that bikes will play a major role in the future of urban transportation. As of June 2014, public bike-sharing systems existed on five continents, including 712 cities. Despite growing interest in the role of bikes in urban transportation, there is little discussion of how the bike will fit into the Internet of Things and the vision of the "connected city." The connected city is an urban area where connected technologies are used to enhance transportation performance and reduce costs. This chapter outlines how biking could be integrated into the Internet of things, potential futures for transportation agencies as bike managers in the connected city, and what potential issues need to be addressed.

Keywords Connected biking · IoT · Urban transportation · Alternate modes

1 Introduction

This section discusses the role that bikes and biking could play in the connected city. Booz Allen Hamilton defines a "connected city" as "*An urban area where connected digital technologies (connected* via *vehicle-to-vehicle [V2V], vehicle-to-infrastructure [V2I], and hybrid [V2X]) are used to enhance transportation performance (in terms of convenience, access, and quality of experience), reduce transportation costs, resource consumption and impact on the environment, and improve the health and well-being of individuals that use transportation in that area. The connected city effectively and actively engages transportation users in transportation decisions and provides real-time feedback on the economic, health, and environmental impacts of*

V. Adams · S. Murari (✉) · C. Round
Booz Allen Hamilton, 8283 Greensboro Drive, Mclean, VA, USA
e-mail: Murari_sudeeksha@bah.com

V. Adams
e-mail: adams_victoria@bah.com

C. Round
e-mail: round_christopher@bah.com

G. Meyer and S. Shaheen (eds.), *Disrupting Mobility*,
Lecture Notes in Mobility, DOI 10.1007/978-3-319-51602-8_18

*their decisions. It also provides transportation agencies with the information on which to base transportation real-time management decisions to maximize mobility, safety, health, and reduce environmental impact, travel time, and cos*t."

There is a growing consensus that bikes will play a major role in the future of urban transportation [1, 2]. For example, as of June 2014, public bike-sharing systems existed on five continents, including 712 cities, operating approximately 806,200 bicycles at 37,500 stations [1, 2]. Major U.S. cities including Chicago have launched ambitious biking programs as part of their future transportation planning.

However, despite growing interest in the role of bikes in urban transportation, there is little discussion of how the bike will fit into the Internet of Things (IoT) and the vision of the connected city. The IoT is the interconnection of uniquely identifiable embedded computing devices within the existing internet infrastructure. The IoT is expected to offer advanced connectivity of devices, systems, and services that goes beyond machine-to-machine (M2M) communications and covers a variety of protocols, domains, and applications. Cisco Systems estimate that more than 50 billion devices will be connected together by 2020. This will transform business, society, government, and how we live and work.

This $100 billion in increased value will be driven by (1) improved asset utilization; (2) improved employee productivity; (3) improved supply-chain/logistics efficiency; (4) improved customer experience; and (5) greater innovation. About 6% of this or more than $5 billion annually will be associated with transportation (see Fig. 1).

The remainder of this paper is divided into four additional sections. Section 1 provides background on the connected bike in the connected city and discusses how biking could be integrated into the IoT. Section 2 describes the concept of the connected, smart, and electric bike. Section 3 identifies potential futures for transportation agencies as bike managers in the connected city. Section 4 identifies potential issues and research questions that need to be addressed.

2 The Future of Biking: The Connected Bike, the Smart Bike, and the Electric Bike

Three related technological changes are occurring that are making the bike a potentially revolutionary surface transportation vehicle for the twenty-first century: the connected bike, the smart bike, and the electric bike. These technologies have the potential to disrupt existing urban personal transportation systems. Figure 2 summarized these technologies. As with connected vehicles, these technologies are combinations of different baskets of technologies.[1] Below we briefly describe each in detail.

[1]Please note the connected and smart bike are in practice very similar. However, we have preserved the distinction because (1) a connected bike is only a subset of the functionality of a smart bike; and (2) it is possible to quickly and easily turn any bike into a connected bike by adding a smart phone whereas a smart bike requires built in sensors or after-market adaptations.

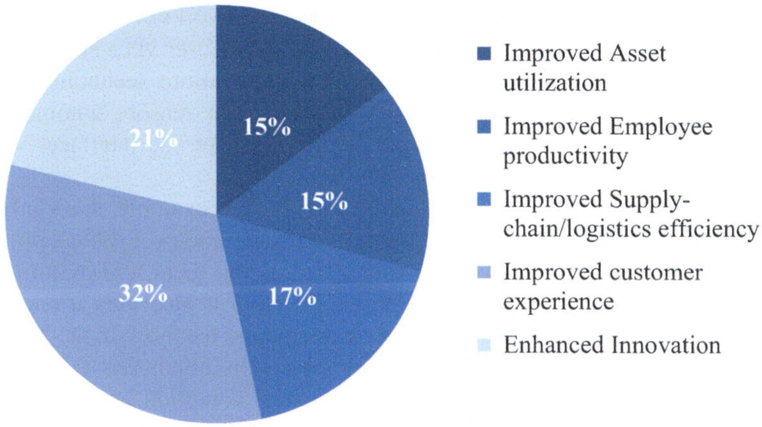

Fig. 1 Distribution of direct and indirect impacts

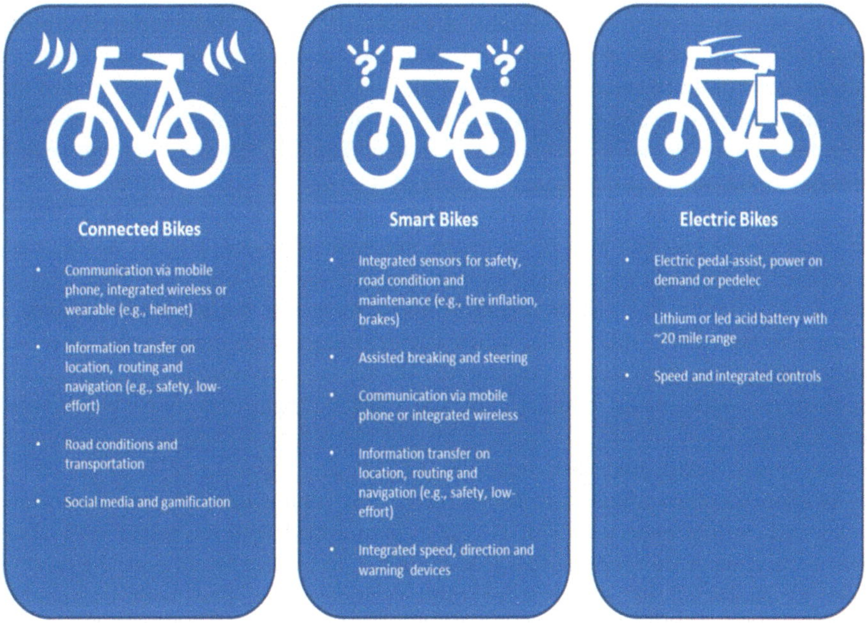

Fig. 2 Three disruptive surface transportation technologies: connected bikes, smart bikes, and electric bikes

2.1 The Connected and Smart Bike

The connected bike is a human-powered, pedal-driven, single vehicle that can transmit, receive, and process data from other vehicles, the infrastructure, as well as other sources (e.g., individuals, traffic management networks).[2] Connected functionality is either built into the bike or is provided by the user via a smartphone or another beacon-style device. A smart bike integrates various technologies (e.g., GPS, rear-view cameras, blindspot sensors, wear-and-tear sensors, automatic suspension, advanced breaking) into the operation of the bike and interfaces with an overall operating control or smartphone.

At present few examples of connected bikes exist. For example, the VanHawks Valour smart bike connects to an iPhone, Android, and/or Pebble app using Bluetooth, allowing riders to plot their routes and giving them turn-by-turn directions (via integrated LED lighting on the handlebars). It also uses a number of sensors to communicate to riders via an integrated haptic feedback in the handlebar grips that alerts riders by vibrating of any object in blindspots (Fig. 2). The VanHawks Valour also collects information on road condition and inclines, and provides information on alternative routes to the user considering traffic, easier and safer routes, while recommending routes that utilize paths and bike-only lanes. In addition, the bike contains theft protection that warns users when the bike is being tampered with and tracks (via GPS) the location of the bike if stolen.

The German Canyon bike contains a variety of sensors that modify the bikes suspension to increase the comfort of ride and increase speed. In addition, an onboard control unit contains a GSM module and GPS that plots the location of the bike, suggests routes, and works with a number of internal sensors that monitor the condition of key elements such as chain and brake pads. The sensors can send a message to the rider's smartphone, warning of the need to replace certain parts, and even connect her to an online ordering facility for spares. A video on how to install the replacement part can also be accessed automatically once the part is obtained. A connected smartphone technology can also be used to trace the bike, and to record performance data. In the event of an accident, the E-Call function sends a message to the rider's smartphone when sensors detect crash damage and a second call to emergency services (if not stopped by the rider) providing location of the accident, the riders name, contact data, and relevant health data (e.g., blood type, drug allergies).

Samsung's Smart Bike also incorporates connected features. For example, its geolocation function connects to social media and friends allowing riders to share routes and cycling experiences. It also comes equipped with advanced safety features such as an Arduino processor that enables it to project lasers onto the road in order to announce its presence to other vehicles equipped with compatible

[2]Please note, throughout this paper we focus primarily on single track, two-wheeled vehicles. However, all of the technologies here can be applied to three- or four-wheeled cycles. These cycles have advantages and disadvantages over traditional two-wheeled cycles. For example, they are safer, more stable and generally easier to pedal. However, they are generally more expensive and take up more room on a bike path leading to overcrowding on heavily used paths and congestion in urban areas.

technology and has a camera incorporated into the bike frame that allows the rider see behind via a handlebar mounted smartphone.

In addition to these other companies, Chinese search engine giant Baidu (the Chinese Google) is developing an operating system for bikes. The Baidu bike (known as the "DuBike") is intended to be a health-tracking system geared for bike riders wanting to improve their workouts. The DuBike monitors riders' health stats, such as calories burnt and heart rate, tracks traffic, and offers map data to find the best routes. Riders can operate the system through voice commands or an onboard display. The DuBike system will also come with social networking functions, safeguards to protect the bike from being stolen, and a battery pack that can be used to recharge smartphones or other devices. One of the great advantages of connected bikes over connected cars in terms of the rate of adoption is the lower cost of connected bikes and the ability to easily modify existing bikes. For example, the VanHawks Valour retails for approximately $1000, which is much less than a connected car. Bikes can be modified to include connected features by adding common apps to a smartphone and including a smartphone mount on the handle bars. Simple after-market add-ons can transform a standard bike into a connected or smart bike. For example, Helios handlebars can be installed on any bike for less than $300. They include integrated LED front and rear lighting, a speedometer, and, via integration with a smartphone, give directions and tracks the bike in the event of theft. Other developers are taking the approach of integrating smart or connected vehicle attributes into the rider's helmet. For example, LifeBEAM's Smart helmet (currently in prototyping) tracks multi-physiological parameters and will provide directions, blindspot warning, and crash detection. The advantage of such a system is that it would allow the user to take connected and smart bike technology with them and not be dependent on a single bike. This makes helmet-based, smartphone only, and other wearables solutions particularly compatible with bike-sharing programs (Fig. 3).

2.2 *The Electric Bike*

Another significant technological development is the growing market for electric bikes (known as an e-bike or booster bike). An e-bike is a bike with an integrated electric motor that can be used for propulsion to either assist the rider's pedal power (i.e., pedelecs) or even propel the bike forward with no rider action. All e-bikes retain the ability to be pedaled by the rider and are therefore not electric motorcycles. E-bikes use rechargeable batteries and the lighter varieties can travel typically between 16 and 20 mph (depending on the weight of the rider and legal restrictions). While not popular in the U.S., they are growing in popularity in Europe and Asia. In Germany they are beginning to cut into the market for conventional bikes, while in China they are largely replacing gasoline powered mopeds

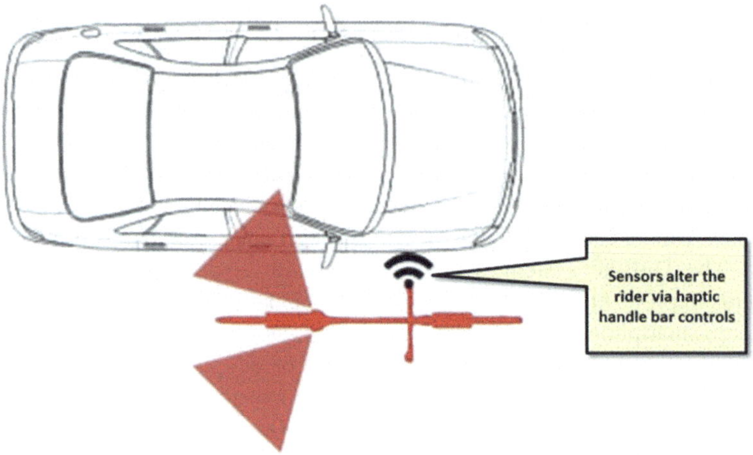

Fig. 3 The VanHawks valour blindspot direction and haptic communication system

such that by 2010 it was estimated that there are approximately 120 million e-bikes in China.[3]

The e-bike has several advantages over traditional bikes. First, by allowing peddle assist it provides a way for older and less athletic people to use bikes and to enter the urban biking system. Second, by requiring less effort it allows individuals who do not wish to change or exert themselves to use urban biking systems. Third, e-bikes allow users to contemplate longer commutes. With a single charge a typical e-bike has a range of 16–18 miles depending on a rider's weight, usage, and terrain. The battery can be recharged at a rider's desk or at a charging station. The potential of e-bikes as a transportation device for older populations is particularly interesting. A number of e-bikes are specifically marketed to this group and may grow in popularity.

The integration of a power source with e-bikes suggests a greater opportunity for developing smart and connected technology. For example, Daimler's Smart e-Bike combines a range of 60 plus miles at a 15 mile per hour peddle assist, dynamic breaking, and full integration with smartphones. Other e-bikes with more connected technology are under development and are likely to arrive on the European and Chinese market in the coming years.

In addition, the e-bike can be expanded to create personal transportation vehicle that can take more than one person. The advantage of such a "velomobile" (i.e., a covered pedal-assisted vehicle) is that they provide increased safety and room for additional passengers and loads, provide cover for inclement weather, and provide increased safety (e.g., greater stability, protective covering, tail lights). However, they have a number of disadvantages. Their size reduces the number of bikes or

[3] J. David Goodman (2010-01-31). "An Electric Boost for Bicyclists". New York Times.

other vehicles that can use a road or bike pathway, they require larger parking spaces, and they can be much more costly that a traditional bike (approximately $5000). However, they offer another alternative for personal transportation. One potential role that vehicles similar to the ELF might provide is to offer a non-car alternative for suburban transportation. The ELF can be used for suburban trips to the grocery store, shopping or to pick children up from day car.

Another interesting development is the so-called "Copenhagen Wheel." The Copenhagen Wheel is a rear bicycle wheel that can be added to a standard bike to transform it into a smart e-bike. The Wheel contains a motor, batteries, multiple sensors, wireless connectivity, and an embedded control system. The Wheel learns how a rider pedals and integrates seamlessly with their motion, multiplying their pedal power between 3 and 10 times. The Wheel is equipped with regenerative breaks such that the Wheel is able to capture energy when braking or going downhill and stores it in the integrated lithium battery pack. All actuation of the Wheel happens automatically via the pedals through sensing and control algorithms. For example, if a rider pedals harder, such as when going uphill, the Wheel feeds more power to the motor to reduce strain on the rider. Using your smartphone with the "Superpedestrian" app, riders can also vary the level of powered assist. The associated smartphone app allows riders to lock/unlock the wheel, choose amongst a menu of customizable rides, and track personal usage statistics including time, distance, calories burned, elevation climbed, and more, all of which can be compared and shared with friends through social media features.[4]

The e-bike can be integrated into bike-share programs. Each bike (known as the "Bycyklen") is equipped with an electric motor and a "vandal-proof" tablet installed in the center of the handlebar that can be used as a GPS or travel guide with tips for what to do in the city. Current bikes are available at 20 locations with 65 more planned. One of the useful characteristics of integrating smart and e-bikes with bike-share programs is that the connected nature of the bikes means that the bike-share program can monitor bike use and institute dynamic price to manage the demand for bikes. For example, currently day users (e.g., tourists) are offered discounts if they return bikes to bike stations in time for peak rush hour demand.

2.3 The Bike of the Future: Connected, Smart, and Pedelectric

One vision for the future of biking is that all three of these technology baskets would be combined. In this vision, the bike of the future would be smart, connected, and would provide an optional pedal assist. The bike would be fully connected into an urban areas traffic management system allowing it to communicate with the traffic

---

[4]Please note, alternative designs for "FlyKly" wheel exist such as the SmartWheel and other products.

control system and the connected infrastructure (Fig. 4). For example, bikes could communicate with traffic lights to provide "bike-only" crossing opportunities before other traffic to improve bike safety. Alternatively by receiving information from multiple connected bikes, a traffic management system could open more lanes to bikes and dynamic control the amount of road way reserved for bikes. Thus, more lanes could become "bike-only" on nice days when demand for cycling is high and fewer lanes could be reserved during inclement weather when demand is low. Similarly, data collected from the routes that bikes take through an urban area could be used to identify new sites for bike-only lanes, identify road conditions and areas with safety or bike congestion problems, or identify new pick-up/drop-off stations for bike-sharing schemes. In the following section we discuss how these new technologies can be used to encourage greater use of biking and the potential role that transportation agencies can undertake to encourage cycling and adoption of these technologies.

3 The Role of the Connected and Smart Bike in the Connected City

3.1 The Role of Connected, Smart, and Electric Bikes in Encouraging Increased Bike Use

Connected, smart, and electric bikes can provide a safer, potentially easier, and more convenient biking experience. These factors address the reasons why individuals do not use bikes in the urban area and show the revolutionary potential of these new technologies; that is, they can increase overall bike usage by encouraging non-cyclists to become cyclists.

While the bike is generally accepted to be a cheap, environmentally beneficial, healthy alternative to the automobile, research shows that the one of the major reasons that individuals do not use bikes in an urban setting is concern for safety, the effort involved and convenience [3]. A study of Portland, Oregon's cycling population shows this. The study breaks down Portland's potential cycling population into four groups (Fig. 5):

"***Strong and fearless***": A group of approximately 1% of the potential cycling population who are in good shape and perceive very little threat from urban cycling (e.g., bike messengers).

"***Enthused and confident***": A group that comprises approximately 5% of the potential cycling population who are in good physical shape and aware of the risks of urban cycling but feel able to deal with these risks.

"***Interested but concerned***": By far the largest groups are those that fall into the "interested but concerned" group (approximately 60%). This group would like make more use of bikes in an urban environment but is concerned about (1) the risks and safety of urban cycling; and (2) whether their health is sufficient to support regular cycling. This group would like to cycle and could be persuaded to cycle if the conditions were correct.

Fig. 4 Alternative approaches to creating a connected bike

"***No-way-no-how***": A group of about 30% that will never use bikes for urban transportation for reason related to safety, health, or their circumstance (e.g., the need to pick up children at daycare, length of commute).

The value of this topology is that it provides a way of thinking about how to encourage increased cycling and what are the issues involved in shifting individuals toward greater participation in cycling. Specifically, the 60% of the population that fits into the "Interested and Concerned" group constitutes the "persuadable" population. As discussed above, the key factors associated with persuading this group to make greater use of bikes are (1) perceived risk and safety; (2) perceived health and effort involved; and (3) more convenient. Connected, smart, and electric bikes

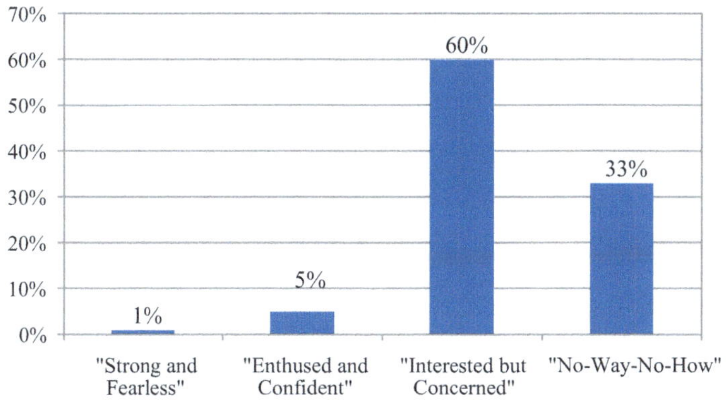

Fig. 5 Portland Oregon characterization of potential urban cyclists

Table 1 Impact of connected, smart, and electrics bikes on encouraging bike use in the connected city

Impact	Connected bikes	Smart bikes	Electric bikes
Improve perception of safety	Provide traffic information to identify safer (e.g., routes with designated bike lanes) or areas with less traffic Communicate directly with connected infrastructure to give cyclists preference (e.g., cycling preference at traffic lights) Communicate directly with traffic management systems to provide real-time on congestion and street conditions—can be used to turn on an off bike lanes or expand biking facilities and infrastructure Preferences Provide anti-theft devices and tracking of stolen bikes	Provide warning and blindspot indicators Integrated directions (e.g., handlebar indicators) Maintenance and functional monitoring (e.g., break functioning) provide integrated day and night lighting at multiple points (e.g., front, rear, side view) provide rear-view cameras provide anti-theft devices and tracking of stolen bikes	Reduces effort and allows greater concentration on road conditions covered cabins, rear/front/indicator lighting can increase safety
Improve perception of effort and health required to cycle	Provide traffic information lower effort routes (e.g., low incline easier rides) Link to health apps and indicate routes by difficulty and effort	Link to health apps and indicate routes by difficulty and effort Adjust bike suspension to optimize riders weight, reach and comfort level	Decrease effort involve—encourages people with longer commutes or biking on difficult terrain (e.g., high inclines)
Improve perception of convenience and access to bikes	Show where bike-share programs and stations are available and how biking might fit to public transport When combined with e-bikes, can identify charging stations		Increases speed allowing faster commutes Allows larger loads and multiple person bikes Covered cabins protection from inclement weather Pedal assist means can be used during the work day without sweating or appearing disheveled

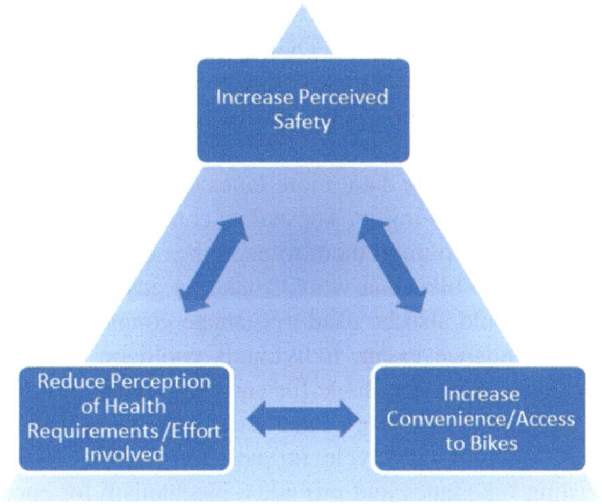

Fig. 6 The triangle of urban biking success

can play a key role in affecting the perceptions of this population. Table 1 shows these potential roles. See also Fig. 6.

A critical element of encourage greater use of these technologies and ensuring that full benefit of new cycling technologies is to link cycling into a city or urban areas traffic management and connected vehicle program. In the next section we discuss the role of transportation agencies in encouraging the use of connected and smart cycling.

3.2 The Role of Transportation Agencies in Increasing Bike Use

By connected cyclists to transportation agencies, new bike technologies suggest a number of potential roles and technologies that transportation agencies can fulfill in encouraging cycling. Below we briefly provide an overview of these technologies.

3.2.1 Dynamic Cycling Infrastructure Management (DCIM)

One of the keys to encouraging cycling in an urban area is the existence of cycle-only infrastructure. Surveys have shown that one of the main factors encouraging individuals to use bikes is the existence of a cycle-only roadways and paths. These reduce individual's perceptions of the risks of urban cycling and can encourage greater participation in cycling. However, cycle-only infrastructure can be costly and interfere with other traffic. Furthermore, use of cycle-only infrastructure varies with the weather (e.g., fewer people cycle in extremes of heat and cold or when it is raining) and other factors (e.g., more people may decide to cycle when construction impedes automobile transportation). Dynamic cycling infrastructure management or "DCIM" offers a solution to this problem.

Using this technology, connected bikes would provide information to a central traffic management system that would use these data to estimate the demand for bike-only roadways or lanes for a particular period (e.g., a morning or evening rush hour). Using these data, more lanes could be converted into bike-only roadways or given over to all traffic. These data could be communicated to cyclists and potential cyclists to encourage them to use bikes or to identify a route where lanes had been converted to bike that would make an easier or quicker route.

This could also be used to manage congestion or air quality. For example, for a high-attendance event, individuals could be encouraged to bike by reducing the number of lanes available for cars. Similarly, in days with heavy congestion or air quality problems, more lanes could be converted to bikes to encourage cycling. This "on-demand" cycle infrastructure would greatly increase the flexibility of existing bike lanes and provide a mechanism for encouraging greater cycle use and improving cycle safety.

3.2.2 Bike-to-Infrastructure Communication

Bike-to-infrastructure (B2I) communications could be used to increase the safety of cycling and the flow of bikes through a traffic signal cycle. For example, smart and connected bikes could communicate to traffic lights when a large number of bikes are inbound. The light could then provide cyclists with a priority or a time when no other vehicles are crossing, increasing the safety of biking at intersections and providing a smoother flow for cyclists through the intersection. By connecting multiple traffic lights into a single system, bike traffic could be efficiently routed through an urban system.

3.2.3 GIS-Disks and the "Last-Mile-Home"

One of the challenges of public transportation is the so-called "last-mile-home" problem. This problem refers to the fact that individuals must typically travel a short distance to reach public transportation. This distance is typically too long to walk and individuals are faced with the prospect of driving to a public transportation station or bus stop, finding parking (and frequently paying for parking) before taking a bus or train. Bikes offer an alternative. However, many individuals do not have or do not use bikes. Bike-share programs can play a role in this problem but the problem is that individuals would need to take the bike home and keep it overnight before returning. This could lead to lost bikes or unreturned bikes pilling up in a particular location.

GIS-disks are cheap (currently about $1.50) and small (less than the size of quarter) disks that can be attached to any item. Attached to a bike they can be used to see the location of that bike at any moment. Currently they are marketed as an anti-theft device. However, combined with a bike-share program they could be used to address the "last-mile-home" problem. Specifically, ride-share bikes could be tagged with these disks and users can be allowed to take the bikes home. They could either return them within 24 h or they could leave then at that location and collection vehicles could pick them up (with an additional fee billed to the rider).[5] This would encourage users to use bike-share programs for the "last-mile-home" and increase the flexibility of bike-share programs.

This could also be extended to create data for new bike stations. For example, if the bike-share system, using data from bikes with GIS-disks, noticed that a large number of bikes were accumulating in one area, they could establish a regular pick-up and drop-off station. Similarly, data collected from GIS disk and other connected bikes could be used by a transportation agency to identify areas that are heavily trafficked by bikes and establish new bikes lanes or bike infrastructure. This could be integrated into DCIM systems to create allow transportation agencies to dynamically manage the flow of bike traffic through a city.

3.2.4 Using Connected Cycle Data to Improve the Cycling Experience and Encourage Cyclists

Data from connected bikes could also be used to improve the cycling experience. For example, using data from connected cycles, transportation agencies could identify bottleneck areas for cycles, high-risk/low-safety areas, road conditions (e.g., potholes), times of day and locations when accidents or congestion occurs, under-traffic or under-served areas, and other data. These data could be combined to improve cycling infrastructure and dynamically manage bike traffic.

In addition, these data could be combine with user data (with the proper privacy and security protections) to increase identify groups that are not using bikes or only use bikes under particular conditions. For example, if an individual only users a bike when particular on nice days, a text message could be sent to her to encourage them to use their bike when a nice day is coming. Similarly, individuals could be notified before they leave home of which bike lanes are open, potential low congestion, low-effort routes, or where there is the potential for high-risk areas.

These data could also be used to identify groups that are not using bikes and target them for communication and outreach campaigns to encourage them to use bikes. For example, if older individuals were identified to be under using bikes, targeted information could be developed to explain the health benefits of bike, recommend low-effort, safe routes, or offer discounts on bike-sharing programs.

One particular use that data could be used for is to encourage women to bike. Women are almost an "indicator species" for the success for urban biking programs [3]. Most available data on urban biking suggests that in areas with little biking

[5]Note, bikes could be provided with a common locking device that users could use to attach the bike to an external structure on their home.

infrastructure, most cyclists are male. The rationale for dominance of men is that women have greater concerns for their safety and security (e.g., concern over poorly lit bike paths) and fulfill most of the daily shopping and childcare responsibilities. Gender parity is only achieved when cycling is perceived to be safe and convenient. Using data from connected bikes, transportation agencies could identify areas of concern to women for remediation (e.g., improved bike path lighting, greater traffic calming) or conduct targeted outreach to encourage cycling.

3.3 Transportation Agency as Bike Transportation Manager

The emergence of new biking technologies suggests that there is an emerging role of transportation agencies as a manager of bike transportation. In this role, the transportation agency could be a facilitator for biking (i.e., encouraging the use of biking, encouraging the growth of support shops and facilities, establishing bike-share programs), a planner and manager of bike infrastructure (e.g., bike-only lanes, protected bike paths, ample bike racks, ramps on stairs, space allocated on trains and buses and well-designed way-finding, traffic calming mechanisms in high bike traffic areas), and a dynamic manager of bike transportation.

To date most of the transportations in leading bike-friendly cities in the world have limited themselves to being facilitators and/or planners of bike infrastructure. The absence of a large number of connected or smart bikes has prevented them from actively entering as a manager. A review of the most recent Copenhagen Index of 80 global cities in terms of the degree that they support biking showed that in 2013 none had developed mechanisms to manage connected cycling [4]. This suggests that the next stage in the evolution of the urban biking experience may be the more from passive manager and facilitator to a more active, dynamic manager.

4 Future Research

Based on the research conducted for this white paper, we have identified a number of potential issues that need to be addressed in future research in this area. These include the following:

What are the potential roles for transportation agencies as managers of bike transportation, what works and what are the main success factors? This research area would review the experience of US and non-city transportation agencies and identify potential roles, policies, and programs that have been undertaken to encourage and create pro-cycling urban areas. This research would identify key lessons learned, assess impact of different programs, identify factors that led to success or failure, and show the business case for different policies and programs.

What are potential future technologies for biking and what is there potential impact on urban transportation? This research area would extend the current research and document potential future biking technologies, assess their viability and potential impact, and identify key enablers and the role of transportation agencies in encouraging their adoption.

What are potential future roles for transportation agencies as managers of bike transportation? This research area would extend the current research and explore potential roles for transportation agencies in new technologies, identify potential activities and requirements that transportation agencies could take to encourage adoption. One output of this could be a maturity model and self-scoring guide for transportation agencies on where they with regard to supporting cycling and preparing for the future. This could be accompanied by guidance and educational documents for transportation agencies to help them create more bikeable urban areas.

References

1. Midgley, P.: The role of smart bike-sharing systems in urban mobility, May Journeys, www.lta.gov.sg/ltaacademy/doc/IS02-p.23%20Bike-sharing.pdf (2009). Accessed 30 June 2016
2. Lerner, W.: No. 1: Future of urban mobility. In: Little A.D. (ed.) Towards Networked, Multimodal Cities of 2050, www.adlittle.com/downloads/tx_adlreports/ADL_Future_of_urban_mobility.pdf (2011). Accessed 30 June 2016
3. Baker, L.: How to Get More Bicyclists on the Road, 21 Sept 2009
4. Copenhagenize.eu: http://copenhagenize.eu/index/index.html (2016). Accessed 30 June 2016

iTRANS: Proactive ITS Based on Drone Technology to Solve Urban Transportation Challenge

Luis E. Ferreras

Abstract Transportation has become the latest field for disrupting innovation, as were telecommunications or computers in the last decades. We are seeing significant advances in the development of autonomous or semiautonomous vehicles and an immense surge of shared mobility services, such as bike-sharing and ride-sourcing services like Uber—all promising to improve the life of the average urban commuter. There are significant obstacles to the deployment and/or full utilization of these technologies, however, which could be mitigated with an additional layer of transportation system monitoring. This chapter introduces the use of unmanned aircraft vehicles (UAV), or drones, as the cornerstone of an advanced and proactive new intelligent transportation system (ITS) called, iTRANS. This new approach will reveal important traffic variables that are currently unpredictable, as well as solve most jurisdictional conflicts of interest, and regulatory constraints. Furthermore, iTRANS is designed to be one of the main tools to complete a successful transition from a dysfunctional transportation system to an optimal linear programming one in which transportation supply and demand is proactively managed through advanced ITS software. This sophisticated ITS system, connected to open software platforms, would gather and integrate all available information from the different modes of transportation, allowing real-time traffic management of the entire transportation system, as a whole. In addition, this chapter addresses the current technologies, urban transportation challenges related to autonomous vehicles, and describes the multifaceted approach of iTRANS and how its application would be advantageous in the deployment of autonomous vehicles. In a nutshell, this provides a systematic approach that twenty-first century engineers could use to create more eco-friendly, affordable, safe, and sustainable transportation environments.

Keywords Unmanned aerial systems (UAS) · Intelligent transportation systems (ITS) platform · Artificial intelligent (AI) · iTRANS

L.E. Ferreras (✉)
Parsons Transportation Group, Orlando, FL, USA
e-mail: luis.Ferreras@parsons.com

G. Meyer and S. Shaheen (eds.), *Disrupting Mobility*,
Lecture Notes in Mobility, DOI 10.1007/978-3-319-51602-8_19

1 Introduction

In today's world, we are going through a new disruptive period in which cutting-edge technologies are transforming the way humans communicate, interact, travel, and create new wealth. On this new horizon, transportation will play an important role in meeting the growing needs of individuals and society as whole. While the impact of disruptive innovations on transportation has not been accurately estimated yet, we know very well the legacy of our current transportation system, which defaults mostly to private vehicles propelled by carbon combustion engines and human control. This system has proved to be extremely complex and optimally unsatisfactory due to human limitations, a large number of unpredictable variables, and the necessity to process a massive amount of information to reach acceptable twenty-first century goals. Moreover, it is well known that the current transportation system also creates negative impacts on climate change, air quality, land use, motorist and pedestrian safety and ultimately, productivity all caused by transportation inefficiencies and gridlocks.

These losses can be seen, for example, in highway capacity ratios, the total surface amount of vehicles occupying a highway surface at optimal speed up to 100 km/h. According to California's Performance Measurement System (PeMS) data, and studies performed at the California PATH Program Institute of Transportation, a typical U.S. highway has a maximum occupancy per lane of 2200 vehicles per hour, representing 5% of total highway surface [1]. This means that the other 95% of the highway surface is used as safety buffer areas between cars, which are necessitated by the limitations of human driver abilities.

Another useful measure to quantify the effectiveness of the main American transportation network is the amount of time lost due to transportation congestions. According to the Texas Transportation Institute of Texas A&M University, in 2014, Americans spent per commuter an average of 42 h in car traffic congestions. This costs the nation, an all-time high of $160 billion dollars [2]. Moreover, focusing on the data provided by the U.S. Federal Highway Administration (FHWA), in 20 major urban areas of the U.S., the daily amount of congestion for 2010 was 4:38 h [3].

The three data sets outline above indicate that the freeway and highway networks in the U.S. are absorbing a lot of resources that could be used more efficiently in the future if a better combination of technology and resources could be found. Fortunately, in the last 10 years, breakthrough technological advances have been made in the fields of autonomous vehicles and Intelligent Transportation Systems (ITS) that could nourish new disrupting approaches to solve this complex problem.

Moreover, new applications based on GPS technology, advances in car design with new features (Active Brake Assistance Systems, Lane Departure Warning, Lane Keeping Systems, Night Vision Systems, Adaptive Cruise Control, etc.) and the mass use of mobile devices are improving our transportation life [4]. What is missing, however, is the ultimate optimization goal by constant location in place and in real time. The dawn and progress of automated vehicles could pave the way to a better transportation system one, in which the combination of high-performance driverless cars and intelligent transportation systems could create a powerful tool to reach new levels of optimization in urban transportation.

Another technological advance that can contribute to transportation optimization is 3D computer visualization, which is based on the combination of powerful cameras with smart algorithms [4]. With the addition of Light Detection and Ranging (LiDAR) information, this information is making it possible for computers on different platforms (autonomous vehicles or autonomous drones) to provide real data on a real-time basis. This point is critical, because it represents our ability to move from an unknown data transportation world into a reliable one. This property, will also allow us to integrate all data information online, and build future computer interfaces that will make all transportation systems optimal for every participant.

All these promising advances combined could be a catalyst to transitioning from a human control-oriented device system to a more efficient automated, self-decision-making vehicle system. This future step would provide transportation engineers with new tools for designing an optimal transportation system. Also, urban planners would benefit from large amount of real transportation data which would help them, design for more efficient urban growth and performance. The transportation benefits of moving from human control devices to self-decision-making machine devices are countless. However, satisfying the population's transportation needs without jeopardizing enormous capital resources is a major challenge. Another main challenge is to make this transition smooth enough to satisfy all transportation agents without penalizing the most vulnerable ones, which has been done during many dramatic transition periods in human history.

2 Related Work

2.1 Related Research

The related research work published up to today on drone technology is based on building mathematical models for drone control, performance, and collaboration between human–machine and machine–machine. The main points that we can underline based on past literature review on drones can be summarized as follows:

- UAV, or drones, have been limited to military purposes since the 1900s due to the large cost of development, since only national strategic purposes could justify its costly development [5]. In the U.S. alone, the Pentagon spent \$284 million on drone technology in the year 2000 [6]. In fiscal year 2016, it is going to spend \$3 billion on this technology [7]. As a result, in the last 15 years, sophisticated drones have reached notorious fame within the civilian population due to their ability to command missions only meant to be conducted by humans not long ago. Now, the symbiosis of human and machine has reached new levels of collaboration. Most recently, this drone technology has been introduced successfully into civilian environments, mainly due to tremendous advances in microelectronics and machine-learning algorithms, which has allowed commercialization of personal and commercial drones at affordable prices.

- Drone performance in flying time is measured by constrains that physics applies on small-size electrical motors with a performance ratio of approximately 50% of efficiency [8]. Drones can be classified into two main groups. The first is the fix-wind group that allows long-time flight and great aerodynamic performance but less capability of hover. The second group named "rotorcrafts", encompasses devices with great hovering capabilities that are usually propelled by small rotors. The most typical of this group are quadcopters because they meet the four degrees of freedom (roll-x, pitch-y, yaw-z, and acceleration).
- Quadcopters based on the outward mounting of the propellers offer great agility with high thrust-to-weight ratio and large angular acceleration [9], making them great candidates to conduct aerial robotic tasks. Studies by different researchers have proven the capacities of these autonomous aircrafts to be optimal tools to complete tasks [8].
- With funding from the Michigan Department of Transportation (MDOT), Colin Brooks from Michigan Tech Research Institute has conducted research demonstrations on the use of UAVs as ITS tools, using tethered blimps for long-term traffic monitoring, thereby providing a near real-time imaginary of traffic conditions [10].

2.2 Literature Review

The literature review on autonomous vehicles, smart ITS systems, and smart cities using sophisticated technologies—which are extensive across the research community—could be resumed on these following points:

- In Europe, the PROMETHEUS Project involved more than 20 manufacturing car companies and important research institutes such as the Argo and VisLab. In the 80s, Dr. Alberto Broggi from VisLab, Dr. Ernst Dickmanns from Navlab, and Dr. Charles Thorpe from VamP Institutes were the pioneers of autonomous vehicle technology. The three of them designed and built self-driving vehicles [11]. VisLab pioneered the development of self-driving car technology based on artificial intelligence and vision cameras while the other laboratories simply used LiDAR, Ladar radars, IR rangers, cameras, and GPS. During the 90s, these research teams built prototypes and tested them in highways at 130 km/h [11].
- In the U.S., and for more than 20 years, the PATH Project at the University of California, Berkeley has developed automatic highway technology that emphasizes collaborative technology between autonomous vehicles and the highways [1]. As a result, a large number of research papers on cooperative vehicle-highway automation systems have been published. Also at Berkeley,

Vason P. Srini at the Wireless Research Center envisioned autonomous navigation in urban environments using wireless technology that communicates with the road, the self-driving car, and the intelligent transportation system [12].

- The most important advance in the U.S., however was sponsored by the U.S. Defense Advance Research Project Agency (DARPA) through its Grand Challenges in 2003, 2004, and 2007. Research teams from all over the world developed the most efficient and reliable autonomous vehicle on different environments at the DARPA Grand Challenges. Also, the industry was able to have a prototype ready to be self-drive in any urban or highway environment without major issues. Afterward, Google decided to hire the winners of past DARPA Grand Challenges. Today, this research team is leading the race on driverless cars due to the extensive experimentation across many real traffic situations. Also, driverless cars testing experimentation has been authorized in Nevada, California, Florida, Michigan, and the District of Columbia always with the supervision of a human driver according to The Center for Internet and Society on Automated Driving: Legislative and Regulation Actions [13].
- The autonomous car brain is based on machine-learning algorithms more than a large amount of programming line code. Machine-learning algorithms need a large amount of trial-and-error experimentation to build the core of their knowledge. In this case, Google cars—with more miles of accumulated experimentation—explain why it is the leading team on this race [14]. However, once this knowledge is built, it is easy for it to be transferred from machine-to-machine or to a fixed communication network that can coordinate other autonomous cars running around that specific area, which will promote the use of "Vehicle to Vehicle" (V2V) and "Vehicle to Infrastructure" (V2I) technologies.
- Another aspect of the driverless brain, is its need for representation of a three-dimensional environment, in order to process all information of the surroundings. Thanks to sophisticated tools such as LiDAR, IR, high-resolution cameras and radars, autonomous vehicles are able to gather all the information needed to build an accurate copy of the space-time environment around the car. Moreover, many researchers have proven that the coordination of V2V and V2I technology collaboration allows the autonomous car to perform much better than in other cases [15].
- The European Roadmap Smart Systems for Automated Driving document [16] describes an extensive number of groups, teams, and governmental entities across the globe doing research on driverless cars. The U.S. and many members of the E.U. (Germany and France mainly) are not the only participants in this technology competitive race. Countries in Asia, such as China, Japan, South Korea, and Singapore, with Australia in Oceania are joining a crowded research field of public institutions and private companies. Based on this information, the success of transforming a promising technology into an industrial reality is closer now than ever.
- In Austria, the Centre of Regional Science at the Vienna University of Technology has identified six components of the smart city [17]. These are a

smart economy, smart environment, smart people, smart living, smart governance, and smart mobility. For many scholars [18], smart mobility implies traffic services with more intelligence and optimization of traffic flow [18] that will combine traditional transportation with new forms of transportation.

- Smart cities are connected and interact based on cutting-edge information and communication technology (ICT). ICT would be part of the core infrastructure of the future smart city concept. This new technology should be able to make new smart transportation mobility thrive. Moreover, advanced ITS systems are necessary to support the concept of the smart city [18]. The idea is to integrate the transportation infrastructure with cars and transportation signalization to optimize city traffic flow.
- Some authors have presented a real-time traffic flow network system for public transportation based on crowd-sensing participation using smartphones as a pivotal tool. The software can represent something close to real-time traffic situation. However, the accuracy of the system depends on the number of people who input data into the system [19].

3 Description of the Transportation Challenge

A simplistic approach to the transportation problem will answer how a person can go from point A to point B using the most optimal mix of transportation resources available. It becomes completely useless, however, when we add in the large number of stakeholders who participate in this "challenge". To increase the complexity of the problem, our transportation system is based on human decisions, thereby increasing the number of variables to be solved and sometimes making public transportation an incomplete solution to the problem. In fact, urban environments in which authorities have invested heavily in public transportation still suffer from private car overuse, which in some cases has persuaded authorities to ban the use of private cars in city downtowns. So far, the automobile is the best individual choice due to its feasibility to simply transport individuals from point A to point B. Moreover, automobiles can be the perfect mix to massive public transportation as first and last mile vehicles. This car flexibility makes researchers consider automobiles an important tool to solve this transportation challenge in urban as well as nonurban environments.

The large amount of resources our societies are spending trying to satisfy their transportation needs, (measured in capital and operational costs), and the negative outcomes calculated in productivity loss, human loss, and environmental damage have convinced authorities to develop research studies to try and find optimal solutions to the "transportation challenge". In past decades major advances in car mechanics, public transportation, UAV technology, and Intelligent Transportation Systems have made some improvements. However, the major problem is still not solved because the imbalance between transportation demands versus the supply available, especially as the demand tends to change faster than supply. These resources can be measured in many ways, but two clear metrics would be freeway and highway capacity and the time lost due to congestion. Both measures reveal the optimal ability of the transportation system to reach transportation needs, as well as the cost society needs to spend. If an optimal solution can be found, more resources can be liberated from this transportation challenge and instead be used on new matters that can contribute to increase the wealth of the country.

3.1 UAV Technology as Advanced ITS Platform

The benefits of using tethered UAV technology as a main ITS tool for our transportation needs may include:

- Ability for long periods of flying time if the UAV is tethered to the ground. By connecting the UAV with the ground, we can provide all power needed on a 24/365 basis. The drone can become a camera in the sky that can monitor real-time traffic conditions.
- A secure and fast communication channel. The idea of a cable connected to the UAV will allow us to avoid any wireless hack attempts from an outsider, plus provide the safety of fast communication, between the tools installed on the UAV and the central command center where all information will be processed and evaluated on a real-time basis.
- Real-time traffic monitoring in large metropolitan areas. Having a cloud of tethered UVAs at certain elevation will allow us to scale up the surface area that ITS can survey, avoiding the need for a large quantity of ITS tools to be installed along transportation infrastructure, thereby providing a real V2I system.
- Ability to have an advanced ITS system in place that communicates with future autonomous technology, which would support it or control any potential issue with this nascent breakthrough technology. Also, this new advanced ITS system can be supported with new V2I technology that would be developed in the near future.
- Potential use of UAVs as Wifi boosters to ensure Internet communication within iTRANS platform and all transportation stakeholders, so there is no missing communication within people, transportation systems, and traffic management software.
- This advanced technology can provide a workable transition from our current transportation model to one in which driverless cars could be the main actor of an autonomous transportation system. Also, we will enhance the sustainability of our societies by producing a significantly reduced air pollution impact and a more environmentally friendly solution, especially in large metropolitan areas. Also, sharing this technology between costumers will demand fewer resources, thereby reducing capital costs for the transportation authorities.

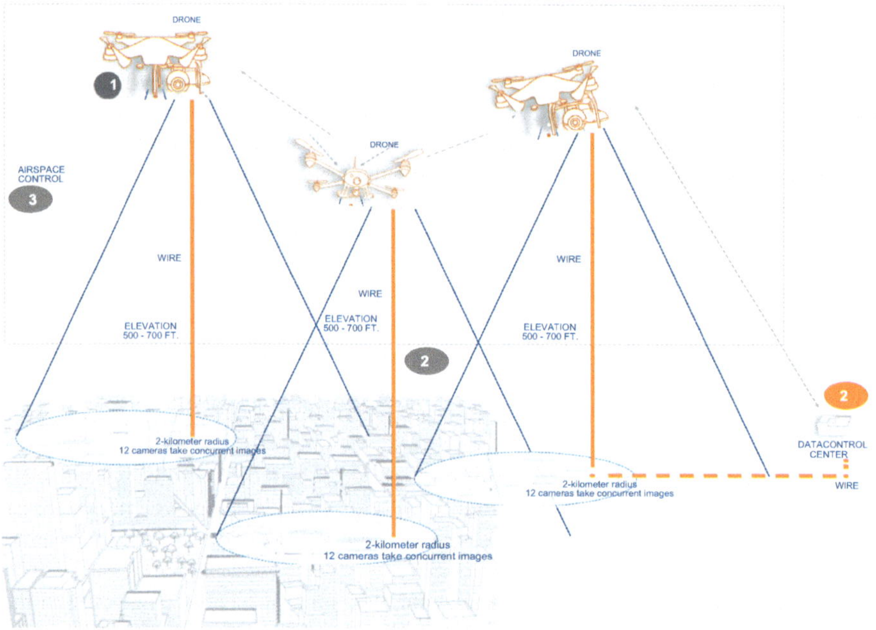

Fig. 1 iTRANS Schematic Design. *1* UVAs cloud to monitor urban traffic in real-time basis. *2* Tether communication within the UVA and the ITS central control to provide power supply, and safe and fast communication. *3* Central control where gathered traffic information is processed and send it back to the car or open software networks

- According to iTRANS theory, the best approach to solve this complex transportation challenge is to adopt a holistic solution, one in which all systems interact together. The general idea is to synchronize all the transportation methods at once into one unique system based on autonomous vehicles with UAVs as a cornerstone, and running by machine-learning algorithms in order to improve it as is used. The autonomous vehicle will be the link of this online decision-making system and the UAV would be the monitoring tool that will control all of the decision processes. The vehicle will provide input data to the system and will receive it as well, adopting the optimal route to go to its destination (Fig. 1).

3.2 *ITS Platform Characteristics*

Furthermore, the autonomous vehicle becomes the connection between all systems by intercommunicating between the autonomous vehicle and other vehicles, or between autonomous vehicles and the ITS in order to provide information about infrastructure status, traffic situations, and the best available route to reach the destination. The communication technologies used in this holistic approach that is based on driverless cars and UVAs are as listed:

- Intercommunication car to car. Cooperative autonomous cars [1]
  - Interchange information regarding location and destination based on GPS location and smartphone device location tools.
  - Ability to share same route and allow cars to drive closer together forming convoys as car trains, increasing highway occupancy.
  - Increase cooperative efforts between driverless cars in terms of better drive performance, car mechanics, or car status using stochastic optimization methodologies.
- Intercommunication car to infrastructure (ITS)
  - Double direction communication car to infrastructure in order to inform about traffic conditions, weather, future destination, and predictable paths. The system will learn about people's transportation needs and can proactively prepare the infrastructures for that.
  - Future traffic status and time delays based on historical data collected. Also the system can introduce machine-learning techniques to avoid past mistakes.
  - Infrastructure status. Devices can be installed in the infrastructure to inform the ITS system and autonomous vehicle about the structure status and condition of the infrastructure.
- Intercommunication car to Internet/Web 2.0
  - Software integration between autonomous cars and smart devices such as smartphones or tablet PCs.
  - Autonomous cars can become portable routers providing access to the Internet. The driverless car passengers will have Internet access during their trip.
  - Smartphone applications to provide driverless rental cars or autonomous taxi vehicles for instantaneous urban transportation. A pool of cars can be driven autonomously along city downtowns with specific lines to pickup and drop-off customers around the city limits.
  - Ability to share cars, thanks to social networks and instantaneous communication. People can ask for a common ride service through a social network. The same autonomous car can share the route with two or more people that live in the same area and have the same transportation needs.
  - All data can be stored in the cloud system, making it available to individual autonomous vehicles to get access to large amount of data to choose the best transportation option online, on time. It can be combined with public transportation services.
  - Machine-learning algorithms will improve the transportation system anytime is used. Optimal feedback interactions and software updates will make the iTRANS software better as much as its application is extended.

- In case of accident, an autonomous car can provide status of the car, and location to emergency services online.

- Intercommunication between car and public transportation management centers.
  - Autonomous vehicles as the first and last miles for public transportation systems will result in an integration of the driverless car into the public transportation system.
  - Autonomous vehicle can have access to public transportation data centers, which can have information on public transportation services, status, location, and occupancy. This information will be critical to a perfect combination between private autonomous cars and public transportation services.
  - Public transportation management centers can account for the number of passengers provided by autonomous vehicles. This way can better adapt the supply to potential daily demand based on historical data and also increase frequencies if it is needed.

These communications have the goal to feed our system with information to better match the demand curve of people's transportation needs to the less flexible supply in transportation resources available at a specific moment on time. This system with all input data will try to alleviate the potential congestion that our urban transportation network deals with every single day. Congestion is defined as the lack of transportation supply services at a specific place and time. Using this holistic approach will make more resources available and provide alternatives to the congested infrastructure that otherwise would be less used.

4 Conclusions

By allocating the best way possible transportation demands with supply transportation resources, this intelligent network system will be able to satisfy people's transportation needs from point A to point B. Sharing routes, changing information between transportation systems, using social network to share cars, and smartphone applications to solicit a momentarily available rental car for short urban transport are some of the beneficial uses of applying autonomous vehicles in all transportation systems.

The research interest so far has been the focus on the application of UAVs as a main tool on advanced ITS systems and autonomous vehicles, but a combination of all technologies need be developed to reach the maximum potential of autonomous vehicles.

However, not only engineers need to work together, but also authorities and industry leaders need to work together in order to establish clear rules for autonomous vehicles, lower the insurance premiums, and liberate resources that can be applicable in other fields. All these characteristics can drive for a more sustainable urban environment, and allow developing nations to have affordable transportation networks without jeopardizing more natural resources.

References

1. Shladover, S.: Cooperative (rather than autonomous) vehicle-highway automation systems. IEEE Intell. Transp. Syst. Mag. 10–19 (2009)
2. Schrank, D., Eisel, B., Lomax, T., Bak, J.: Urban mobility scorecard. Texas A&M Transportation Institute and INRIX, August 2015. http://d2dtl5nnlpfr0r.cloudfront.net/tti.tamu.edu/documents/mobility-scorecard-2015.pdf, accessed March 1, 2016
3. U.S. Department of Transportation: Federal Highway Administration 2010. 2010 Urban Congestion Trends, Enhancing System Reliability with Operations. http://ops.fhwa.dot.gov/publications/fhwahop11024/index.htm. Accessed 11 Jan 2012
4. Ilas, C.: Electronic sensing technologies for autonomous ground vehicles: a review. In: The 8th International Symposium on Advanced Topics in Electrical Engineering, IEEE, Bucharest, Romania, 23–25 May 2013
5. Amukele, T.K.,; Sokoll, L.J., Pepper, D., Howard, D.P., Street, J.: Can unmanned aerial systems (Drones) be used for the routine transport of chemistry, hematology, and coagulation laboratory specimens?, PLoS ONE (2015)
6. Ian, G.R.S.: The rise of the predator empire: tracing the history of U.S. drones, understanding empire, 2014. https://understandingempire.wordpress.com/2-0-a-brief-history-of-u-s-drones. Accessed 1 Mar 2016
7. Center for the Study of the Drone: Drones in the defense budget, Bard College, 2015. http://dronecenter.bard.edu/drones-in-the-defense-budget. Accessed 1 Mar 2016
8. Floreano, D., Wood, R.J.: Science, technology and the future of small autonomous drones. Nature **521**, 460–466 (2015)
9. Mueller, M.W., Hehn, M., D'Andrea, R.: A computationally efficient motion primitive for quadrocopter trajectory generation. Institute for Dynamic Systems and Control, ETH Zurich, IEEE (2015)
10. Brooks, C., Roussi, C., Colling, T.: Unpaved roads assessment. Michigan Technological University (MTU), characterization of unpaved road conditions through the use of remote sensing. www.mtri.org/unpaved, 1 Mar 2016
11. Urmson, C., Whittaker, W.: Red, Self-Driving Cars and the Urban Challenge, pp. 66–68. IEEE Computer Society, March–April 2008
12. Srini, V.P.: A Vision for Supporting Autonomous Navigation in Urban Environments, pp. 68–77. IEEE Computer Society, Dec 2006
13. Asaro, P., Calo, R., Walker Smith, B.: Autonomous driving. center for internet and society, Stanford University, http://cyberlaw.stanford.edu/wiki/index.php/Automated_Driving:_Legislative_and_Regulatory_Action. Accessed 1 Mar 2016
14. Driverless Car Market: Top misconceptions of autonomous cars and self-driving vehicles. www.driverless-future.com. Accessed 26 Feb 2016
15. Cheng, X., Yang, L., Shen, X., D2D for intelligent transportation systems: a feasibility study. IEEE Trans. Intell. Transp. Syst. **16**(4) (2015)
16. Dokic, J., Muller, B., Meyer, G.: European roadmap smart systems for automated driving. European Technology Platform on Smart System Integration, April 2015
17. Albino, V., Berardi, U., Dangelico, R.M.: Smart cities: definitions, dimensions, performance, and initiatives. J. Urban Technol. (2015)
18. Chun, B.: A study on intelligent traffic system related with smart city. Int. J. Smart Home **9**(7), 223–230 (2015)
19. Farkas, K., Feher, G., Benczur, A., Sidlo, C.: Crowdsending based public transport information service in smart cities. IEEE Commun. Mag. 158–165, Aug 2015

Mobilescapes: A New Frontier for Urban, Vehicle, and Media Design

Lino Vital García-Verdugo

Abstract What if we could transform traffic chaos into mobile landscapes? This paper introduces a new discipline, mobilescape design, at the intersection of urban, vehicle, and media design. Its catalyst is the capability of driverless vehicles to redefine urban spaces. In addition to offering improvements in safety, traffic circulation, and pollution levels, autonomous driving could sublimate the chaos of traffic into meaningful dynamic environments out of moving or static vehicles. The opportunities are endless. Creating meaningful situations out of mobile arrays can transform the interaction with local spaces, opening up new avenues in the arts, design, and business. The main goal of this paper is thus to raise awareness of the true potential of autonomous driving to fully exploit its capabilities in cities.

Keywords Autonomous driving · Vehicle design · Urban design · Smart environments · Interdisciplinary

1 Introduction

"In the roads which are or shall be within the city of Rome, or within the limit of continuous habitation, no person [...] shall be allowed in the daytime [...] to lead or drive any heavy wagon; except where it shall be requisite, for the sake of building [...] or carrying out some public work" [1].

Little has changed in the relationship between cities and mobility in the last 2000 years. This excerpt from the Lex Julia Municipalis, a legal code issued in 45 BC, endures as a pertinent comment on the contextualization of transport within the embodiment of community life that cities represent. Mobility has always been considered a necessary evil: a daily phenomenon that has to be minimized, subdued by the role of urban master planning. However, trends toward unprecedented scales

L.V. García-Verdugo (✉)
Automotive R&D Consultant (Rucker Lypsa) and Independent Design Researcher, Barcelona, Spain
e-mail: lino@mobilescapes.tech

G. Meyer and S. Shaheen (eds.), *Disrupting Mobility*,
Lecture Notes in Mobility, DOI 10.1007/978-3-319-51602-8_20

of urbanization might foster a reevaluation of Vitruvian urban conceptions. Daily headlines about pollution levels, safety and traffic congestion illustrate the paradox in which the intrinsic dynamism of urban life both creates settlements and negates their most basic habitability.

Industry pushes forward with technical developments, notably electric and autonomous vehicles, to reduce the problem of private cars (the demons of urban transport). However, car-centric approaches fail to take account of the 'wicked' problem at hand: the use of smarter 'wagons' might merely postpone the inevitable degradation of the fabric of urban spaces.

Wider approaches illustrate how cars establish an unwieldy combination of individual convenience and public discomfort by design. Obvious aspects, such as pollution, are merely components of the general destruction of urban life, which also includes the wasting of urban space. Moreover, spatial effects represent a vicious circle that generates saturation by parked and moving vehicles. For instance, 30% of congestion is produced by drivers looking for parking spaces [2]. Unlike those associated with architecture, the negative effects of vehicle design are not relegated to bad practice, constituting by-products of current approaches. Cars are conceived as commercial products, showroom eye-catchers with little concern for contextual integration [3]. An one-dimensional approach such as this would be acceptable if it were not for the scale and number of cars in urban spaces.

Nevertheless, automotive omnipresence is also indicative of a latent potential. The timely confluence of the increasing necessity of mass urbanization and new vehicle technologies offers the opportunity to reconsider the role of urban dynamics in the articulation of cities. This article will introduce a new concept, mobilescapes: the generation of immersive and interactive urban environments through the integration of vehicle, urban, and media design. It will focus on an approach to the dynamic layer juxtaposed with fixed urban contexts, and how to harness its power to project new experiences and conceptions of the city.

2 Spatial and Temporal Considerations in Urban Mobility

A brief analysis of two traditional concepts, space and time, will help to initiate an approach to the dynamics of mobility as a design issue.

Space has been considered as creative material since the dawn of humanity, and was formalized in architectural practices. The materiality of the city is a product of the confluence of function and communication [4]. Even the role of space in the definition of social interactions [5] is a theme that informs a range of disciplines, from urban design to context-aware computing [6]. There is a fundamental humanist dimension to architecture. However, this cannot take account of the significant influence of cars on architectural environments: vehicles affect both internal and external spaces [3]. Maybe the ephemeral nature of this influence is partly the reason for the omission in this design context: compared with this gap in design practice, the literature on spatial design and its practice is amply developed elsewhere.

In contrast, the dynamics of cities are mostly unaddressed. Spaces are not merely static images, as expressed in an architectural drawing. Interactions that occur between places, time, and energy imply the understanding of urban rhythms [7]. Understood as constituting elements of urban environments and social interactions [5], dynamics are less established than static spaces as a design theme. This situation has a negative effect on the definition of urban hubs as embodiments of community life, which is particularly worrying as the megacity emerges: the social and service dynamics motivating the agora, the generative isolation of urban space from natural 'chaos' [8], also become the invisible cancer that is slowly transforming it into meaningless chaos.

Paradoxically, there is a striking contrast between the ephemeral nature of time as a design material and the ultimate confrontation of the city dweller with its effects. From a perceptual perspective, time appears as a mere flow, background noise, rather than a succession of objective realities [9]. Moreover, traffic flow lacks any inherent value, in comparison with other types of flows. While running water has been an evocative resource in the creation of spatial masterpieces such as the Alhambra [10], to the casual observer vehicle traffic appears as a low-resolution and pointless succession of objects, such that the attentional impulse needed to disrupt it, to perceive time passing [9] is missing until the observer is confronted with it. The flow of traffic represents risks (accidents) and annoyances (noise) that attract attention. It is certainly a daunting picture, a flowing urban reality, emerging as merely a disturbance to the observer's inner dynamics. Moreover, this fast-paced reality invades and divides urban spaces, creating separations between bystanders, vehicle occupants, and local areas.

From this problematic viewpoint, significant advances in multimedia technologies have offered individuals an escape route from their environment. Spaces have degenerated from social scenarios to locations for even more meaningless flows (bits per second), where dislocated interactions take place; time has stopped in the sea of media, the timeless time of Castell [11], preventing any possible change in the current decadent approaches to urban design.

Rather than being a merely abstract exercise, mobility occupies a significant amount of an individual's lifespan. Beyond simplistic approaches that deny the role of private mobility, there is a latent need to address both the private allure of cars and their equally ephemeral and imposing presence in local neighborhoods [3]. In particular, this consideration is a timely investigation into a foreseeable future of smarter automotive examples. These vehicles offer controllable space and time [12] in the chaotic urban environment of uncontrolled flows; and their appeal will presumably grow in relation to a number of current issues:

- *Pollution*: The 'bubble' cars created can reduce the dose and exposure to air pollution [13], minimizing its effects on health [14]; they also protect the passenger from noise and distract from the chaos of the surrounding traffic flow with attractive interior design.
- *Safety*: Cars offer an unequaled level of protection against traffic and attacks (allowing travel through dangerous areas).

- *Traffic*: Autonomous cars will mean a better use of time spent on the road and independence for users who have not been able to access private cars before.

Furthermore, autonomous driving could reignite the role of private cars in uncontrolled urban sprawl [15].

The inevitability of car use in megacities reinforce the role of the vehicle as an interface, a generative force that, combined with new technologies could redefine perceptions: a new approach to question the role of the vehicle in its context vehicle, to avoid the 'smarter wagons' syndrome referred to earlier.

At a time when industry goes so far as to introduce somewhat unsettling humanoid robots [16] to respond to daily needs, the exploration of the role that could be played in the positive enhancement of our environment by existing elements that are already smart and dynamic certainly seems pertinent.

3 Approaching Mobilescapes

This study pursues the creation of malleable urban realities by rationalizing vehicles as parts of wider entities. It introduces a crucial notion: the infusion of meaning into urban traffic: a movable reality that is meant to contribute. Just as architecture constitutes a fixed embodiment of community life, the design of mobilescapes will address the materiality of mobility (understood as the aggregation of vehicles) to extend that embodiment to a dynamic dimension.

Unlike most of the technological developments supporting this vision, the use of vehicles to generate urban environments constitutes a 'wicked' design problem [17], which requires an interdisciplinary approach with respect to conventional vehicle design [18]. The definition of this approach poses a big challenge: how to articulate chaos while maintaining an exploratory approach.

In social life, functional objects are immediately transposed into the realm of meaning. Clothes are not just body coverings but communicate a particular style (formal dress, for example); cities are not mere shelters but readable and writable environments [4]. Traffic, however, only operates in the functional realm. From a communicative perspective, the aggregation of transport solutions results in a pointless succession of modern 'knights in shining armor' [19]. Harnessing the creative power of traffic will imply the creation of a structure that can integrate new and traditional functions with the codes of communication that they can generate.

In this sense, the use of related design codes is tempting but inappropriate, since it limits the capacity to explore new creative realms outside the current range of disciplines [20]. Semiotics, however, can offer a nonpartisan approach to avoid hindering future evolution. While the dynamism and presence of mobilescapes as a medium offer great potential, the emitter and receptor of this communication, in a world of varied stakeholders and big data, might not yet have been identified.

Semiotics works with codes as systems of significant units and rules of transformations and combinations. A fairly crude, but useful, approach can help to identify significant units of mobilescapes with vehicles. The most important point in the creation of a new realm is the definition of its new codes. For Eco [20], real creation happens according to the following sequence:

1. Identification of new needs
2. Translation into functions that will be primary (the inherent function) and secondary (the symbolic function)
3. Formalization of the functions in new designs

The rest of this paper will follow this route to offer an initial approach to the design of mobilescapes.

4 Apprehension of Latent Needs in Dynamics Environments

From the initial analysis of spatiotemporal considerations, the exploration of mobilescapes requires a practical understanding of the elements of community life that go unnoticed in the daily dynamics within which mobility operates. The modulation of space and time is twofold. It affects the experiences of actors and observers [7], or occupants of vehicles and city dwellers; and it might work as a context in the making or as a substratum for additional media.

In the consideration of the city as embodied community life, there is a distinction between embodiment itself (the present reality at any given moment) and a state of time-dependent need (ranging from inner, social, to outer needs). It is the adequacy of current embodiments in relation to need that will be the spark igniting the mobilescape generation.

4.1 Realm (Time-Sensitive, Space-Based)

The realm ultimately constitutes the canvases on which mobilescapes will operate (like land in relation to architecture). The challenge is approaching a canvas that is fluctuating, both in material qualities and in time. Mobilescapes have to operate within a layer that is free of any planning except technical traffic control. This realm, as a perceived environment, is articulated in both material and immaterial dimensions.

The material dimension relates to the intrinsic affordances [21], the enablement of primal possibilities in environments related to traffic. These can be articulated by the position of observers (immersive, approachable or contemplated realms) and vehicle speed (static or quasi-static, low-speed zones, high-speed networks). Depending on whether observers are occupants of a vehicle or bystanders, material realms tend toward certain sensory channels (i.e., high-speed immersive environments will emphasize visual inputs).

The immaterial dimension relates to the alteration of intrinsic perception. Realities such as ubiquitous media make mere physical descriptions obsolete [22], as they imply multisensory perception and range from intrinsic (pure material perception), through mediated (spatial interventions such as modulated lighting or acoustics), to semantic (pure messages isolated from location) inputs [23]. They also introduce rates of variation [23]. For practical purposes, immaterial conditions can vary from zero (nondistracting variations) to medium (perceived changes) to high (attention-grabbing changes).

The combination of both dimensions creates a dynamic realm that offers a perceptual baseline, imposing its own tension and quality of approachability to synchronize with the inherent tension of urban dynamics.

4.2 Dynamics of Urban Life (Space-Affecting, Time-Based)

Current realms appear as the result of, and conditions for, the development of urban dynamics, considered from the perspective of the functions, relations, and actors involved.

Functional dynamics relate to rhythms of generation and the demand for, for example, energy, information, transport (goods and people), and shelter. Some of these rhythms affect the spaces related to social life by occupying them (i.e., streets) or conditioning their perception (i.e., attention diverted from street to media consumption). They represent both needs and opportunities.

Relational dynamics refer to the required quality of social life at any given moment. Ranging from sightseeing, social activities, and individual activities to moments of pure introspection, the social character of spaces changes drastically in time.

In terms of actors, a space can be used by different types of individuals, each with a specific set of needs. Demographic aspects such as age, income level or occupation can significantly affect the nature of individual needs in an urban context. The possibilities a park offers to a child on a summer afternoon are very different to those it offers to a homeless person on a winter night.

4.3 Toward a Synthesis of Latent Needs

In terms of both realm and needs, there are side effects in community environments. Individual needs will affect and be affected by others. The key motivation for mobilescapes may be the purely social dimension, where one life affects others beyond mere social interaction. It also equates to the humanist dimension of architecture in its pure form, transcending commercial needs into those that are a result of human coexistence.

Moreover, the introduction of autonomous urban vehicles will add new dimensions in terms of actors and the articulation of spaces. The minimization of the need to concentrate on driving and wayfinding will unlock new needs with respect to urban spaces, while the introduction of smart mobile objects will redefine the relationship of people to their environment.

The main object will be to manage the tensions generated by both the realms and the dynamics of everyday needs. The ancient Greek agora, or meeting place, was the first step: the establishment of social cohesion by separating it from uncontrolled nature. However, this separation does not imply a hieratic construction. A lack of understanding of these dynamics can lead to the degeneration, or abandonment, of public spaces due to their lack of adaptability for the conditions of daily life.

What, then, would be the functions that could harness these dynamics? Rather than getting lost in a mass of variability, this precise variability emerges as the raison d'être for the incorporation of temporal articulations.

5 The Functions of Mobilescapes

The constantly changing needs of urban hubs motivate an infusion of new functions to make use of the physical and ephemeral influence of vehicles. This influence places transport solutions in a key position between the materiality of architecture and the mutability of media. It should allow both the articulation of fixed urban environments and the embodiment of media in local environments. This vision highlights the dual nature of mobilescapes as a motive (dynamic spaces) and a substratum for additional motives. Mobilescapes represent the sublimation of traffic chaos into meaningful interactive environments.

This new definition poses two questions. What are these new functions? How will these new articulations be made accessible for city dwellers?

For the creation of spaces and perspectives, motives and substratum articulated by mobilescapes, the primary and secondary functions will need to be defined. Primary functions denote immediate utility [20], new uses of transportation. From the emergence of mobilescapes as a dialogue between transport, context, and user, secondary functions will relate to the connotations of particular embodiments [20]. They will embody every aspect of the ways that primary functions connote specific modes of acting upon urban environments. In comparison to the functional realm of primary functions, secondary articulations encompass a wider and more subtle sensory spectrum.

Crucially, the new functions that respond to latent needs of urban dynamics will have to be implemented for audiences who have not already experienced them. How will their validity be ensured, particularly in the realm of individually adopted options? Aristotle's 'Poetics' asserted that to generate new messages, there should be elements of redundancy in relation to current ones [20, 24]. The code containing the new functions should thus relate to existing codes such as those of architecture and multimedia interactions.

The following paragraphs present a brief outline of the new functions involved.

5.1 Primary Functions

Primary functions start from the consideration of vehicles as communicators and mediums for other communications. Primary functions will appear as a response to both the intrinsic and the latent capabilities of traffic, and as differentiations from related disciplines such as architecture and media design. To survive uncertain scenarios, they will need to be variable [20]. Thus, to enable engaging dynamic environments for heterogeneous groups of stakeholders inhabiting different realms, vehicles should provide:

- *Awareness*: mobilescapes should take account of both the qualitative and the quantitative influences of locations and contexts to comply with changing sets of necessities.
- *Interactivity*: multimodal interactions between vehicles, city dwellers, contexts, and pervasive computing will enable the creation of highly articulated environments.
- *Motion control*: the consideration of movement should transcend mere displacement from A to B in order to generate new spaces and perspectives by harnessing the ephemeral characteristics of transport flows. To use Laban's analysis [25], vehicles can provide outer movement (displacement) and inner movement (expressiveness). These would be used in relation to both the modulation of spaces and the perceptions of the occupants.
- *Active space segmentation*: the generation of spaces (internal and external) by the presence of vehicles should be articulated by 'membranes', rather than by traditional panels [18]. Using the example of digital imaging, current vehicle design uses 1-bit surfaces (passive transparent or opaque panels) implying a fixed configuration of inner and outer spaces; 8-bit surfaces would represent articulated relations capable of responding to context (divide, filter, or communicate). An augmentation of the material qualities of vehicle surfaces will also offer a vast resource for design, with the use of layering [18].
- *Dynamic spatial relations*: Active membranes should also embrace superposition and simultaneity in both vehicles and context. Designs are not limited to single units, but also include arrays. This function opens a truly creative realm where the combination of (static and mobile) vehicles will generate authentic mobilescapes.
- *Digital–physical integration*: The dynamism of mobilescapes will offer an unequaled resource for finding correspondence between the mutability of media and the fixed nature of urban design. They offer both dynamic spaces and

perspectives to be used as mediated resources or supports for immersive multimedia experiences.

Depending on specific realms and needs, these functions will work across multisensory spectra.

5.2 Secondary Functions

Mobilescapes will truly crystallize in their secondary functions. Primary functions should allow a modulation of space and time to respond to specific needs within changing realms, but it will be the subtleties in the way they materialize that will constitute social communication: secondary functions will connote a particular way of operating within the urban fabric. As demonstrated before in the contexts of fashion and architecture, these connotations are keys to social life. In the same way, that the individual nature of traffic flow in different global cities already informs the experience of both the inhabitant and visitor, the modulations introduced by mobilescapes will communicate too.

The presence and plasticity of mobilescapes will introduce a vast creative field where existing codes (i.e., architecture, dance, cinema, video games) can adopt these new functions to both address needs (modify urban tempos and spaces) and project disruptive experiences. It will be the unexpected combinations, rather than the superposition of functions, that will multiply the potential of mobilescapes as a creative resource.

Examples abound. Focusing on the relations between stakeholders, mobilescapes can range from experiences designed by leaders in multimedia technology and practice to embodied social platforms. In terms of mediated environments, mobilescapes could create embodied information. For instance, many cultures have learnt to identify birds flying low with approaching storms; the interactivity and motion control of smart vehicles could create a similar metaphor, where weather changes are announced with distinctive motion patterns. In terms of urban spaces, imagine the complexity and mystery of coordinated translucent vehicles, performing a dance of digital fog in any of the dehumanized plazas of our contemporary cities, a "space that is veiled for our eyes, but transparent to action" [26]. These ephemeral qualities would be a great resource to infuse the qualities that are valued in natural landscapes.

Again, the creation of robust new codes implies the adoption of open secondary functions to incorporate the element of unexpectedness intrinsic to human creation. Further definition would deny the exploratory nature of this study.

In the same way that the agora was the beginning of the humanization of environments in opposition to nature, mobilescapes can enable a humanized control of spatiotemporal dynamics in the city: traffic and multimedia harnessed as realms of social life.

6 A Route Map for the Development of Mobilescapes

How might mobilescapes emerge from existing knowledge about vehicle design and control? This section will briefly outline the general implications for the materialization of the new concept. Nevertheless, the role of this knowledge is not merely instrumental. Technologies such as autonomous driving will not only enable mobilescapes, but might also trigger unexpected developments in its evolution. Unlike enduring concrete or stone, the elements that mobilescapes can use are fairly recent: communication technologies, new materials, and control engineering are opening new frontiers on a daily basis. Taking interactions as an example, how will people coexist with smart cars that are able to 'see' and act accordingly? Or how will they live in mazes of vehicles that can go from private to transparent at a single click? Understanding how to materialize new functions with current technology is as important as detecting additional functions resulting from the interaction with technology itself.

6.1 Vehicle Definition

Nowadays, vehicles are conceived as commercial end products based on complex architectures and purpose-designed industrial processes. This conception limits their communicative realm to certain high-speed dynamism and stylistic connotations, which results in clumsy and imposing presences in an urban context [3]. The generation of mobilescapes implies two disruptive changes in vehicle ideation: the vehicle as mobile canvas and as space modulator [18]. While both changes still rely on automotive technology, they introduce two fundamental reconsiderations of standard developments: specific vehicle dynamics and surface articulations.

6.1.1 Vehicle Dynamics

For the design of mobilescapes, vehicle dynamics are an important consideration both as a communicative motive and as a substratum for media. Moreover, considering the role of vehicles as mobility solutions, these qualities must be aligned with intrinsic automotive functionality.

While architecture can enrich the quality of the spaces created by vehicles, the fixed nature of its creative realm makes it notoriously oblivious to a fundamental aspect: vehicle weight and its distribution. This is a familiar and fundamental concept in racing car design [27], for example. Weight not only conditions vehicle dynamics, it also affects the perception of these dynamics. In contrast with a bicycle, where movement is modulated by the rider's weight, a conventional

Table 1 Perception of vehicle dynamics both as motive and substratum

Parameters		Dynamics perception
Weight value	High	Restriction to Labanian strong and sustained movements [25] A, mobile room'
	Low	Full spectrum of expressive movement [25] Risk of 'breaking the fourth wall' when working as a medium
Center of gravity	Close to occupants	Embodied dynamics: vehicles as extensions of occupants
	Far from occupants	Perceived separation of vehicle occupant Complex perceptions of transitory movements
	High	Amplified perceptions of movement
	Low	More subtle perceptions of movement
Polar moment of inertia	High	Restriction to subtle yaw changes
	Low	Full spectrum of expressive movement [25]

enclosed vehicle weighs at least five times more than a single occupant. Hence the perception of movement will rely heavily on its own qualities. Three aspects to consider will be weight value, the location of the center of gravity, and the polar moment of inertia.

Weight value in itself already conditions the dynamic experience a mobile space can provide: low weight at speed might make occupants aware of both increased responsiveness and instability, in comparison with users' experience of a heavier vehicle.

Perception will affect both the expressive qualities of movement (i.e., understanding the vehicle as a dancer, or as a mobile 'nest') and its qualities as a substratum (contributing to 'breaking the fourth wall', drawing the observer's attention away) (Table 1).

In addition, other systems will affect its dynamic capabilities, mainly the chassis and power train. Within chassis systems, steering will condition changes in direction, ranging from conventional front-wheel steering to four-wheel systems, allowing omnidirectional capabilities and turns around the vehicle's own axis. Suspension layouts are defined with respect to weight distributions and can alter changes provoked by weight transfer (pitch, roll, and yaw response at the limit) with the location of roll centers, and anti-squat and anti-dive geometries. More advanced active systems simplify mechanical designs and can even add their own dynamic perceptions, inclining the vehicle with actuators. Finally, the power train provides the realm of dynamism a vehicle can project and modulates its transitions. They intervene with motors and transmissions, and in the case of electric motors can greatly condition weight balance with the position of battery packages.

6.1.2 Design Articulation of Vehicle Space

The mass characterized above by its dynamic qualities should ultimately take form by responding to the dynamics of urban environments and needs. New technologies enable time-based adaptability over the rigid transparent–opaque dichotomy of conventional cars. However, not everything is visual: these 'membranes' are also part of the technical architecture of vehicles. Rather than prescribing a particular configuration, this text highlights the dialectic reactions between functions, technique, and style that generate new purposeful typologies [28].

Taking vehicle functions as a starting point, the membranes intervening in mobilescapes should also create protection from the following:

- Hazardous mechanisms, the classic genesis of design [29]
- Disturbances (weather, noise, vibrations, smell)
- Accidents and attacks (i.e., theft, tampering)

The ideal partition will be the membrane [18], a malleable element that can serve both as a modulator of contextual elements and as a substratum for added media. The automotive functions will intervene, conditioning and adding new elements to final designs. These automotive functions can range from qualities such as acoustic isolation to whole structural elements, such as those required for passive safety. Contemporary state-of-the-art materials and electronics generate vast numbers of possible configurations. In terms of visual aspects, the optical qualities of panels can be varied while embedding displays and a variety of electronics. Structures can incorporate aesthetic considerations in their design, and electric power trains allow greater creative expression with their simplified layouts [18].

From a stylistic point of view, the available technologies will enable disruptive designs, but creative expression will mostly be limited by user acceptance. Futuristic concepts might have to be tamed to respond to known codes of what a vehicle 'should' look like, fostering interesting dialogues between the revolutionary functions of mobilescapes and automotive culture. Of course, this culture can be used as a guiding force, or be overtly criticized, but the generation of new vehicle proposals should not ignore its importance.

6.2 Control Systems

This text has mostly dealt with the role vehicles play and should play in the creation of urban space. However, both the realm they would operate in and the technology enabling the new functionalities belong to control engineering. In parallel with the perceived dimension, control engineering articulates the functionality of a city as a whole down to the inner workings of any remotely smart vehicle. Mobilescapes are created to intervene in lived environments, but should be integrated with the structures of functional control. This implies a consideration of the different levels of engagement: cities, areas, and local interactions with end users.

Nowadays, worldwide megacities cannot be understood without taking account of the massive amount of data they generate, and their functional control depends on it significantly. From past times of investment in infrastructures to cover the transport needs of cities, current trends seem to show a swift movement toward the optimization of related data management and sharing [30]. Systems such as these will integrate the control systems of traffic police with urban management in order to manage peak hour traffic flow in densely populated areas [31]. As in any network system pursuing control optimization, it is easy to foresee a trend toward fully computerized and interconnected urban operations (especially with the introduction of fleets of electric vehicles). In this future scenario, mobilescapes will have to navigate not only roads but also communication networks. This will imply the need for an adequacy to the defined protocol interfaces, regulatory commands for traffic, and also access to data feeds that will be provided by the city.

On a closer scale, mobilescapes will imply the direct interaction between a network of smart vehicles, humans, and other devices. Research from space exploration missions will be a valuable reference for this. Their multiagent systems establish a cooperative environment to relieve humans of tasks that can be performed by robots [32], a similar situation to that found in autonomous driving. Agent models (i.e., vehicles, humans, smart buildings) and specialized middleware can systematize interactions and scalability in the generation of smart dynamic environments.

Ultimately, the middleware should extend its accessibility to end users, making some of the functions enabled by mobilescapes accessible to app developers. Who knows what could emerge from the introduction of mobilescapes to the universe of these common technologies, in terms of urban life?

Needless to say, the control systems of the vehicles would be affected by the introduction of mobilescapes. Depending on the nature of the articulations (additional to conventional transport or the generation of pure dynamic environments), the introduction of mobilescapes would vary from corrections to control inputs (i.e., those imposed by lane departure avoidance systems in conventional cars) to separate states in the finite-state machines governing the actuation of the vehicle.

7 Conclusion

Given the daily reality of traffic, the notion that cars passing by can be transformed into dancers or mobile architecture might sound preposterous, a mere theoretical exercise. However, the daily ubiquity of this itself represents the motivation of this quest. Traffic flow constitutes a burden, but also a huge opportunity; in the same way that the use of a well-designed space or a responsive vehicle can affect not only its functional aspects, but also the user's mood, harnessing the power of daily dynamics could redefine life in urban environments.

The proposals in this article are presented as a response to the static traffic gridlocks that have existed since Roman times. However, practical implementation does not have to involve the whole proposal: vehicle and traffic planners might be influenced by a number of specific parts of it.

Despite the eccentric nature of this idea, the reality is that technology is slowly enabling this alternative reality. Materials and computers can help to liberate us from the burden of meaningless tasks such as driving in order to reorientate lives toward social and individual development. Moreover, legislators have already shown a predisposition to include frameworks that help to articulate the dynamics of urban traffic in vehicle-type approval processes [33].

What is missing is an ambitious vision, a disruptive departure from known solutions and disciplines that truly breaks down the boundaries between architecture, vehicle, and media design to rehumanize urban environments.

Obviously there are risks. Mobilescapes can easily become instruments of persuasive control, but the same has been said of traditional and new media; despite this reality, these new technologies have also enabled their own realm of social life that transcends location. It might even be the moment now to use technology to relocate societies too, challenging the frontiers between traffic networks and social spaces, and urbanizing 'blank' areas with ephemeral configurations of vehicles.

Both the dynamic articulation of cities and the integration of media deserve the exploration of new interdisciplinary fields.

References

1. Hardy, T.: Roman Laws and Charters. Clarendon Press, Oxford (1912)
2. Shoup, D.: Parking data and analytics: what can your parking spaces tell you? https://eu-smartcities.eu/sites/all/files/docs/best-practice/Streetline%20Parking%20Data%20%26%20Analytics%20whitepaper%20-%20LowRes.pdf. Accessed 27 Feb 2016
3. Vital, L.: Towards a new discipline: the design of urban vehicles. In: Spinney, J., Reimer, S., Pinch, P. (eds.) Designing Mobilities. Routledge, London (2016)
4. Barthes, R.: Semiology and the urban. In: Leach, N. (ed.) Rethinking Architecture: A reader in cultural theory, 165–172. Routledge, London (1997)
5. Hall, E.: The Hidden Dimension. Anchor Books, New York (1969)
6. Greenberg, S., et al.: Proxemic interactions: the new Ubicomp? Interactions. **XVIII**(1), 42 (2011)
7. Lefevbre, H.: Rhythmanalysis: Space. Time and Everyday Life, Continuum, London (2004)
8. Ortega y Gasset, J.: The Revolt of the Masses. W. W. Norton & Company, New York (1932)
9. Merleau-Ponty, M.: Phenomenology of Perception. Routledge Classics, New York (2002)
10. Rabbat, N.: The palace of the lions Alhambra and the role of water in its conception. Environ. Des. J. Islamic Environ. Des. Res. Centre **2**, 64–73 (1985)
11. Castell, M.: Materials for an exploratory theory of the network society. Br. J. Sociol. **51**, 5–24 (2000)
12. Urry, J.: Automobility, car culture and weightless travel: a discussion paper. http://www.lancaster.ac.uk/fass/resources/sociology-online-papers/papers/urry-automobility.pdf. Accessed 28 Feb 2016
13. Tesla Motors, Model X.: https://www.teslamotors.com/modelx. Accessed 28 Feb 2016
14. Gorham, R.: Air pollution from ground transportation. www.un.org/esa/gite/csd/gorham.pdf. Accessed 28 Feb 2016

15. Anderson, J., et al.: Autonomous Vehicle Technology: A Guide for Policymakers. RAND Corporation, Santa Monica (2014)
16. NTDTV, Robot Department Store Receptionist Starts Work in Tokyo. https://www.youtube.com/watch?v=s95ZGQKl96k. Accessed 28 Feb 2016
17. Rittel, H., Webber, M.: Dilemmas in a general theory of planning. Policy Sci. **4**, 155–169 (1973)
18. Vital, L, µcar: Multidisciplinary development of an electric vehicle typology for the city http://ethos.bl.uk/OrderDetails.do?uin=uk.bl.ethos.577716. Accessed 28 Feb 2016
19. McLuhan, M., McLuhan, E.: Laws of Media: The New Science. University of Toronto Press, Toronto (1988)
20. Eco, U.: La Estructura Ausente: Introducción a la semiótica, Editorial Lumen, Barcelona (1974)
21. Gibson, J.: The Ecological Approach to Visual Perception. Psychology Press, New York (1986)
22. McLuhan, M.: Counterblast. Rapp & Whiting Limited, London (1970)
23. McCullough, M.: Ambient Commons: Attention in the Age of Embodied Information. The MIT Press, Cambridge (2015)
24. Aristotle.: Poetics. Hackett Publishing Company, Indianapolis (1987)
25. Salazar, N.: Rudolf Laban and topological movement: a videographic analysis. http://epubs.surrey.ac.uk/789620/6/__homes.surrey.ac.uk_home_.System_Desktop_rudolph%20laban.pdf . Accessed 28 Feb 2016
26. Bachelard, G.: The Poetics of Space. Beacon Press, Boston (1969)
27. Costin, M., Phipps, D.: Racing and Sports Car Chassis Design. B. T. Batsford Ltd., London (1967)
28. Rossi, A.: The Architecture of the City. The MIT Press, Cambridge (1982)
29. Maldonado, T.: El Diseño Industrial Reconsiderado. Gustavo Gili, Barcelona (1993)
30. Barbaresso, J., et al.: USDOT's Intelligent Transportation Systems (ITS) ITS Strategic Plan 2015–2019. US Department of Transportation (2014)
31. IISI Group, Advanced Urban Traffic Management System. www.iisigroup.com/en/products/tra-autms.html. Accessed 28 Feb 2016
32. Kaskiris, C., et al.: Mobile agents: a ubiquitous multi-agent system for human-robotic planetary exploration. http://people.ischool.berkeleyedu/~kaskiris/papers/ISSHS2005.pdf. Accessed 28 Feb 2016
33. European Commission, Proposal for a European Parliament and Council Regulation. Regulation (EU) No. /2010 of the European Parliament and of the Council on the approval and market surveillance of two- or three-wheel vehicles and quadricycles, European Commission, Brussels (2010)

The manufacturer's authorised representative in the EU is Springer
Nature Customer Service Centre GmbH, Europaplatz 3, 69115 Heidelberg,
Germany. If you have any concerns regarding our products, please
contact ProductSafety@springernature.com

Printed and bound by CPI Group (UK) Ltd, Croydon, CR0 4YY
30/09/2026
02225651-0011